MIRROR | MIRROR

MIRROR MIRROR

A History of the Human
Love Affair with Reflection

MARK PENDERGRAST

BASIC

BOOKS

A Member of the Perseus Books Group

NEW YORK

Published by Basic Books,
A Member of the Perseus Books Group

Hardback first published in 2003 by Basic Books
Paperback first published in 2004 by Basic Books

Designed by Brent Wilcox

Books published by Basic Books are available at special discounts for bulk
purchases in the United States by corporations, institutions, and other
organizations. For more information, please contact the Special Markets
Department at the Perseus Books Group, 11 Cambridge Center, Cambridge
MA 02142, or call (617) 252–5298, (800) 255–1514, or e-mail
special.markets@perseusbooks.com.

Library of Congress Cataloging-in-Publication Data
Pendergrast, Mark.
    Mirror mirror : a history of the human love affair with reflection /
Mark Pendergrast.
        p.   cm.
    Includes bibliographical references and index.
    ISBN 0-465-05470-6 (HC) ISBN 0-465-05471-4 (pbk)
    1. Mirrors—History.   2. Reflection (Optics).   3. Reflecting
telescopes.   I. Title.
TP867.P46   2003
535'.323—dc21
                                                          2003002544

04 05 06 / 10 9 8 7 6 5 4 3 2 1

To my parents,
Nan and Britt Pendergrast,
mirrors for those who seek peace and
justice in a difficult world.

The world is full of fools, and he who would not see it should live alone and smash his mirror.

ANONYMOUS FRENCH PROVERB

Strange, that there are dreams, that there are mirrors.
Strange that the ordinary, worn-out ways
of every day encompass the imagined
and endless universe woven by reflections.

JORGE LUIS BORGES

They gave us things like solid water, which were some-times brilliant as the sun and which sometimes showed us our own faces. We thought them the children of the Great Spirit.

CHIEF CAMEAHWAIT OF THE SHOSHONE,
August 1805, upon receiving mirrors from
the Lewis and Clark expedition.

Mirrors symbolize reality, the sun, the earth, and its four corners, its surface, its depths, and all of its peoples.

CARLOS FUENTES

CONTENTS

INTRODUCTION

Every morning there you are again. It's a ritual that humans perform daily, something so commonplace that we hardly notice it. Perhaps you're a little bleary-eyed, but that's you in the mirror, all right, maybe with a toothbrush in your mouth or a washcloth in your hand, trying to reorient yourself for another round in life's everyday affairs. Like most people, you've become so accustomed to this morning routine that you rarely think about it. Yet it's almost unique in the animal kingdom, because the ability to recognize the creature in the mirror as *you* seems to be limited to the higher primates and, perhaps, dolphins and elephants. Other animals see only a rival or a friend.

Mirrors are meaningless until someone looks into them. Thus, a history of the mirror is really the history of looking, and what we perceive in these magical surfaces can tell us a great deal about ourselves—whence we have come, what we imagine, how we think, and what we yearn for. The mirror appears throughout the human drama as a means of self-knowledge or self-delusion. We have used the reflective surface both to reveal and to hide reality, and mirrors have found their way into religion, folklore, literature, art, magic, and science.

Humans have been intrigued with mirrors since prehistoric times. The ancients—Egyptians, Indians, Chinese, Mayans, Incas, and Aztecs—buried their dead with magical metal or stone reflectors, to hold the soul, ward off evil spirits, or allow the body, before taking the final trip to the afterlife, to check its hair.

Because a round mirror can both reflect the sun and become a miniature imitation of it, early metal reflectors came to be associated with sun gods. At the same time, however, mirrors as secular objects were used to apply cosmetics, foreshadowing thousands of years of people peering into the "flattering glass."

Yet the magic of mirrors remained. Scryers (gazers into reflective surfaces) in the Middle Ages used them to look into the mystic future; in this way, mirrors served as a portal to the divine or demonic. Magicians manipulated them to create illusions to impress kings and commoners.

And from the earliest times, mirrors have also been used for scientific applications. According to legend, Archimedes used mirrors to set fire to Roman ships during the siege of Syracuse, and the controversy over whether or not this feat was possible led eventually to modern solar ovens and generators. Concave mirrors made early lighthouses possible, and the reflecting telescope changed our view of the universe. Today, huge mirrors permit us to peer back through time billions of years— into the most distant regions of space—and lightweight gossamer optics will allow us to look even farther. Some envision using giant, orbiting mirrors to manage the earth's climate.

Thus, the story of mirrors is also the story of light, that mysterious medium that acts simultaneously like a wave and a particle, imposes a speed limit on the universe, and in a sense *is* the universe, at least according to Albert Einstein. Yet no one really knows what light is. As if these mysteries were not enough, visible light is only one octave in the spectrum that ranges from mile-long radio waves to high-energy bursts of gamma rays. After World War II, our ability to explore the universe dramatically expanded as scientists figured out how to make unusual mirrors to reflect most of those wavelengths. That story, too, is part of the mirror saga.

The glass mirror industry, since its inception in the Middle Ages as a secret Italian guild, followed by the seventeenth-century French industrial espionage that broke the monopoly, has grown to huge proportions. The common glass mirror also had an unforeseen and revolutionary impact on Renaissance literature and art, rendering them more realistic, secular, and sexy.

With the advent of cheap industrialized glass and modern methods of applying reflective material to it, mirrors have become common objects even in the poorest homes. They have been used creatively by architects and home decorators, and in the twentieth century glittering mirrors helped transform the United States into a pleasure-seeking, vain, celebrity-driven society. Psychologists, advertising men, police, and voyeurs peer at us through one-way mirrors. Now more than ever, mirrors are ubiquitous reminders that the study of mankind is man and woman.

Mirrors ushered in the earliest human civilizations, and now they point us into the future—while simultaneously allowing astronomers to peer ever farther back into time. The history of mirrors covers a vast territory, from the creation of the universe (perhaps along with alternate mirror universes), to the first hominids, to the Hubble Space Telescope, and beyond. The cast of quirky characters looking into and manipulating mirrors is equally diverse.

When I first conceived of this project, I knew it would be interesting, but I didn't realize how many facets it would contain or where it would take me. I examined ancient Egyptian mirrors in the Louvre, walked through the Hall of Mirrors at Versailles, looked at myself in an Aztec divining mirror at the British Museum, stumbled through a century-old mirror maze in Lucerne, buried myself under books and manuscripts in various archives and libraries, visited a French nudist colony (few mirrors, as I suspected), lay on my back to see the world's largest kaleidoscope in a silo in Upstate New York, looked at myself as I really am (not flipped right-to-left) in a "True Mirror" in Manhattan, clambered to the top of a new 300-foot-diameter radio telescope in rural Green Bank, West Virginia, gazed down into the vast pool of the 200-inch mirror on Mount Palomar, and lived at a Vedanta monastery while tailing John Dobson, the extraordinary missionary of amateur-telescope mirror-makers.

As I type this sentence, I am looking into my own eyes in the hinged "PC mirror" attached to the side of my computer monitor. This device is sold by a New York company primarily as a sales tool for telemarketers—if you smile winsomely, people are more likely to buy your product. I put the mirror on my computer, however, not to sell something but to remind me of my own humanity. Right now, I see a man in his early fifties, graying around the temples, who needs a haircut and who, though he hates to admit it, looks a bit like Woody Allen. I am confessing all this because the book is not only a history of mirrors but also, like all books, a reflection of a particular person and his experience. Although I will not reappear in person until the final chapter, readers should bear in mind throughout that I am lurking somewhere behind these words.

An overarching theme in this book is that as human beings we use mirrors to reflect our own contradictory nature. On the one hand, we want to see things as they really are, to delve into the mysteries of life. On the other hand, we want the mysteries to remain mysteries. We yearn for definitive knowledge, yet we also revel in imagination, illusion, and magic.

The German poet Rainer Maria Rilke may have been right when he wrote, "Mirrors: still no one knowing has told / what your essential nature is." In J.K. Rowling's Harry Potter fantasy novels, the Mirror of Erised shows us "the deepest, most desperate desire of our hearts." In a way, all mirrors are like that. Ultimately, what we see in them depends on what we bring to them.

Chapter 1

THE MIRROR OF THE SOUL

If I ask if all be right
From mirror after mirror,
No vanity's displayed:
I'm looking for the face I had
Before the world was made.

WILLIAM BUTLER YEATS

THE SCENE: AN AFRICAN savanna after a torrential seasonal rain. Water still drips from the leaves of the scattered trees and seeps into the roots of the tall grass. In the peaceful aftermath of the storm, animals gather to drink in the temporary pools, lapping and slurping with their tongues. At one pool, however, a creature standing upright on hind legs leans over, preparing to scoop water with its hand. But it stops, furrowing its brow in curiosity.

The hominid has noticed how in the still pool the baobab tree magically appears to grow down into the water. Now, he sees a fellow creature looking back at him, hand cupped, ready to drink. Is it an enemy? The hominid bares his teeth. So does the man in the pool. He grunts and hits at him, but the image disappears in a splash.

He dips his hand and drinks, then sits back and contemplates the scene before him. The ripples gradually settle. He smiles at the beautiful reflection of the tree, then leans over again to see his silent fellow creature. He, too, is smiling. Perhaps he is not an enemy after all.

The man frowns; so does his reflected companion in the pool. He sticks out his tongue; they both do. They touch their noses, show their teeth, pull their ears, wink simultaneously. He understands, on one level, at least. They are the same, yet they are different.

Such was probably the first mirror, as humans evolved from apes and developed self-consciousness. Of course, such a fable is a simplified version of what took evolution millions of years to accomplish. According to paleontologists, our forebears stretch back some 18 million years. *Homo sapiens* appeared only some 200,000 years ago.

Our sapient ancestors could think abstractly, use tools, create art. The magnificent paintings of deer, horses, bison, and other animals at the Chauvet cave in southern France were created 32,000 years ago. Physically and psychologically, we haven't changed much since then, despite our enormous technological progress. "Remember that the Cro-Magnon people are us—by both bodily anatomy and parietal art—not some stooped and grunting distant ancestor," observed paleontologist Stephen Jay Gould. "Large, widespread, and successful species tend to be especially stable. . . . Human bodily form has not altered appreciably in 100,000 years."

We all know what it is like to be human, yet we do not know, and the history of the mirror is intimately tied to that questioning ambivalence. We think, and so, as René Descartes posited, we self-consciously exist. But this self-awareness leads to questions. *Who is doing the thinking? Who am I? Am I that image in the mirror? How do I fit into the universe? What is beauty, and why does it move me? What is love, and why am I so obsessed with thoughts of sex? And by the way, how does my hair look? Do you think my nose is too big?*

We are a curious species, and so we are always turning the next corner, wondering what is over the horizon, opening Pandora's box. That same curiosity leads us to the mirror to gaze deeply into our own eyes in search of answers. It appears that only a few animals—higher apes, man, perhaps dolphins and elephants—have the mental capacity to realize that they are looking at their own reflections. That capacity for self-consciousness is apparently fundamental to the human experience, connected to self-awareness, logic, and empathy.*

---

*The ability to recognize oneself in a mirror has profound implications that will be explored fully in the final chapter.

"Come along, child, have no fear," a French educator writes in a message to students about to enter the Chauvet caves. "He who walked about here . . . was your fellow creature, your brother. . . . He is also your mirror and your memory."

The Ka in the Mirror

It is impossible to pinpoint when humans first created an artificial mirror. Initially, they probably gazed into bowls of water, then made the logical connection between still waters and other flat, reflective objects. As nomadic hunter-gatherers, Stone Age people learned to work rocks into weapons, so it is not surprising that the earliest artificial mirrors archaeologists have discovered, dating from around 6200 B.C.E. at Çatal Hüyük (near Konya, Turkey), were made of polished obsidian, a natural black glass created during volcanic eruptions.

Other candidates for the first man-made mirror are a slab of selenite, with traces of wood around it that may have been a mirror frame, and a disk of slate. Both were found in El-Badari in Egypt and date from around 4500 B.C.E. A reflective piece of mica pierced with a hole, presumably for suspension from a wall, was also found in Egypt from the same period.

The mining and working of metals marked the European Copper, Bronze, and Iron Ages (ca. 4000 B.C.E. to 1 B.C.E.). The scientific quest—the human desire to explore, explain, and transform the world through logic and experiment—arguably began with pottery, then metallurgy. Copper was a pliable substance. Tin, too, was easily worked. At some point, someone combined the two metals, discovering that the resulting bronze alloy, using mostly copper and some tin, was stronger and less subject to corrosion. Thus with the first great civilizations and cities came the bronze sword, efficient warfare—and many more mirrors.

The earliest copper mirrors were found in Iran and date from around 4000 B.C.E. Other copper mirrors, found in an Egyptian grave dating from the period of the First Dynasty of the pharaohs (ca. 2900 B.C.E.), were shaped like upside-down pears with handles. They probably came in trade from elsewhere. Because the Egyptians were so obsessed with death and the afterlife, however, and because they preserved their possessions in tombs, we know more about their mirrors than about those of other cultures.

The typical ancient Egyptian mirror was essentially flat (a few were convex or concave), polished on both sides, and slightly elliptical (wider than high) with a sharp metal tang at the bottom that fit into a handle made of wood, stone, ivory, horn, metal, or clay. The highly polished surface was protected with cloth, animal hide, or woven rushes. In the tomb of Tutankhamen, there was a mirror in its own custom-fitted wooden box, embossed with sheet gold and inlaid with colored glass, carnelian, and quartz.

Generally made of copper until around 2100 B.C.E., then of bronze—and sometimes gold or silver—Egyptian mirrors were both secular and religious objects. They were often used for such familiar purposes as applying makeup. The Egyptians' elaborate cosmetics probably first developed as a defense against the fierce sun, moisturizing the skin and protecting against glare. But it is clear from paintings and carvings that Egyptian men and particularly women spent a great deal of time working on their appearance, applying makeup of yellow, green, black, and red. Priests used mirrors to see while they shaved their heads; others fixed their hair or wigs. In addition, the Egyptians were susceptible to eye diseases and probably used their mirrors to examine their eyes.

The mirror's primary religious connection was to Ra—the most powerful deity, the omnipresent African sun—and the mirror was his symbol brought to earth. In Egyptian sculpture and painting there is always a round sun-mirror atop Ra's falcon head. Even the mirror's elliptical shape imitated the rising or setting sun, stretched sideways as it refracted through the atmosphere.

Egyptian mirrors also were associated with Hathor, the goddess of love, fertility, beauty, and dance. Hathor was usually represented as cow-headed, her horns enclosing a sun-mirror-disk, and she was identified as the eye of the sun god. Perhaps this is why some Egyptian mirror representations have magical eyes painted in their center. Mirror handles sometimes show Hathor as a lithe nude, and dancers depicted in tomb paintings frequently hold mirrors. In the "Erotic Papyrus" (ca. 1300 B.C.E.), a naked woman with splayed legs masturbates atop a pointed cone while looking into a mirror to apply makeup.

The *ankh*, the Egyptian symbol of life, looks like a mirror—it is egg-shaped, with a *T*-handle attached at the small end. The long name for mirror is *ankh-en-maa-her*, meaning something like "life-force for seeing the face," and was shortened to "see-face." On a typical coffin lid, the goddess Hathor (this time with a lovely human face rather than a cow's head) holds *ankhs* that look very much like mirrors. In addition to their "see-face" names, Egyptian mirrors were also given religious titles such as "the divine," "that which is in eternity," or "the truth."

The Egyptians believed that each person had a double called a Ka, which represented a person's essential genius, energy, and identity, as well as a Ba, the soul or consciousness, usually shown as a bird. The elaborate mummification of the body and other funeral practices were designed to preserve both Ka and Ba. The Ka, like its former body, required food for energy, which is why the Egyptians brought food and drink regularly to the tombs. The Ba flew off to heaven during the day, but at night it reunited with the mummified body. The deceased thus became identical to the sun-god, who rose each day and, like Osiris, died every night only to be reborn at dawn.

Mirrors were an essential element in tombs. In her book *Ancient Egyptian Mirrors*, Christine Lilyquist describes a tomb scene at Thebes where a girl presents ointments and a mirror to the deceased with the statement: "To thy Ka. It has made thee, namely the House of Morning, thou being living . . . vigorous like Ra every day." The Egyptians may have believed that the mirror helped preserve the Ka, the double discovered in the mirror's depths, and allowed it to make a transition to another life.

Thus mirrors are frequently depicted on the wall paintings directly before the face of the deceased, or in his hand, beneath his chair, or in his coffin. Although they were more elaborate in noble burial sites, mirrors are also found in very simple graves, even with children—some of these cheaper "mirrors" are made of painted wood.

The Egyptians also understood some scientific uses of mirrors, redirecting sunlight down into pyramids to provide light for workmen in the dark tombs. One papyrus even relates how a magician replaced a severed head during a seance, apparently using a mirror to create an optical illusion. Thus the Egyptians employed all the main themes associated with mirrors—religion, cosmology, vanity, beauty, sex, death, magic, and science.

Golden Reflections for the Lady of Uruk

Another ancient civilization flourished in the Fertile Crescent, which nestled between the Tigris and Euphrates Rivers as both wend their way to the Persian Gulf. Around 4500 B.C.E., at Tell al-Ubaid on the Euphrates, a tribe settled and created an agricultural community. The Ubaidians knew how to make clay bricks, plaster walls, mosaic decorations, turquoise beads, and copper mirrors.

The Ubaidians left no written records, but their descendants, the Sumerians (3000 B.C.E. and later), who invented cuneiform writing, left clay tablets and archaeological evidence. We know from these that they were practical traders who valued the art of metalworking and supported a thriving trade in tin, which came over mountain passes from inner Asia or by sea. The cuneiform archives contain a number of recipes giving the amounts of copper and tin to be used for bronze.

By 2000 B.C.E., the *tibira*, or metallurgist, was a prized specialist in the cities of Uruk and Ur. "The list of metals used in the foundry of the smith," observes historian Samuel Noah Kramer, "includes almost all those known at the time: gold, silver, tin, lead, copper, bronze." The cuneiform texts frequently mention mirrors, mostly of copper and bronze. One tablet refers to repairs to a golden mirror belonging the "the Lady of Uruk."

Sexually explicit Sumerian myths featured mirror metaphors for excellence or beauty. In an exchange between Inanna, the goddess of love, and her hairy husband, Dumuzi, she cries in passion, "Rub it against our breast, my sweet! . . . My one worked on by a skilled metal worker!" He cries out, "May you be a shining mirror! . . . Come with the sun, stay with the sun!"

The Sumerians wanted to understand the world in which they found themselves. How could such joy and life coexist with suffering and inevitable death? They created numerous gods, and through various methods of divination they sought ways to learn (and sway) the future. Along with examining animal entrails and studying the heavens, they looked into a kind of mirror—a bowl of water, usually with floating oil—to see visions.

One God, Many Mirrors

To the west of Sumer lay the Syrian Desert and the Arabian Peninsula, home to Semitic nomads even in the time of the Ubaidians. For thousands of years, the Semitic tribes periodically infiltrated, conquered, assimilated somewhat, then departed Mesopotamia again for the desert. Around 1850 B.C.E., a Semite named Abraham, apparently fed up with the soft lifestyle and belief in multiple gods, left Ur with his wife, Sarah. The patriarch of a tribe that came to be known as the Hebrews or Jews, Abraham was typical of the nomads who provided a key link between the civilizations of ancient Egypt and Sumeria.

Abraham's great-grandson Joseph, sold into slavery by his half-brothers, eventually wound up in Egypt around 1700 B.C.E. There, because of direct access to his singular God, he could correctly interpret dreams, locate stolen goods, and see the future in the water reflections of his magic silver goblet. As a result, he thrived as the right-hand man of a grateful Semitic Hyksos pharaoh.

Some 450 years later, however, the Hyksos no longer ruled. According to the Bible, the resident Jews had been reduced to brick-making slaves when their leader, Moses—another magician with strong ties to God—inflicted various plagues on Pharaoh (probably Rameses II), who finally let the Jews go, along with "articles of silver and gold [and] whatever they asked for." Apparently among the goods they requested were mirrors (or perhaps, as some scholars have conjectured, they were expert mirror-makers). In Exodus 38:8 we read: "The bronze laver, with its bronze base, was made from the mirrors of the women who served at the entrance of the meeting tent."*

The Hebrews invaded and conquered Canaan, renaming it Israel, but for centuries they were subject to marauding tribes from the east and west. Throughout their turbulent history, the Jews struggled to main-

---

*Later, these same women serving at the entry of the meeting tent offered sexual intercourse to visitors, perhaps as ritual prostitutes alluded to in Genesis.

tain their identity. They also continued to admire themselves in mirrors, to the dismay of their prophets. "Because the daughters of Zion are haughty," Isaiah stormed, "and walk with necks outstretched, ogling and mincing as they go, their anklets tinkling at every step, the Lord will do away with the finery of the anklets, sunbursts, and crescents; the pendants, bracelets, and veils; the headdresses, bangles, cinctures, perfume boxes, and amulets; the signet rings and the nose rings; the court dresses, wraps, cloaks, and purses; the mirrors."

When Elihu scolds Job for daring to question God, he says, "Stand and consider the wondrous works of God! Do you know how God . . . makes the light shine forth from his clouds? . . . Do you spread out with him the firmament of the skies, hard as a brazen mirror?" Like many other ancient peoples, the Hebrews thought that the sky was a literal dome of heaven, and when this ceiling became as hard as metal, it prevented rain from falling.

Jewish folklore incorporated mirrors into magical thinking, often as a method of securing love. In Joshua Trachtenberg's book *Jewish Magic and Superstition*, we learn how to arouse passion by writing the name of the beloved three times on the back of a small mirror before holding it up in front of two copulating dogs to capture their image. Then, get your intended to glance into the mirror, exciting her through the magic power of the sexual act, fixed in the mirror.

Jewish scribes believed that they could improve weak eyes by taking a break from the scrolls and staring into a mirror. And when a Hebrew died, mourners would cover his mirrors or turn them to the wall to protect his soul—trapped in the mirror from reflection during life—from being carried off by demons or from haunting the household.

Trade, Empires, and Etruscan Art

By 1000 B.C.E., humans were making mirrors all over the world. Traders such as the Phoenicians and Etruscans sailed the waters of the Mediterranean and beyond, carrying news, goods, and customs with them. Many cultures modified the traditional bronze Egyptian mirror to create their own versions—though most of these were round rather than elliptical. The Phoenicians' successful if often amoral trade in slaves, mirrors, and other goods declined only gradually as they survived successive incursions by the Assyrians, Babylonians, and Persians.

The Persian Empire ruled all lands between the Nile and the Indus at its height under Darius the Great around 500 B.C.E. Nobles from conquered nations trekked to Persepolis, the newly created capital city, with mounds of silver, gold, and jewels for Darius, who sat in purple robes on his golden throne, surrounded by incredible luxuries, including the best bronze and silver mirrors to reflect his accomplishments.*

Many of those mirrors were probably made in northern Italy, where the Etruscans, who by 600 B.C.E. had grown enormously wealthy through trade, mining, and agriculture, made exquisite bronze and silver mirrors, slightly convex, with remarkable line engravings on the back. Many look as if they could have been drawn by a modern graphic artist. Although heavily influenced by the Greeks, who were making their own mirrors by that time, the Etruscans' art was far freer. Like the Egyptians, they apparently believed that a happy afterlife required the proper grave furnishings. In underground tombs cut from the coastal region's soft volcanic stone, the Etruscans re-created the home of the deceased, including beds, tables, stools, candelabra, gold brooches, earrings—and mirrors, which served as receptacles for the soul. The Etruscan word for soul, *hinthial,* also means "image reflected in a mirror."

The backs of many Etruscan mirrors depict sexually charged scenes where partially clothed men and women gossip and flirt. On one, a nearly nude seated man (a cloth is draped over his legs) embraces a naked young woman on his lap. To their right, a woman stands and stares at them. She is also naked, wearing only a pair of sandals and a necklace with a half-moon pendant. On the couple's left, a fully clothed girl holds a mirror and stares into it with a distinctly unhappy expression, perhaps trying to block out the lovemaking by looking into the mirror (or she may be holding the reflective surface toward the lovers so they can see themselves).** The inscription identifies the lovers as Mexio and Fasia.

---

*Under Xerxes (485–465 B.C.E.), however, the empire began to slip away after the Greeks defeated the Persians, and Alexander the Great eventually destroyed it completely in 331 B.C.E.

**"Among the Etruscans," Greek visitor Theopompus noted around 380 B.C.E., "Etruscan slave girls wait on their masters naked.... [Etruscans] adore sex and have intercourse sometimes with others watching."

The Etruscans also made small, portable mirrors with hinged covers, quite similar to modern compacts. The inner surface of the cover was concave and probably used to direct light onto the face, or it may have been used as a magnifying mirror. The engravings on the cover often show Dionysus, the god of wine, and Eros, the god of love, with a muse playing a lyre. Another popular cover depicts Odysseus and Penelope with the dog Argos. Some show a satyr and maenad dancing, while still others feature Athene fighting a giant or Hercules vanquishing a lion. Even heroes needed maternal support, however. One mirror shows the mighty Hercules sucking from his mother's breast while shocked attendants look on.

Greek Self-Examination

Like the Etruscans, the Greeks made mirrors primarily of bronze, a craft they learned from the Minoan culture on Crete. By the time the city-states emerged around 700 B.C.E., the Greeks had created a pantheon of gods who dwelled on Mount Olympus and about whom they told wonderful stories. In some of these myths, mirrors figure prominently.

Consider poor Medusa, one of the gorgeous Gorgon sisters. When Medusa slept with Poseidon in one of Athene's temples, the infuriated goddess changed her into a winged monster with glaring eyes, huge teeth, a protruding tongue, terrible claws, and hair of hissing serpents. Anyone who looked directly at her turned to stone. To kill Medusa, the Greek hero Perseus needed magical aids provided by the still-vengeful Athene. Most important, Athene lent him her bronze shield, which he used as a mirror to look at Medusa without being turned to stone. It is probably a good thing that Athene guided his sword, because eye-hand coordination in a mirror can be quite difficult.

In another myth, Narcissus, an exceptionally beautiful young man, frustrated the woodland nymphs and lustful males by remaining aloof. To punish the disdainful youth, Nemesis made Narcissus understand what it was like to have an unrequited passion by having him fall in love with his own unattainable image. When he bent to drink in a still pool, he saw a beautiful young man whom he took to be a water spirit. Ovid quotes Narcissus in *Metamorphoses*:

But a thin film of water parts us. He is eager
For me to hold him. When my lips go down
To kiss the pool, his rise, he reaches toward me.
You would think that I could touch him—almost nothing
Keeps us apart. Come out, whoever you are!

But every time he brought his lips near to kiss his beloved or plunged his arms into the pond for an embrace, the spirit fled. Eventually, Narcissus figured it out. "I know the truth at last. He is myself! I feel it, / I know my image now. I burn with love / Of my own self; I start the fire I suffer." Even so, he could not tear himself away from his reflection. Narcissus pined away and died, turning into a beautiful flower that droops over as if looking into a pool.

Greek oracles made ample use of mirrors. Ailing Greeks consulted "an infallible oracle" devoted to the goddess Demeter, as the Greek writer Pausanias noted. "They [the keepers of the oracle] tie a mirror to a fine cord and let it down, judging the distance so that it does not sink deep into the spring, but just far enough to touch the water with its rim. Then they pray to the goddess and burn incense, after which they look into the mirror, which shows them the patient either alive or dead." At a different oracle, this one for Apollo, "the water shows to him who looks into the spring all the things he wants to behold." Yet another oracle site featured a wall mirror that reflected the viewer only dimly or not at all but in which "the actual images of the gods and the throne can be seen quite clearly"—perhaps an early optical illusion.

Pythagoras, the mystic mathematician who died around 475 B.C.E. a few years before the birth of Socrates, also reportedly possessed a magical mirror, which (according to legend) he held up to the moon before reading the future in it. For Pythagoras, numbers were the soul of the universe, and abstract math, music, and astronomy were sacred. Perhaps in his magic mirror he contemplated an orderly universe in which, he believed, the world progressed by the interaction of contraries, pairs of mirror-image opposites.

In the dialogues of Socrates, Plato recounted the parable of the cave, in which he portrayed humans as akin to prisoners chained since birth in an underground cavern, able to see only projected shadows, which they take for reality. If suddenly freed to come into the light of day, they would be unable to cope with their new reality right away. "At first it would be easiest to make out shadows," said Socrates, "and then the images of men and things reflected in water, and later on the things themselves." In like manner, Socrates and Plato asserted that our illusory reality is only the reflection of a greater, abstract Goodness that lies in a hypothetical upper world beyond the mirrorlike dome of the sky. Even though he considered this world a mirror-illusion, however, Socrates urged his followers to study themselves in mirrors in order to make sure their faces did not reflect dishonorable thoughts or deeds, apparently assuming that they could monitor their inner reality by their outer appearance.

Most Greeks didn't worry about such deep matters. Instead, they used mirrors to admire themselves and fix their hair, as we see in numerous illustrations on vases, urns, and friezes. Many scenes containing mirrors show women with their children and husbands, listening to music or getting dressed. Sometimes the mirror hangs on the wall. While women are usually shown gazing into handheld mirrors, men, too, examine themselves. Thus Eros, the god of love, admires himself in the metal.

Greeks also made mirrors with stands, usually with female "caryatid" figures holding up the round reflective disks. In a typical mirror from the fifth century B.C.E., a woman lifts her gown delicately with her left hand, apparently about to step into water. In others, the lady has her hands on her hips, is arranging her hair, holds castanets, or is looking into a tiny mirror. Around the rim, many mirrors featured doves, flowers, fruits, rabbits, or flying horses—all associated with Aphrodite, the goddess of love. While female figures predominated as mirror handles in Greece itself, the mirrors made by Greek craftsmen in southern Italy during the same period featured strong nude men, presumably athletes. Greeks gave valuable mirrors as votive offerings to various gods, and mirrors were frequently placed in tombs.

"Barbarian" Celts, "Civilized" Romans

To the north of Greece and Italy, the Celtic tribes thrived throughout France, Germany, and Great Britain. They were, as one Roman writer put it, "fair and of ruddy complexion; terrible from the sternness of their eyes, very quarrelsome, and of great pride and insolence." The Celts were fierce warriors whose priests, called Druids, sometimes sacrificed humans to their gods—a practice that had been abandoned by their southern neighbors in favor of animal sacrifice. The Celts were also skilled metalworkers, making iron swords and, eventually, bronze mirrors. At first, they imitated the Greek and Etruscan mirrors they received in return for salted meats, leather, and textiles. But eventually the Celts created a distinctive artistic style of intricate whorls and curling tendrils that decorated their swords, jewelry, and mirror backs. They revered the human head, which they believed contained the soul after death (and which accounts for their head-hunting and tendency to drink out of skulls), and they undoubtedly valued mirrors as magical repositories of the head's image.

Attacked by the Celts from the north and the Romans from the south, the Etruscans gradually lost power, but the Romans survived the Celtic sacking of Rome around 390 B.C.E. and created an empire that lasted until 476 C.E. Having observed the Etruscans' hedonistic lifestyle, the Romans outdid them in sumptuous feasts, luxurious quarters, and huge public entertainments, including chariot races and gladiatorial combats. The Romans were cultural sponges, adopting many of the gods and ways of conquered peoples such as the Greeks, Etruscans, and Egyptians.

Wealthy Roman women spent amazing amounts of time striving to be beautiful, partly because beauty was associated with virtue and fertility. "She must these days use powders, pomades, paints," observed Lucian. "Each chambermaid, each slave carries one of the essential objects for the toilet. One holds a silver basin . . . another a water pot, still others the mirror." Each morning, a slave girl brought her mistress a bowl of scented water to rinse off the night's facial cream, a mixture of flour and milk. Then, after brushing her teeth and rinsing her mouth with a breath sweetener, she soaked in a scented bath before receiving an oiled massage. Following that, the *ornatrix*, her dresser, fixed her elaborate hairdo, perhaps using a *calamistrum*, a hot iron, to produce curled ringlets. Often, her hair was dyed or she wore a wig. The *ornatrix* applied white powder, then rouge on the cheeks and lips, and black kohl, the ancient Egyptian favorite, on her eyelids. Finally, the lady dressed and put on rings, bracelets, necklaces, and other jewelry.

All the while, of course, she checked herself out in her mirror. The Roman artisans mass-produced their mirrors, abandoning fine engravings on the backs for simple concentric circles. They used a hot-tinning process to give the bronze a hard, silvery-white reflecting surface. Even poorer Romans and servants usually had some kind of mirror, keeping a sponge with powdered pumice handy for polishing it, but the aristocracy preferred mirrors of silver.

Roman men, too, were vain, curling their hair in front of a mirror and worrying about going bald. Large metal mirrors hung along the walls of the public baths, according to Seneca, who complained: "We think ourselves poorly off, living like paupers, if the walls [of our baths] are not ablaze with large and costly mirrors." The paranoid Emperor Domitian had his gallery lined with polished reflective stone (probably white marble or selenite) so that he could see what was happening behind his back.

Many Romans, particularly in the more decadent days of the empire, took to heart the hedonistic philosophy of the Greek Epicurus, becoming gluttons for food and sex. The insides of some elegant drinking vessels were cut into many-faceted mirrors, so that the imbiber could see multiple images of himself getting drunk. At banquets, guests reclined on couches for elaborate meals, surrounded by wall paintings, mosaics, and sometimes full-length metal mirrors. Perfumed doves might be released to flutter around the room. After the meal, once the wives and children had left, some male guests would linger to dally with courtesans.

One wealthy, dissipated Roman named Hostius Quadra took the art of the orgy to an extreme, installing large concave metal mirrors in his chambers so that they enlarged whatever they reflected. The repulsed Seneca voyeuristically described the scene:

> He was vile in relation not to one sex alone but lusted after men as well as women. He had mirrors made . . . in which a finger exceeded the size and thickness of an arm. These, moreover, he so arranged that when he was offering himself to a man he might see in a mirror all the movements of his stallion behind him and then take delight in the false size of his partner's very member. . . . Mirrors faced him on all sides in order that he might be a spectator of his own shame. . . . And, because he could not

watch so attentively when his head dipped in and clung to his partner's private parts, he displayed his own doings to himself through reflection.

"Sometimes shared between a man and a woman, and with his whole body spread in position for submitting to them, he used to watch the unspeakable acts," Seneca continued, adding that "no night is deep enough to conceal" the acts that the author had just revealed in such loving detail.

It is possible that Hostius and his magnifying mirrors were a figment of Seneca's pornographic imagination, but there is no question that other Romans also connected mirrors and sex. The fourth-century poet Claudian pictured Venus readying herself for a meeting with Cupid in a chamber covered with mirrors, "so that whichever way her eyes turned, she could see her own image"—and, presumably, their subsequent coupling as well. If other Mediterranean peoples had employed mirrors to reflect on the soul, for Romans they mainly provided reflections of the self.

While these large erotic mirrors were made of metal, the Romans also learned to produce small glass mirrors. Around 100 B.C.E., Syrian craftsmen near Sidon discovered that they could dip a long, hollow metal tube into a batch of molten glass, retrieve a glob on the end, and blow it into shape. Elegantly shaped hollowware now became relatively quick and easy to make, and mass production became possible by blowing glass into molds. Within the well-organized Roman Empire, this revolutionary new method spread quickly. Glass plates, bottles, cups, mosaics, and fake jewelry flooded the market. "Sidon was once renowned for its glass factories," Pliny the Elder observed.* "Glass mirrors, among other things, were invented there."

These small convex pocket mirrors were produced by blowing a thin glass sphere, then pouring hot lead into it to coat the inside. When broken and cut, these made adequate mirrors for household use or magical amulets, and they have been found in graves all over the Roman Empire. Because Roman glass was generally discolored, wavy, and full of bubbles and lines, the mirrors were far from perfect, but the thin glass compensated for these faults to produce a relatively good reflection.

---

*A workaholic with an encyclopedic curiosity, Pliny died of asphyxiation while investigating the eruption of Vesuvius in 79 C.E.

With the fall of the Roman Empire, the art of convex mirror-making apparently went into decline throughout most of Europe until the twelfth century, though the art was apparently kept alive in Near Eastern countries. Mirrors made of silver and bronze, which predominated among the nobility, were too expensive for peasants. For the next few centuries, mirrors in the Western world, as well as those in Asia and the Western Hemisphere, would remain essentially the same—usually made of metal or stone, and used primarily by the elite because of their expense and difficult production.

Cosmic Mirrors of the Orient

To the east, ancient mirrors were occasionally connected with sex, but usually they were more concerned with everyday functions, religious symbolism, or magic. In the Indus Valley, citizens of Moenjodaro and Harappa admired their elaborate headdresses in copper and bronze mirrors with simple handles, though the archaeologist Stuart Piggott complains that the people suffered from "standardization and an almost puritanical utilitarianism."

The same cannot be said of an ornate silver mirror, dating from around 700 B.C.E., found in Kazakhstan. Engravings on its gold-molded back show a lion biting the back of an ox, heroes attacking a winged griffin, and other scenes from nature and mythology. This round mirror, with two holed knobs on the back for strings, was created by the nomadic Scythians who roamed the steppes, part of the Indo-European hoards who periodically invaded India, China, and the Mediterranean. In another burial tumulus, at nearby Issyk, the so-called Man of Gold warrior was buried with an elaborate tunic with more than 9,000 gold plaques depicting snow leopards, tiger heads, deer, trees, and mountains, along with a small bag containing a mirror, perhaps to ensure that his soul passed unharmed into the next world. The Scythians were probably influenced by Siberian tribes to the north, who had been making round, slightly convex bronze mirrors since 1500 B.C.E.

The earliest mention of a Chinese mirror, from 673 B.C.E., refers to a "girdle mirror of the queen's"—an indication that women carried toilet mirrors even then. These early mirrors may have derived from the

Siberians or Scythians. But it was only during the period of the Warring States (475–221 B.C.E.) and the subsequent Han Dynasty (206 B.C.E.–200 C.E.) that the Chinese really mastered mirror-making—a mastery that would last for some 2,000 years.

The earliest Chinese mirrors may have been polished jade, followed by iron, then bronze. The first bronze mirrors were extremely thin and delicate, becoming thicker and more durable with time, and were almost all round, with a single pierced knob on the back through which a silk cord or ribbon could be run. The Chinese also made stands to hold these round mirrors on dressing tables. Even the earliest examples are remarkable works of art, cast in clay molds (later using the lost wax method), smoothed with chisels and scrapers, and then brilliantly polished with *xuanxi*, a mixture of tin, mercury, alum, and deerhorn ash. Chinese emperors and nobility could admire themselves and adjust their elaborate headdresses in convex surfaces that allowed them to see their entire head even with a mirror as small as 2 inches across.

"In making these mirrors," one ancient commentator asserted, "[the craftsmen] put into them the vital essence of creation, therein following the fundamental principles of the universe so that they compare in brilliancy with the sun and moon, and communicate the will of the gods, thus defending us from evil spirits, and curing our diseases." The Chinese often gave mirrors as gifts on special occasions, and their possession was considered essential to ensure the Chinese ruler's access to ancestral wisdom.

The circular mirrors were emblematic of the universe, which was thought to be round, like a parasol-sky. A few mirrors were square, as the earth was supposed to be, but more often a square was cast on the back inside the round universe-mirror. These mirror backs were exquisitely decorated with dragons, phoenixes, plants, flowers, fruits, insects, and birds, but they were also symbolic of the complementary male and female principles of yin and yang and the Four Spirits, which ruled over time and space. One mirror-maker inscribed: "Commanding the four directions are the [Blue] Dragon at left and [White] Tiger at right. The Red Bird and the Dark Warrior are in tune with the forces of *yin* and *yang*." Because they represented the totality of existence, these were called "cosmic mirrors."

The round mirrors were also emblematic of the glorious sun and the luminous moon. "Its inner purity is shown in perfect illumination," wrote one mirrorsmith on his product. "Its light is the image of the sun and moon." A Chinese poet described one mirror: "On the front it emits a light as of four jewels. / Seen from a distance it might be taken for a suspended moon." Many such mirrors were decorated with sunburst patterns on the back. Well before the Christian era, the Chinese had learned to make concave burning mirrors called *yang-suei* that could, as one early text put it, "draw fire from the sun." The casting of such sacred mirrors, to light sacrificial fires, had to be done at midnight on the solstice.

Like the Greeks, the Chinese had a vast array of mythological figures, which appeared on the backs of their mirrors. On one such mirror, for instance, Hou-i, the divine archer, received the pill of life from Hsi-wang-mu, the Queen Mother of the West, and then brought it down to earth. Like Hercules and Gilgamesh, Hou-i was famed for his heroic deeds, such as battling the Speckled Serpent, the Great Boar, and the Celestial Wolf. These myths explained various astronomical phenomena and natural cycles and were all subjects for the mirrorsmith. "If you carry this mirror," one inscription reads, "you will see the great divinities."

The Chinese inscriptions illustrate universal human aspirations for happiness, wisdom, prosperity, health, and longevity. "May you see the light of the sun, and live in endless joy!" one exuberant mirror proclaims. "May you enjoy lasting fortune. May you enjoy wine and food. May you enjoy freedom from worry," reads a Han mirror from 113 B.C.E.

Many fine Chinese mirrors were interred with their owners at death, presumably to give light to the dead. Several hundred iron mirrors were entombed with the Prince of Wei in 295 B.C.E. In another Chinese tomb, the corpse's head was equipped with a wooden box covered with metal mirrors on the inside. *Hu hsi ching*, "heart-protecting mirrors," were often placed on the dead person's chest. Unlike the Egyptians and Etruscans, the Chinese did not usually try to recreate life conditions in the tomb but believed that because the soul wandered the earth after death, innumerable spirits and demons had to be propitiated in various rituals.

Other mirrors were not buried but were passed on for generations, as some inscriptions indicate: "May [the mirror be] cherished by your sons and grandsons for a long time." Others were intentionally broken in two, so that a husband and wife could each keep half, symbolizing their love and faithfulness even when apart. They could rejoin the pieces when reunited to see themselves as one. In some tombs, the husband and wife were each buried with a mirror half. "Looking at the light of the sun [in the mirror], let us forever not forget one another," an inscription reads. Another tender message says, "The autumn wind rises; my mind is sad, for long I have not seen you."

Other huge, probably mythological mirrors called *chau-ku-pau* (or "the precious mirror that would illuminate the bones of the body") were supposed to allow people to see their interior organs and thus somehow cleanse their innards. One such mirror, kept in a grotto in a cliff face, was reputed to be more than 10 feet square and could reflect the "five viscera" of a human being. Some otherwise normal Chinese mirrors appear to make similar claims. "This mirror's light shows the man inside," says one. Another, which purports to reveal the gallbladder, also claimed to "fathom the hidden and the subtle. Its clarity and luster put the pearl to shame, and the moon to pale."

Still other magical mirrors, called *t'ou kuang chien*, cast a reflection that showed the image on the back, as if the light had penetrated the metal. This effect was an artifact of the polishing technique, which caused imperceptible irregularities on the mirror surface that corresponded to the raised pictures on the back. Such magic mirrors were so strong that one lover's inscription assured his beloved: "In no way would I use it to unveil your hidden desires. Verily, I only wish to lay bare my own heart." Or perhaps the mirrors were magical only because they saw everything as it was: "The bright mirror, which reflects the figure, knows people's feelings," one inscription reads.

Some mirrorsmiths turned their inscriptions into a form of advertisement. "A fine mirror it is, my making of it!" boasted one such inscription around 20 C.E. "I can show the ageless immortals up above." Another explained that he used the best metals: "Good copper is being mined at Danyang. Mixed with silver and tin, the alloy is clear and bright." Several fine craftsmen even named themselves: "Mr. Tu has made a precious and marvelous mirror; there has never been such a one in the world."

During difficult times, such as the last decades of the Han period, when rebellion and warfare engulfed the country, mirrors were more popular than ever, and not just for the elite. The masses demanded mirrors with supernatural protective powers, even if the workmanship was often shoddy. A mirror worn on the back was thought to defend against evil demons, which would become visible when their image was reflected in a mirror. Chinese soldiers wore them on their breasts going into battle. When a person was near death, his empty coat with a mirror on it was sometimes hung on a fresh bamboo branch and carried around in the hope of enticing the departing soul to reenter the mirror and thus save the patient's life. Even when mirrors broke, they were thought to be magical, so they were often ground up and ingested with more palatable fare as medicine.

The Japanese initially imported Chinese mirrors, perhaps as early as 250 B.C.E. In 238 C.E., in a formal exchange of gifts, the Chinese emperor Ming-Ti gave silks, gold, pearls, and 100 bronze mirrors to the Japanese empress Miyako. Eventually, however, the Japanese craftsmen formed an honored guild and learned to make their own, including magic mirrors, retaining the Chinese circular pattern with a pierced knob in the back. They created their own style of decoration, however, sometimes attaching little bells to the outside rim. Some of the mirrors, up to 3 feet in diameter, were much larger than their Chinese cousins.

Many Japanese mirrors were dedicated to the Shinto sun goddess, Amaterasu-Omikami, the Heaven-Shining-Great-Deity. Amaterasu was once so outraged by the behavior of the ruler of the Nether World that she retired to the Rock Cave of Heaven, casting the earth into darkness. The other gods tried to lure her out by dancing, lighting fires, and reciting liturgies, but nothing worked. Finally, they made a mirror of metal taken from the sacred mountain and told her that it was "spotless and indescribably beautiful, as though it were thine own august person. Pray open the cave door and behold it." When she did and paused to admire herself, the gods grabbed her and led her out of the cave.

Later, when Amaterasu sent her grandson down to Japan, the Cradle of the Sun, she gave him the mirror, telling him to "reverence it as if reverencing us and rule the country with a pure luster such as radiates from its surface." And so such a mirror was kept in the sanctuary in the Imperial Palace in Edo (now Tokyo) until it was removed to a shrine at Ise, where it remains to this day.

By the seventeenth century Japanese mirrors had become more secular. Popular prints depicted prostitutes and actors using metal mirrors to apply makeup before they performed, in one way or the other. Mothers gave marriage mirrors to their daughters, usually depicting a pair of faithful cranes, a tortoise, pine, and bamboo, all symbols of long life and happiness. These mirrors became valued heirlooms that held ancestral spirits with whom the owner could supposedly commune. Men and women wore small mirrors on their sleeves in order to check their looks, and some mirrors had handles. As in many other cultures, the Japanese mirror was often associated with women. An old saying went: "The mirror is the soul of a woman just as the sword is the soul of the samurai."

Peruvian Sun-Catchers

In the Americas, mirrors were created by the descendants of those who crossed the land bridge during the last ice ages or, as some have argued, floated across the sea. While nomadic tribes spread throughout the hemisphere, two areas produced a series of high civilizations: the Mexican-Guatemalan complex, known as Mesoamerica, and the Peruvian Andes. Both American cultures produced magical mirrors that were central to their belief systems. "It is plain," writes the anthropologist Nicholas Saunders, "that for many Central and South American Amerindians the reflected image stands for the soul or essence of the person who looks into the mirror." In addition, Saunders believes that such mirrors allowed shamans to connect with "parallel spirit worlds."*

Oddly, even though both peoples became master metallurgists, Amerindian mirrors were primarily made of stone for most of the pre-Columbian era. Many were of pyrite (known commonly as fool's gold, a metallic sulfide with a flat crystalline surface), others of hematite or magnetite (forms of iron ore), anthracite (a form of hard coal), mica (an aluminum silicate that forms in thin reflective sheets), obsidian, or

---

*Although most ancient mirrors in the Western Hemisphere were found in Central or South America, there were also slate and iron-ore mirrors in use by North American tribes.

slate.* Because such rocks are more fragile than metal, only a few have survived.

The earliest American mirrors archaeologists have discovered are "jet mirrors" of anthracite dating from around 1500 B.C.E., found in the Peruvian highlands and coast. We have only fragments, each with one side polished, but it appears that they were both square and circular, ranging up to 5 inches.

The rugged Peruvian landscape yielded a variety of mysterious civilizations, ultimately dominated by the Inca by the time the Spanish arrived in the early sixteenth century. Because none produced written records, we can only guess at their way of life from the archaeological record. The Chavin civilization, whose artifacts often feature a ferocious jaguar, thrived from 850 B.C.E. to 300 B.C.E. before inexplicably disappearing. They produced numerous anthracite jet mirrors, most now in pieces. We do have one highly polished Chavin-era mirror of hematite, however, nearly 3 inches round, with two hand-drilled holes, presumably for a cord to hang around the neck.

The Moche people maintained an empire in the northern Peruvian coastal desert between 200 B.C.E. and 800 C.E., building cities and a huge brick pyramid to the sun called Huaca del Sol in the Moche Valley. Several Moche pyrite mirrors have survived, set into wooden frames with handles. The carved back of one round mirror shows a round-eyed face surmounted by a hat with a cat's head in the front. A square mirror, probably made around 1 C.E., has a copper frame and handle with twenty separate birds perched around the mirror's edges. The reflective surface is a mosaic of straight-edged pyrite pieces carefully fitted together like a puzzle.

Although the Moche mirrors may have been reserved for ritual or religious purposes, it is quite likely that they also saw daily use, even though the mosaic reflections would have been imperfect. From the paintings on the prodigious amount of Moche pottery that has survived, we know that the women bobbed their hair, plucked their eyebrows, rouged their cheeks, painted their lips, and applied mascara, so they presumably ap-

---

*The Mesoamericans may have used liquid mercury mirrors as well, as gourds containing mercury have been found in ancient grave sites in Mexico, Guatemala, and Honduras.

preciated mirrors. The men wore face paint along with elaborate costumes and headdresses. The Moche also created erotic pottery, depicting every conceivable form of sexual intercourse, including threesomes. There is no evidence, however, that they, like their Roman contemporary Hostius Quadra, admired their sexual prowess in their mirrors.

The sparse archaeological record in Peru suggests that each civilization probably made similar stone mirrors. The Huari people in the southern highlands, for instance, produced a lovely round pyrite mirror framed with blue polished stone with a trapezoidal handle that can stand on its own. Two cats look at the observer from the top of the frame. The Chimu, whose kingdom reached its peak in the fifteenth century before falling to the Inca, also made ornate wooden frames for handheld mirrors. One features a husband and wife holding hands, standing atop the round mirror—just as they presumably appeared within it.

By the time of the sun-worshipping Inca, who thrived from 1200 C.E. until 1532 C.E., many mirrors were still made of pyrite, but they were also sometimes made of copper, bronze, silver, or *tumbaga*, an alloy of gold and copper. In 1526, when Francisco Pizarro sent his pilot, Bartolomé Ruiz, from Colombia to explore the southern coast, Ruiz seized a flat-bottomed Inca log boat loaded with trade items, including silver-framed stone mirrors.

The Inca created vast stone public works and food storehouses, practiced sporadic human sacrifice, and instituted a hierarchical system in which hair length denoted social status—the shorter, the better. Also, the bigger the ear lobe and its ornament, the higher the status.

Garcilaso de la Vega (1539–1616), a mestizo (mixed-blood) whose mother was an Inca princess, wrote a fascinating two-volume history of Peru in which he explained that "they worked with instruments of copper and brass mixed together" while creating gorgeous golden objects in imitation of plants, animals, and gods. Yet the Inca never invented scissors, and they used thorns as combs. "The mirrors used by the women of the blood royal were of highly polished silver," de la Vega observed, "the ordinary ones of brass."

He asserted that "the men never looked in a mirror—they held it as shameful and effeminate." Given their evident vanity and resplendent dress, however, Inca men surely sneaked glances in mirrors—indeed, on their left wrists they wore bracelets called *chipanas* with small concave mirrors attached for starting fires. De la Vega quoted a young Inca boy as saying, "If the Spaniards, your fathers, had done no more than bring us scissors, mirrors [presumably European glass backed with tin], and combs, we would have given them all the gold and silver we had in our land."

During the great feast of the sun, a priest started the sacrificial fire with a large polished mirror, "a highly burnished concave bowl like a half orange," de la Vega wrote. "It was placed against the sun and at a certain point where the rays reflected from the bowl came together, they placed a piece of well-carded cotton, [which] was quickly fired." Such a bronze concave mirror was found at Machu Picchu, the sacred mountain-shrouded Incan city.

Aztec Smoking Mirrors

Far to the north, the Olmec civilization—the forerunner of the Maya, Zapotec, Mixtec, Toltec, and Aztec—thrived in the coastal plains near the Gulf of Mexico from 1800 B.C.E. to 200 B.C.E. All of these Mesoamerican civilizations shared certain characteristics, including human sacrifice, cultic ball games, stepped pyramids, hieroglyphic writing, and stone mirrors.

In Olmec myth, kings descended from offspring of a jaguar-human mating. One of their sculptures shows a jaguar being with a mirror chest pendant, copulating with a human woman.* The Olmecs made pyrite and obsidian mirrors in San Lorenzo, their first capital, located in southern Veracruz. These were used (among more mundane functions) as ritual portals to another world seen in visions, probably aided by eating hallucinogens made from a local toad's flesh.

"The Olmecs lived in a world of spirits and invisible masters," wrote Muriel Porter Weaver in her classic text *The Aztecs, Maya, and Their Predecessors*. "When hunting, fishing, and planting, the Olmec de-

---

*The jaguar, a dominant predator, was chosen as an object of worship and identification with shamans, but other mysterious creatures of the night were also associated with mirrors. In an ancient Panamanian burial site, for instance, there was a gold mirror frame in the form of a bat god that originally had a pyrite mirror set in its chest.

stroyed something of nature, and this disruption had to be compensated through ritual and offerings." Those offerings frequently included human sacrifices to the jaguar fire-god who created the sun. The shaman-rulers probably identified themselves with the predatory jaguar with its mirrorlike eyes. Behind a cat's retina is a light-reflecting layer of cells made of zinc and protein called the *tapetum lucidum*, which acts as a mirror, reflecting light back to the retinal cells. That is why they can see so well at night and why their eyes reflect an unearthly glow. "As jaguars see with their naturally mirrored eyes," writes Nicholas Saunders, "so shamans see with the aid of mirrors."

The Olmecs' cultural base shifted to La Venta, a small, swampy island near the Gulf of Mexico, from 900 B.C.E. to 400 B.C.E. There they created concave mirrors of magnetite, ilemite, and hematite, up to 10 centimeters in diameter, pierced with two holes for hanging around the neck. "No verbal description can convey the remarkable technical and artistic quality of the La Venta mirrors," wrote the archaeologist Jonas Gullberg, who first examined them upon their discovery in 1955. The mirrors are not simply concave but nearly paraboloid, focusing sunlight efficiently. Though nearly three millennia old, some are still so well polished that they can start fires, and they also reflect enlarged faces like a modern makeup mirror. "They have a gracefulness, dignity, and perfection that makes it hard to think of them as incidental or even only ornamental," Gullberg observed.

Little female clay figurines found at La Venta have tiny pieces of polished hematite hanging around their necks. "The impressiveness of [one such] figure lies in its realism," wrote the archaeologist Philip Drucker, who worked with Gullberg, "of which the delicately captured smile is but one feature. One is struck by the thought that it must be a portrait, carved by a master craftsman." These mirrors were probably worn by shamans, priests, or the noble elite, and they may have been used, as the archaeologist Gordon Ekholm put it, "to reflect the rays of the sun in a darkened room for divinatory purposes." They may have also been used for self-examination.

But the most likely use was for the Promethean task of lighting sacred fires. "I repeatedly made fire in pieces of dry, rotted wood at a time of between twenty and thirty seconds," Ekholm wrote about one such ancient mirror. "Thus the making of fire could have been the primary use of these mirrors, and what could have been more magical or wonderful than to take fire from the sun!"

The fire sun-god of the Olmecs became God K, or the Flare God, of the Maya, with a forehead mirror punctured by a smoking torch or cigar. The Highland Maya interred their royal dead along with their retainers—men, women, and children all sacrificed for mass burial—and lavish offerings of pottery, jade, obsidian, and pyrite mirrors. Many such mirrors, sometimes more than a foot in diameter, were made of mosaic pieces, carefully crafted to create one reflective surface, which was glued onto a slate or elaborately carved wooden back.

In addition to those carried on the chest or headdress, many Mayan mirrors were worn at the small of the back, with the reflective surface facing out. No one is sure why, but it is likely that the Mayans, like the Chinese, thought the mirrors could ward off unseen evil spirits or protect their backs during warfare.

The anthropologist Karl Taube points out that Mayan mirrors were associated with numerous symbols, such as the sun, human eyes or faces, flowers, butterflies, fire, pools of water, spider webs, shields, caves, or passageways—all emblematic of communication with the supernatural world where the priests sought answers to the most vital questions. "These ancient mirrors expressed a rich body of esoteric lore," Taube writes, "much of it still present among . . . contemporary peoples of Mesoamerica," particularly the Huichol tribe. In addition, circular mirrors sometimes appear on the abdomens of figures as the earth's navel, symbolic of life and its mysterious connections.

Their vase paintings indicate how important mirrors were to the Mayans. Rulers are frequently shown sitting cross-legged and staring into mirrors, often held by assistants or dwarves. Sometimes they are clearly examining their appearance, as in one picture where a mirror is used so that a ruler can see how the painted decoration on his back is progressing. In others, men dance with mirrors or gaze intently into them, sometimes while smoking a drug through a tube.

The Toltecs and Aztecs worshipped the god Tezcatlipoca, whose name means "Smoking Mirror" and who had a mirror instead of a right foot. This dark god thrived in the village culture before the arrival of the Aztecs, according to Cottie Burland and Werner Forman, authors of the chilling history *Feathered Serpent, Smoking Mirror*. "In the twelfth and thirteenth centuries, the whole of Mexico was made up of small groupings of tribal societies. . . . The temple priests continued to feed the gods with human sacrifice. Most of the hearts torn from the breasts of the victims were offered to the great Smoking Mirror."

The Aztecs, whose brief ascendancy began in 1325, brought such sacrificial worship to its apex. Their priests—painted from head to toe with a black ointment—fell into possibly drug-induced trances while gazing into black obsidian mirrors, seeing pictures that revealed the future and the will of the gods. That future was uncertain, because it predicted the return of Quetzalcoatl, the bird-lord of healing and magical herbs, who would one day come back to overthrow Tezcatlipoca and the Aztec ruler.*

When Hernando Cortez arrived in 1519, he was taken to be the returning god, partly because he wore flashing, reflecting spectacles that made magic mirrors of his eyes, just as the god Tezcatlipoca's skull-like mask featured convex pyrite eyes. One of Cortez's men later described a huge figure of Tezcatlipoca as having a "countenance like a bear, and great shining eyes of the polished substance [obsidian] whereof their mirrors were made."

Among the treasures Cortez sent back to the king of Spain that year were "a mirror placed in a piece of blue and red stone mosaic-work, with a feather stuck to it . . . , a mirror with two faces; a mirror with a figure of guastica; . . . a round mirror like the sun; a mirror with the head of a lion; a mirror with the figure of an owl."

Even after the Spanish conquest, mirrors remained important in Mexico, with many obsidian and glass mirrors incorporated into Catholic Church decoration there, as in Peru, where visitors to Cuzco can enter Santa Clara's church (construction began in 1558) and see tiny mosaic mirrors reflecting light everywhere. Neither did their magical properties entirely disappear from Latin America. A folklorist visiting Mexico in 1883 noted that special mirrors had "the power of reflecting the past and future. . . . There is scarcely a village in Yucatan without one of these wondrous stones."

---

*According to myth, Quetzalcoatl originally fled because Tezcatlipoca held a mirror up to him. Horrified that he had a human face and therefore a human destiny, Quetzalcoatl got drunk, fornicated with his sister, then fled in shame.

Chapter 2

MAGIC VISIONS

*Therefore have I brought it to the window of thy senses, and
doors of thy imagination.*

THE ANGEL URIEL SPEAKING TO JOHN DEE
FROM THE SHEWSTONE, APRIL 6, 1583

*Do not infest your mind with beating on
The strangeness of this business.*

PROSPERO, IN *THE TEMPEST*,
BY WILLIAM SHAKESPEARE, CA. 1611

BY THE TIME CORTEZ, wearing his shiny spectacles, was mistaken for a returning god by the Aztecs, mirrors were reflecting human faces around the world. Even more remarkably, they were universally connected with religious practices and attempts to delve into the mysteries of life, including magical divination in dark reflective surfaces.

One of the Aztec mirrors sent back to Europe ended up in the possession of Dr. John Dee, a renowned scholar, mathematician, philosopher, and adviser to Queen Elizabeth I. His "Shew-stones," as he called his polished obsidian mirror and magic crystal balls, were to lead him into a search for the ultimate truths of the universe. His story, told at the end of this chapter, provides a fitting culmination of several millennia of human interaction with mirrors and marks a crucial moment in history, when magic and science, which had existed in uneasy alliance within the mirror frame, split from one another. In this chapter, we will explore religion and the occult; in the next, science.

Demon-Haunted Worlds, Sacred Metaphors

Beginning in terror and evolving toward wisdom in an effort to explain the world and give it meaning, shamans and theologians looked to mirrors. Every human culture has valued supposedly powerful objects—especially mirrors, crystals, and other reflective talismans—to control evil and preserve the soul. The Chinese believed that demons avoided mirrors because they would be made visible in them. The Aztecs also used mirrorlike surfaces to ward off evil spirits, placing a bowl of water with a knife in it at the entrance to their homes, so that a spirit looking into the water would see its soul pierced by the knife and flee.

But mirrors could also be frightening because of their power to capture an image. People feared that their souls might stray into the reflective surface and never come out. Thus, some cultures believed that covering a mirror in the house where someone had died prevented the soul of the living from being carried off by the ghost of the deceased.

According to James Frazer in *The Golden Bough*, the fear of souls being captured in mirrors was widespread: "It was a maxim both in ancient India and ancient Greece not to look at one's reflection in water, and . . . the Greeks regarded it as an omen of death if a man dreamed of seeing himself so reflected. They feared that the water-spirits would drag the person's reflection or soul under water, leaving him soulless to perish." Similar beliefs explained why a sick person should avoid looking into a mirror for fear that the loosely held soul would take flight into it. Infant souls were particularly susceptible to harm, so folklore warned parents to keep children under a year of age away from mirrors.

Because mirrors were so powerful that they could capture souls, breaking them was thought to bring bad luck in numerous cultures. In China, for instance, it meant that the owner would lose his best friend. The Romans believed that breaking a mirror would cause seven years' bad luck (they thought a person's health changed in seven-year cycles).

"Among some peoples," wrote Wallis Budge in *Amulets and Talismans*, "the belief is common that 'the little man of the eye,' *i.e.* the figure seen in the pupil of the eye, can leave a man and enter another person and do harm. . . . Of all the things which have driven man in all ages to invent and to use magic, the most potent is the 'Evil Eye.'" According to Plutarch, who wrote around 100 C.E., it was possible for people to injure themselves by staring into a mirror. For similar reasons, the only way to destroy the mythical basilisk—whose gaze killed—was to hold a mirror up to it. A witch's eye, on the other hand, reputedly did not reflect a beholder's image.

Such fears and superstitions may have been the original religious impetus, but man also has an innate sense of the sacred, a feeling that life's meaning consists of more than brutish survival. The great religions and their founders sought to lift humans out of their myopic rut into a perception of the universal and divine, frequently using mirrors as metaphors. In a Hindu parable, for instance, two seekers hear, "He who has known the Self and understood It obtains all the worlds and all desires," so they look for the Self in a pan of water. The reflection they find there is the mutable self, though, not the universal Self they seek.

Some Buddhist mirror parables are humorous. A prostitute demanded money from a young man who told her that he had "diverted, enjoyed, and amused" himself with her in a dream the previous night. The wise Buddha figure ruled: "The fee should be paid by the merchant's son . . . in just the same fashion as he consorted with her." He had the young man place the money in front of a mirror and told the woman to take her payment from the mirror. This story was a popular way to teach the Buddhist notion of illusory reality.

Chinese philosopher Hua-yen taught that each element of the universe, from a grain of sand to the sun, contained within itself every other element. The concept puzzled most people until teacher Fa-tsing set up a demonstration around 700 A.D. He placed a brilliantly illuminated statue of Buddha in the center of ten bronze mirrors, arranged so that viewers saw an infinity of reflected Buddhas in each mirror, receding until they were too tiny to perceive.

Taoists sought to adjust to the natural as well as the supernatural world. "The still mind of the sage," the *Tao Te Ching* states, "is the mirror of heaven and earth." For one of its sects, Feng Shui, a life force called *chi* had to be directed properly for good health and happiness. Mirrors played a vital part in reflecting *chi* in the right path and deflecting harmful energy.

Because light was sacred to Zoroastrians, mirrors appeared in their art, architecture, and writings to represent self-reflection and divine knowledge. During Noruz, the celebration of the Iranian new year at the spring equinox, mirrors still play a major role.

The Jewish mystic Solomon Ibn Gabirol spoke of "the souls, close packed, / Peering in mirrors, [who] hope these may reflect / God's image glimpsed," and Christian mystic Meister Eckhart wrote: "The soul contemplates itself in the mirror of Divinity. God Himself is the mirror, which He conceals from whom He will, and uncovers to whom He will. . . . The more the soul is able to transcend all words, the more it approaches the mirror." Muhyi 'd-Din ibn 'Arabi, a Mohammedan, wrote of man's resemblance to God, and vice-versa: "God is the mirror in which thou seest thyself, and thou art His mirror in which He contemplates His names."

As translated in the King James Bible, Saint Paul wrote: "For now we see through a glass, darkly; but then face to face: now I know in part; but then shall I know even as also I am known" (I Corinthians 13:11–12). But the Elizabethan "glass" is misleading. What Paul actually meant was: "Now we see indistinctly, as in a mirror; then we shall see face to face." Referring to poor-quality mirrors of metal, he meant that our view of the world (and of ourselves) is flawed in comparison with the overwhelming knowledge and love of God.

The Mayan creation myth in the *Popol Vuh* eerily echoes St. Paul's image of a poorly reflecting mirror in which humans see reality only dimly. The four earliest humans were all-wise and all-seeing, which alarmed the gods. "What should we do with them now?" one asked. "Their vision should at least reach nearby, they should at least see a small part of the face of the earth." Another suggested, "We'll take them apart, just a little, that's what they need." So that's what the gods did to the four ancestors: "They were blinded as the face of a mirror is breathed upon. Their vision flickered. Now it was only when they looked nearby that things were clear."

Striving to See More

The Mayan tale of the fall of man echoes the Garden of Eden story in Genesis, as well as myths from many other cultures. Humans suffer an apparently universal feeling of loss and longing, a sense that once, in a distant past, we were wiser, more peaceful, longer-lived. We were more like gods. But we somehow lost our way. To become *seers* again—the word literally means "those who see"—visionaries resorted to magic mirrors. In Europe, this ancient practice was called scrying, from *descry,* "to see something difficult to make out." Scryers peered into dark mirrors to see what ordinary mortals could not.

Actually, scryers stared into reflective surfaces of all kinds—bowls of water, ink, oil, mirrors, crystals, swords, fingernails, bones, even fresh animal livers. *Catoptromancy* is the official term for divination in mirrors. By staring fixedly in the mirror or other bright object, the mediums put themselves into a kind of trance in which they could see the past, present, and future. Through these visions—which frequently included an auditory component as well—they tried to bridge the gap between their limited knowledge and the wisdom available to their ancestors.

Not everyone had the innate ability to scry. Children around the age of seven or eight were likely subjects, as were pure-of-heart virgins. So were imaginative men so highly strung that they trod a tortuous course between sanity and madness. The scryer frequently had to pray, fast, and abstain from sexual intercourse before attempting to look into the magic mirror. Elaborate incantations, drawings of magic circles and hexagrams, candlelight, and special treatment of the mirror helped to create a heightened atmosphere of tense expectation.

The historical and geographical extent of scrying is astonishing. The ancient Egyptians, the Sumerians, the Hebrews, and the ancient Chinese practiced scrying. Prepubescent Vedic Indian girls could see the future in a mirror or spoonful of water. The Persian Magi—from whom the word *magic* derives—used magic mirrors. The tenth-century Persian poet Firdausi described a scrying session in "the cup that mirroreth the world":

He took up the cup, and gazed.
He saw the seven climes reflected there,
And every act and presage of high heaven. . . .
In that cup the wizard-king
Was wont to see futurity.

Greeks and Romans peered into magic mirrors, crystals, and waters to gain supernatural knowledge. Roman scryers were called *specularii,* after *speculum*, the Latin Word for "mirror." Both stem from *specere*, "to look," as (appropriately) does *speculation*. So, too, the Aztecs and Incas sought enlightenment from their stone mirrors, and archaeological evidence suggests that their predecessors had passed the practice on to them. Virtually every culture has practiced the art of scrying: Mongolians, Siberians, Japanese, Tahitians, Gypsies, Australian aborigines, Zulus, Congolese, Ethiopians, Papuans.

Most early Christians, too, believed in scrying, although they were conflicted about the practice, since it smacked of paganism. They also feared that demons would appear instead of angels. Saint Hippolytus, inveighing against heresies around 200 C.E., was unusual in warning primarily of fraud rather than demons. He explained that some scryers employed a large magic cauldron with a crystal bottom; hidden underneath were actors "invested with the figures of such gods and demons as the magician wishes to exhibit." Reflective fish scales were glued on the blue ceiling to imitate stars.

After the conversion of the Roman emperor Constantine in 312 C.E., the Christians gradually turned from the persecuted into persecutors. At a synod held by Saint Patrick in the early fourth century, a canon ruled that any Christian who believed that a lamia (a monster with the body of a woman) could be seen in a mirror would be excommunicated.

Jesus himself had discouraged belief in miracles, emphasizing that it was people's faith that had made them well (though he combined earth and spit to make a mud to cure blindness, and he cast out demons). The organized Christian church went much further, trying to substitute its own rituals and prayers for folk superstitions, intent on suppressing and supplanting every other belief system.

At the same time, however, early Christians allowed properly sanctioned miracles. They believed, for instance, that when the pure of heart looked into a particular well at Bethlehem, they would see a magical star. In his prayer for victory over paganism, Saint Patrick's invocation sounded similar to a scryer's chant:

I bind to myself today
The power of Heaven,

The light of the sun,
The brightness of the moon.

Scrying persisted in occult religious traditions such as neo-Platonism, Gnosticism, Hermeticism, Kabbalism, and alchemy, regardless of Christian attempts to stamp them out, and they thrived for centuries as an underground movement, profoundly influencing those who sought to plumb the mysteries of the universe, including many Christians. All of these mystical traditions stressed a transcendent One who could only be reached through ritual magic, a complex hierarchy of angels and demons, and a virtuous life of contemplation.

In *The City of God*, written in the early fifth century, Saint Augustine swung the weight of the Christian church firmly against scrying. He labeled all magic as "entangled in the deceptive rites of demons who masquerade under the names of angels." Nonetheless, the practice of scrying continued, even within the church. Christians did, after all, believe in angels and demons, and mirror-gazers claimed to communicate with them.

In the twelfth century, the Christian scholar John of Salisbury recalled how a priest ordered him and another child to look into polished basins, then to stare at their fingernails smeared with holy oil and to report any ghostly shapes they saw. John saw nothing and so was spared the ordeal of being a child scryer. "The *specularii* [scryers] flatter themselves," John wrote, "that they immolate no victims, harm no one, often do good as when they detect thefts, purge the world of sorceries, and seek only useful or necessary truth." Yet John knew child scryers who had gone blind from the enforced staring into mirrored surfaces.

In the thirteenth century, European scholars began to translate works from Arabic that were to spawn the Renaissance. Along with many important scientific works, they translated the works of Picatrix, who regarded magic as a superior branch of science. He provided an important alternative perspective to the gloomy Christians who denounced magic as demonic. According to Picatrix, the scryer should be chaste, or at least refrain from intercourse prior to a magic session, when he should put himself in an "expectant and receptive" mood. His views influenced European scryers, giving them pride in their exalted but furtive work.

Forty Devils with Their Imps

Regardless of church dogma, scrying remained firmly imbedded in folk practice and belief. As collected by the Grimm brothers, the story of Snow White, for instance, hinges on the evil queen's scrying talents. "Mirror, mirror, on the wall," she begins her invocation, and the infuriatingly honest magic mirror reveals that her despised stepdaughter, living with seven dwarves, is "fairest of them all," even though the queen thought she had eaten the young woman's lungs and liver. This story probably originated in the Middle Ages, but by that time European folktales contained a wild mixture of Hindu, Arabic, and Hebrew material as well, so it is impossible to know the exact provenance of this particular mirror tale.

Similarly, the story of Reynard the Fox, related in various versions and languages from the twelfth century on, involved a mirror "of such virtue that men might see therein all that was done within a mile." The *Gesta Romanorum,* a popular medieval collection of folktales, told the story of a knight who was in Rome on his way to the Holy Land, when a good magician accosted him: "This day you are a son of death unless you have my help, for your wife has made arrangements to kill you!" Only scrying in a magic mirror saved him and killed the evil wizard who was having an affair with his wife.

In Russia, mirrors helped peasant girls determine whom they would marry. The most common method was to go to a bathhouse or deserted hut on a dark night with a torch and mirror. Placing the mirror opposite the open door, a girl might see the image of the spouse appear in the mirror at midnight. Sometimes a group of village girls would form a circle around the mirror and the chosen girl, who would chant, "Come forty devils with your imps from under the tree stumps and roots." Sometimes the future spouse would appear in the mirror, but often the Devil would come instead.

In England, belief in the supernatural was widespread. One Bishop Baldock complained in 1311 that "some pretend to invoke spirits in [finger]nails and mirrors, in stones and rings, and pretend that these spirits give signs and responses." In *The Canterbury Tales,* Geoffrey Chaucer's Parson objected to the "horrible swearing of adjuration and conjuration, as done by these false enchanters or necromancers in basins full of water, or in a bright sword."

Inquisitorial Fires and a Skeptic

The Renaissance (ca. 1300–1600) was a period of intellectual, religious, scientific, and occult ferment and change. As the Roman Catholic Church struggled to maintain control over an ever-changing world—suffering its own Great Schism, followed by the Protestant Reformation—the popes sanctioned the Inquisition to root out heretics and witches. As scrying became more popular as a way to penetrate the mysteries of the universe, it also became more dangerous.

At the same time that inquisitorial fires consumed heretics, however, magic continued to flourish. Around 1350, *The Stone of the Mountain*, a book attributed to Philip I, son of the French king, featured a virgin on a mountaintop in an Eden-like garden, surrounded by attentive philosophers. She held in her hand "the mirror of human life," presumably the stone of the book's title. Another legendary figure, the Christian priest-king Prester John, purportedly consulted a marvelous mirror—guarded by 12,000 soldiers, reachable only by climbing 125 steps—in which all plots against him were revealed.

The longtime popularity of scrying attracted a few skeptics. Nicolas Oresme (1323–1382), a French theologian, mathematician, and translator of Aristotle, and later Bishop of Lisieux, provided a more nuanced psychological critique of scrying. He attributed "marvelous power" to the human soul, which acted most strongly when people went into a trance state. This explained why boy scryers' vision was affected and their "spirits so disordered," even to the point of blindness. Oresme also noted the astonishing changes in the face of a magician during his conjurations and invocations: "He scarcely seems the same person, while his mind also appears to be alienated." Add fastings, special diets, and a solitary lifestyle, and it was no wonder, Oresme wrote, that scrying produced results—but the visions derived not from invoked demons but from "delusion, imagination, [and] an abnormal state of mind and body, terror, and illusion." That explained why only the scryer saw visions, whereas others attending saw nothing unusual. Oresme added that some magicians made "mathematical illusions" by using hidden mirrors. Despite such a rational approach, Oresme also believed that a black cloud obscured a mirror whenever a criminal looked into it.

Meanwhile, the Jewish magical tradition of Kabbalah and divination thrived. Although Deuteronomy clearly forbade wizardry and scrying, it was mentioned in the Talmud, and it was perfectly acceptable even for pious Jews to practice it on the Sabbath.

New Worlds

By the end of the fifteenth century, these trends—burning at the stake, magical practices, and embryonic scientific rationalism—were exacerbated by two events: Gutenberg's invention of the printing press, and Columbus's encounter with North America, both of which opened up new worlds of possibility for adventurous spirits. The rush to claim colonies and find new trade routes helped broaden provincial perspectives and encouraged scientific advances, including much mirror use. But the printing press had the more immediate effect.

In 1438, Johannes Gensfleisch Gutenberg started a mirror-making business in Strasbourg, selling small metal or glass mirrors to religious pilgrims who believed they could thereby capture the reflection of a saint's relics—a poor person's way to bring home holiness. In 1444, Gutenberg returned to Mainz, where he used his expertise in metal-working and the concept of mirror images to create the first printing press. By 1455, he had completed his monumental printing of the Bible.

Subsequent books poured off the new presses, permitting a vast expansion of knowledge as well as nonsense. In 1486, the *Malleus Maleficarum*, or "Hammer of Witches," written by priests Heinrich Kramer and James Sprenger, was published. Enormously influential, it explained how to identify and interrogate witches. As a result, the Inquisition and its fires heated up, though it was fortunate for wizards that Kramer and Sprenger focused primarily on witches; the two priests were obsessed with "all manner of filthy delights" enjoyed by women, whom they blamed for killing cattle, making men impotent, and causing miscarriages. Although the authors expressed mild disapproval of scrying, they admitted to employing magicians themselves when they felt they had been bewitched.

"The years 1500 to 1600 undoubtedly were the century of magic," Colin Wilson observes in *The Occult: A History*. The times nourished a flamboyant sort of freelance intellectual and rogue. In 1501, for instance, an Italian magician calling himself Mercury surfaced in Lyon, claiming to surpass all the occult sciences of the ancient Hebrews, Greeks, and Latins. He presented a magic mirror, made under favoring constellations, to the appreciative French king.

Born in Cologne in 1486, Henry Cornelius Agrippa, fluent in eight languages and an omnivorous reader, seemed destined for great things, but wherever he went—all over Europe—he clashed with priests, whom he regarded as ignorant and narrow-minded. Agrippa consulted a magic mirror in which "the dead seemed alive," as one witness asserted. Fantastic tales swirled around the magician. Agrippa supposedly showed the Earl of Surrey his mistress in a mirror, and another time he conjured up Tully out of the reflective depths to give an oration.

In 1510, Agrippa completed a three-volume treatise, *On Occult Philosophy*, a grab bag of lore written in a mystical, overheated style. In defending the magical use of mirrors, Agrippa focused on the power of the imagination, denying that his scrying had anything to do with sorcery or the Devil. "The fantasy, or imaginative power, has a ruling power over the passions of the soul," he wrote, "when these are bound to sensual apprehensions." Near the end his life, a disillusioned Agrippa wrote *On the Vanity of Sciences and Arts*, in which he attacked all of man's puny efforts to acquire knowledge, whether from magic or science. Nonetheless his occult writings, as well as his legendary magical feats, remained influential after his death in 1535 at the age of forty-nine.

Twelve years after Agrippa's passing, another wandering wizard-physician settled down to become one of the most famous scryers of all time. Born in 1503 of Jewish parents who converted to Christianity, Michel de Nostredame was known simply as Nostradamus. In 1547, Nostradamus commenced scrying, wearing a ceremonial robe, holding a magic wand, and staring into a bowl of water atop a brass tripod. Visions came to him, as did entire quatrains of prophetic poetry, which he published in 1555.

The French aristocracy loved the mysterious messages. Catherine de Medici summoned Nostradamus to her Parisian court because one of his poems might be interpreted to predict the death of her husband, King Henri II. There, the scryer cast horoscopes and interpreted moles on the exposed bodies of various noble clients. In 1559, when Henri II died a lingering death from a jousting wound, Nostradamus was acclaimed a true clairvoyant, and Catherine continued to consult him until his death in 1566. Although she generally strove for tolerance and compromise during the bloody civil wars between Catholics and Protestants, Catherine was known as the "queen-witch" because after Nostradamus died she looked into her own magic mirror.

John Dee: Renaissance Man

In England, another queen also struggled to maintain power amid scheming factions, assassination plots, and foreign intrigue. Like Catherine, Queen Elizabeth had her favorite occult adviser, complete with magic mirror. But John Dee was far more than a mere "conjurer," as public rumor labeled him early in his career. He was an expert in astronomy, mathematics, musical harmonics, optics, cartography, navigation, geography, cryptography, medicine, theology, law, literature, and history. A child prodigy, Dee entered Cambridge University in 1542 at the age of fifteen. During the next four years, he later recalled, "I was so vehemently bent to study, that for those years I did inviolably keep this order; only to sleep four hours every night."*

In 1548, having received his master's degree, Dee went to the University of Louvain, near Brussels, where he studied civil law and mathematics and gained a reputation as a budding genius. As his fame spread, "diverse noblemen (Spaniards, Italians, and others) came from the Emperor Charles V's court at Brussels to visit me at Louvain," Dee recalled, including Sir William Pickering, the English ambassador, whom Dee tutored in logic, rhetoric, math, and astronomy. In return, Pickering gave Dee a large concave mirror that produced extraordinary optical illusions.

In 1550, Emperor Charles V offered the twenty-three-year-old Dee a well-paid position as a "mathematical reader," but he declined, returning to England to seek his fortune with the Protestant regime. Dee took a position as tutor in the household of John Dudley, the Duke of Northumberland, the real power behind the throne held by twelve-year-old Edward VI.

---

*For ease in reading, Dee's spelling has been modernized.

When the sickly Edward died in July 1553, his Catholic half-sister, Mary Tudor, became queen. Two years later, John Dee was arrested on charges of "calculating," "conjuring," and "witchcraft." His real crime was political: He had befriended Elizabeth, Mary's half-sister, who was arrested a week after Dee. Elizabeth and Dee survived their imprisonments to remain friends (on a very unequal footing) for the rest of their lives.

In July 1558, Dee published his first major work, the *Propaedeumata Aphoristica* (Preliminary Aphoristic Teachings), intended as a scientific introduction to astrology and astronomy. He believed that the stars and planets influence events on earth through rays of visible light as well as invisible rays and that "they come together especially in our imaginal spirit as if in a mirror, show[ing] themselves to us, and enact[ing] wonders in us."

Four months after the book's publication, Queen Mary died, and Elizabeth Stuart asked Dee to consult the stars for the most propitious day for her coronation. He chose January 15, 1559.

Dee yearned for royal recognition and at least some monetary security, but the frugal Queen Elizabeth kept her philosopher perpetually dangling, though she dispensed lavish compliments, limited favors, and occasional cash. In 1566, Dee moved into his mother's sprawling house at Mortlake, set on the River Thames 8 miles southwest of London. There he created a lair Merlin would have envied, with outbuildings holding 40,000 books, scientific and magical apparatuses, and several laboratories bubbling with alchemical experiments. His home served as a magnet for intellectuals, nobility, and adventurers of all stripes.

In 1570, Dee wrote his *Mathematical Preface* to the first English translation of Euclid, part of the rediscovery of ancient Greek science. His long essay was an ode to math and its uses in subjects ranging from architecture, commerce, and music to astronomy, mechanics, and magic. Dee praised "the infinite desire of knowledge, and incredible power of man's Search and Capacity," urging bold and mathematical ventures. He discussed submarines, for instance, and asserted (before Galileo) that light and heavy objects fall at equal speed to the ground.

For Dee, mathematics was, next to theology, "most divine, most pure, most ample and general, most profound, most subtle," particularly in its highest usage, "lifting the heart above the heavens, by invisible lines, and immortal beams meeteth with the reflections, of the light incomprehensible: and so procureth joy, and perfection unspeakable." Like Pythagoras and Plato, John Dee believed that pure mathematical forms provided a perfect mirror in which to contemplate the ultimate reality, which existed above earthly matters.

The closest thing to this purity was the study of optics, which Dee called Perspective, the study of "all Radiations Direct, Broken, and Reflected," which concerned "all Creatures, all Actions, and passions, by Emanation of beams performed." By understanding how mirrors worked, Dee noted, we could understand "why so sundry ways our eye is deceived and abused" through mirror tricks and other optical illusions. Dee described the mirror he had received from Sir William Pickering. If you lunge at it with a sword or dagger, "you shall suddenly be moved to give back . . . by reason of an Image, appearing in the air," attacking you in return. "Strange, this is, to hear of: but more marvelous to behold."*

Dee ended his essay with a mysterious passage about the Archmaster, who through his "doctrine Experimental" can accomplish things "so unheard of, so marvelous, & of such Importance." What could this be? "The chief Science, of the Archmaster (in this world) as yet known, is another (as it were) OPTICAL Science: whereof, the name shall be told (God willing) when I shall have some (more just) occasion, thereof, to Discourse."

Dee was referring obliquely to the art of scrying. By 1570, he had almost certainly been using this "optical science" in order to supplement his other quests for knowledge. Dee believed that certain gifted scryers could see and converse with God's angels. He tried a string of scryers, eagerly writing accounts of the sessions, only to be disappointed and disillusioned with each in turn. Dee yearned, as he wrote in the *Mathematical Preface*, for "things Intellectual, Spiritual, eternal, and such as concern our Bliss everlasting: which, otherwise (without Special privilege of Illumination, or Revelation from heaven) No mortal man's wit (naturally) is able to reach unto." And so Dee continued his search for the right scryer, his key to the "special privilege of illumination."

---

*Queen Elizabeth once arrived at Mortlake with an entourage of nobility and "willed me to fetch my glass [mirror] so famous, and to show unto her some of the properties of it, which I did," Dee recalled, "to her Majesty's great contentment and delight."

A Model for Prospero

As a young man in the 1550s, Dee—an expert in mathematical navigation and astronomy—had advised navigator Richard Chancellor in his attempt to find the Northeast Passage to the Pacific. Instead, after discovering the land of endless night, Chancellor abandoned ship and traveled through "extreme and horrible" cold to Moscow, where he established trading relations with Ivan the Terrible. As a consequence, the Muscovy Company was founded back in London and was given a royal monopoly over northern exploration.

Dee's enthusiasm for exploring other new worlds remained high. In fact, the two parallel attempts to expand human knowledge—spiritually and geographically—were directly linked through the Gilbert brothers (Humphrey, Adrian, and John) and navigator John Davis. A curious 1567 document, now in the British Library, describes "Certain Strange Visions or Apparitions of Memorable Note" by "an experimental Magician." It recounts the efforts of one H.G. and his scryer, John Davis. Thus, it would appear that Humphrey Gilbert was attempting to find answers in magic mirrors, too.

It isn't surprising that Gilbert sought anonymity, however, because scrying was technically illegal. Anyone who appeared to possess extraordinary knowledge, such as John Dee, was suspect. Mathematical and scientific pursuits were considered evidence of devil-dealing. As a consequence, Dee was plagued by such rumors all of his adult life, and it was he who partially inspired Marlowe's Faust (1593) and Shakespeare's Prospero (1611).*

In spite of official laws against such practices, Queen Elizabeth and her ministers believed in the efficacy of scrying and witchcraft. Thus, in 1577, when a new comet threw the court into a panic, Queen Elizabeth summoned Dee, who spent three days calming her fears. In return, she

---

*Shakespeare's works are filled with mirrors, magic, and witches. In *Macbeth*, for instance, the three Weird Sisters fill their bubbling cauldron with grotesque items before cooling it to a mirrorlike surface with baboon blood. When Macbeth asked them, "What is't you do?" they answered, "A deed without a name," but it was obvious that they were scrying, producing apparitions. Their last vision included scrying within scrying, when a king appeared holding a mirror that showed the entire Stuart line.

promised to protect her philosopher against any who would "unduly seek [his] overthrow" because of his "rare studies and philosophical exercises."

During that intense time at court, Dee probably advised Sir Francis Drake on his forthcoming circumnavigation of the world. He also met and wooed Jane Fromonds, a young lady-in-waiting. On February 5, 1578, Dee married her, recording it in a diary he had begun the year before. He was fifty-one; she was twenty-three. On July 13, 1579, Jane Dee gave birth to their first child, Arthur (named after King Arthur).

In 1580, John Dee entered a scheme with Sir Humphrey Gilbert, his brother Adrian, and John Davis, who proposed to "discover and settle the northerly parts of Atlantis, called Novus Orbis"—in other words, to colonize North America. In return for Dee's advice and support, Sir Humphrey offered the philosopher ownership of most of Canada and all of Alaska. Unfortunately, Sir Humphrey drowned three years later when his ship sank returning from America, and the plans came to naught. Something else came out of Dee's interests in exploration, however.

A week after Dee signed the agreement, the queen rode to Mortlake and "willed me to resort to her Court" at Richmond, a few miles upriver. There, on October 3, he delivered his manuscript, entitled *Brytanici Imperii Limites*, giving Elizabeth a scholarly rationale for colonial conquest through the direct tracing of her legal right to newly discovered lands back to King Arthur. He urged the creation of a royal navy and (coining a phrase) envisaged a glorious "British Empire." Such matters were very much on Elizabeth's mind, since Sir Francis Drake had just returned from his four-year voyage around the world, "richly fraught with gold, silver, silk, pearls and precious stones" pirated from South America.

How Pitiful a Thing Is It, When the Wise Are Deluded

It is possible that among his lesser treasures, Drake brought back a highly polished black obsidian mirror, used for scrying by Aztec priests, and that either he or William Hawkins, one of his crew, subsequently gave it to John Dee. In his diary, Dee recorded a visit from Hawkins on June 17, 1581. Then, in two cryptic entries, Dee wrote on July 29, "The glass gone," and the next day, "Another glass given." It would appear that he traded mirrors with someone, and perhaps this is when he came into possession of the mysterious black mirror, along with the information that the Aztecs had used it to communicate with their gods. Or Dee could have gotten it during his time in Europe from a Spanish noble, eager to impress the young English genius by dispensing interesting trifles from the booty of Cortez. At any rate, the mirror now rests in the British Museum, where it keeps its secrets.*

By this time, John Dee had begun to consult scryers more assiduously than ever. On March 8, 1582, a new scryer arrived at Mortlake. The rather unbecoming young man introduced himself as Edward Talbot, which Dee later discovered was an alias for Edward Kelley. Kelley limped and wore a cap to cover one ear, which had been cropped for some previous transgression. But his appearance didn't matter. Dee quickly assessed him "a learned man." After supper, Kelley offered to "further my knowledge in magic . . . with fairies," Dee wrote in some horror in his diary. He wanted no commerce with magic, fairies, or demons. Angels were what he sought.

Kelley was a quick study. When he returned two days later, he explained to Dee that he was only trying to trap him, to see if he had "any dealing with wicked spirits." Dee stressed that he did not practice what was "vulgarly accounted magic" but "confessed myself long time to have been desirous to have help in my philosophical studies through the company and information of the blessed angels of God." With that, he pulled out "my Stone in the frame (which was given me of a friend)," and they commenced a scrying session, which Dee called "Actions." After a quarter of an hour of earnest prayer, "he had sight of one in the Stone" who identified himself as Uriel. In this first session, Uriel assured Dee through Kelley that he would live "an hundred and odd years" and that the archangels Michael and Raphael would also be visiting.

That afternoon, in a second session, Uriel warned that an evil spirit named Lundrumguffa was lurking in the house, seeking to destroy

---

*It is possible that this Aztec mirror, first identified as having belonged to Dee in a 1748 catalog of Horace Walpole's collections, was not really owned by the Elizabethan philosopher, for he never mentions it specifically. Dee's fascination with optics, odd mirrors, and scrying, however, plus the circumstantial evidence, lends credence to its identity.

Dee's daughter, Katherine, who was eight months old. To reinforce the validity of the warning, Uriel informed Dee that the demon had maimed his shoulder the previous night. (Dee had, of course, told Kelley that he had awakened with a sore shoulder.)

The next day, a figure appeared in the scrying stone in a purple robe, "all spanged with gold," but Uriel whipped him and stripped off his robe to reveal Lundrumguffa as "all hairy and ugly." Of course, Dee actually saw and heard nothing except Kelley's dramatic recitation, but he was utterly convinced. Uriel threw the demon into a great pit and then all was well. "My scryer saw an innumerable company of angels about him." Then Michael, who sat in a chair with a sword, spoke:

> Go forward: God hath blessed thee.
> I will be thy guide.
> Thou shalt attain unto thy seeking.
> The world begins with thy doings.
> Praise God.

It is little wonder that Dee was seduced. Here were the answers to his prayers, delivered through compelling scenes and in resounding biblical-sounding prophecy. Dee apologized for keeping the angels so long, adding: "But, for my part, I could find in my heart to continue whole days and nights in this manner of doing: even till my body should be ready to sink down for weariness."

Dee was hooked, and over the next year and a half Edward Kelley's hooks would only sink deeper, even though there were early warning signs that all was not right. Kelley proved to be a volatile, difficult houseguest. He declared that the archangel Michael had ordered him to marry, "which thing to do, I have no natural inclination." Kelley, twenty-seven, reluctantly married Joan Cooper, nineteen.

Jane Dee, who was Edward Kelley's age, became increasingly uncomfortable with the situation. Her husband was clearly falling under Kelley's spell, and the scryer himself was attracted more to his master's wife than to his own. One can hardly blame Jane for being disturbed at the scryer who had taken over her husband's life. It was also affecting their sex life, since Dee, in his efforts to please the angels, promised "to forbear to accompany with my own wife, carnally: otherwise than by heavenly leave and permission."

It didn't matter how outrageous Kelley's antics were; Dee's faith could not be shaken. At one point, Kelley observed a "tall well favored man" in the Shewstone who looked suspiciously like Dee himself. The man said, "How pitiful a thing is it, when the wise are deluded." Yet Dee still believed.

Behold, You Are Become Free

When Polish count Albert Laski appeared at Mortlake in the summer of 1583, a sprightly new angel named Madimi appeared, "like a pretty girl of seven or nine years of age." Madimi and other angels suggested that they follow Laski back to Bohemia, and on September 21, 1583, the entire Dee entourage—including three children now (Roland was nine months old)—and Count Laski sailed for the Low Countries. Dee took his holy scrying table, his mirror and crystals, and 700 books.

For the next three years, Dee and Kelley bounced between Prague, Krakow, and Trebon Castle, where the wealthy William of Rosenberg gave them shelter. During the Actions, angels directed Dee to confront Emperor Rudolf II in Prague with his sins. With the remarkable self-possession and courage of the righteously deluded, John Dee spent an hour alone with the Emperor on September 3, 1584.

> I began to declare that all my lifetime I had spent in learning: but for this 40 years continually, in sundry manners, and in divers countries, with great pain, care and cost, I had from degree to degree sought to come by the best knowledge that man might attain unto in the world. And I found (at length) that neither any man living, nor any book I could yet meet withal, was able to teach me those truths I desired and longed for.

And so he resorted to magic mirrors and crystals, which God's "holy angels, for these two years and a half, have used to inform me."

One might think that Rudolf would conclude that Dee was a complete crackpot at this point, but the emperor was intrigued. A melancholy, inscrutable man, Rudolf II was fascinated by all forms of knowledge; his court was famed not only for attracting the best scientists and artists in the world but also for Rudolf's interest in the occult and alchemy.

Then Dee delivered his message. "The Angel of the Lord hath appeared to me, and rebuketh you for your sins. If you will hear me, and believe me, you shall triumph. If you will not hear me, the Lord, the God that made heaven and earth (under whom you breathe and have your spirit) putteth his foot against your breast, and will throw you headlong down from your seat." Incredibly, Rudolf told Dee that he believed him and that "another time he would hear and understand more."

Later, the angels led Dee to upbraid Poland's King Stephen in similar fashion. Yet it wasn't his prophetic chiding that got Dee in trouble; instead, the threat came from the papal nuncio, who suspected Dee of heresy and wanted him to turn over his books of angelic Actions. Kelley staged a dramatic burning of all the Action diaries. The scryer, a sleight-of-hand artist, apparently pulled a classic switch, having prepared an identical bag that he kept under the table. Later, the books miraculously (to Dee) reappeared in the garden.

The trick didn't keep the Pope from pressuring Rudolf into banishing Dee and Kelley from his kingdom, but Rosenberg intervened and got permission for them to stay with him at Trebon Castle, where Kelley began to spend most of his time on alchemical experiments. He convinced Dee as well as many others that he really was producing gold, though his sleight of hand (or "juggling," as the Elizabethans put it) probably accounted for the miraculous results.

That year, 1587, was to bring a dramatic climax to the Actions. On Friday, April 17, Kelley took up scrying again, looking into the Stone, where he saw a swiftly turning globe on which was written: "All sins committed in me are forgiven. He who goes mad on my account, let him be wise. He who commits adultery because of me, let him be blessed for eternity and receive the heavenly prize."

The next day, Madimi appeared. No longer a sweet young girl, she had matured into a voluptuous young woman. "Madimi openeth all her apparel," Kelley reported, "and showeth herself all naked; and showeth her shame also." Madimi gave a speech on free love: "Behold, you are become free. Do that which most pleaseth you. For behold, your own reason riseth up against my wisdom."

Kelley reported that he saw four heads—his, Dee's, and their wives'—on a white pillar and that Madimi brought down a half-moon on which was written, "Nothing is unlawful which is lawful unto God." Madimi declared that there should be "unity amongst you," then disappeared. Dee interpreted the unity as "after the Christian and godly sense," but Kelley took it as sexual and "utterly abhorred to have any dealing with them farther."

Dee prevailed upon Kelley to ask whether Madimi meant "carnal use" or "spiritual love and charitable care and unity of minds." The answer: "*I speak of both.*" Dee couldn't believe it. "The one is expressly against the commandment of God: neither can I by any means consent to like of that doctrine. . . . Assist me, O Christ. Assist me, O Jesu. Assist me, O Holy Spirit."

Kelley then read what appeared in the Stone on a white crucifix: "*If I told a man to go and strangle his brother, and he did not do it, he would be the son of sin and death. For all things are possible and permitted to the godly. Nor are sexual organs more hateful to them than the faces of every mortal.*"

At 2 A.M. that morning, John Dee told his wife in bed, "Jane, I see that there is no other remedy, but as hath been said of our cross-matching, so it must needs be done." She wept and trembled in his arms for a quarter of an hour. "I pacified her as well as I could," he wrote, "and so, in the fear of God, and in believing of his admonishment, did persuade her."

On May 3, the four of them signed an agreement to carry out "this most new and strange doctrine" despite "all our human timorous doubting," and they promised to promote "amongst us four a perfect unity and Christian charity with incomparable true love and friendship, imparting and communicating each with other, of all and whatsoever we have or shall have hereafter during our lives."

On May 21, there was only one diary entry: "*pactum factum*" (pact fulfilled). During an Action the following day, a man on a white horse asked Kelley, "Was thy brother's wife obedient and humble to thee?" Kelley answered, "She was."

Nine months later, Jane Dee gave birth to a boy. No one discussed the child's possible paternity. They named him Theodore Trebonianus Dee, meaning "Gift of God at Trebon."

Final Journeys

Although the two couples attempted to fulfill their commitment to practice "incomparable true love and friendship," it didn't work. Dee could never admit that the angels were figments of Kelley's imagination or that his scryer had deliberately set out to seduce Jane. As Kelley's alchemical experiments appeared to thrive, Kelley's reputation soared as Dee's dwindled. On March 11, 1589, Dee and his family left for England, arriving on November 22.

Kelley remained in Bohemia, was made a baron by Rudolf II, and acquired a castle, nine villages, and two houses in Prague. Kelley died in 1597 of injuries sustained after he jumped out of the window of a castle, trying to escape prosecution for one of his schemes.

Meanwhile, John Dee struggled to survive. He returned to Mortlake to find his library looted and his laboratories trashed. Two months later, in February 1590, Jane gave birth to a girl. Incredibly, the Dees named her Madimi, after the lewd angel who had ordered the cross-matching. In addition to his six children, Dee had to support a houseful of servants. Although Queen Elizabeth received him kindly within two weeks of his return, she repeatedly failed to help him.

In 1605, Jane Dee and two of the children died of the plague. Only three of Dee's eight children now survived. Suffering from painful kidney stones, Dee moved back to Mortlake, where his daughter Katherine cared for him and he resumed Actions with his scryer before Kelley, Barthilmew Hickman. On July 17, 1607, the eighty-year-old Dee wrote in his diary that "Barthilmew and I talked of divers of my doings with Kelley," and he unlocked his chest to pull out his old scrying Stone.

The angel Raphael soon appeared to Hickman in the reflective surface. The angel pledged "to serve thee at all times, when thou art placed in thy journey." Dee was prepared to trust the spirits hovering in his magic mirror one more time. "Thou shalt . . . take a long journey in hand, and go where thou shalt have all these great mercies of God performed unto thee." Dee would finally be given "the secret knowledge and understanding of the philosopher's stone." Great things were in store for him overseas. "They have and do make a scorn of thee here in this thy native country," Raphael observed, but he noted that the same had been true of Jesus.

And so Dee prepared for his last journey, but the spiritual creature's message was apparently metaphorical. Dee died in England on February 26, 1609.

John Dee's Legacy

John Dee resorted to the occult only in a sincere effort to plumb the mysteries of the universe. He tried "natural magic," which we now call science, but it was not enough for him, so he was seduced by the supernatural. Dee's life story offers a tragedy worthy of Shakespeare, his contemporary. In another era that rewarded true scholarship, he might have been an honored optician, physicist, astronomer, or mathematician. Or he may have turned to mysticism without the need for a scryer.

In Dee's own time, the world seemed a ferment of change and discovery, with new wonders revealed almost daily. Many signs, including a nova that appeared in Cassiopeia in 1572, were interpreted by astrologers as an indication that the world would soon end. Thus, it was not surprising that an angel in the mirror told Dee, "New worlds shall spring of these. New manners: strange men: the true light, and thorny path, openly seen. All things in one."

Yet Dee's path remained thorny all of his days, and he never found "all things in one." He was one of the last intellectuals for whom occult and scientific mirrors reflected the same light of truth. Yet Dee's mixed legacy—and his mirrors—also helped lead the world into scientific advances that would revolutionize our views of the universe. Dee stood at a historic crossroad where magic and science were finally to split apart.

Chapter 3

FIELDS OF LIGHT

And God said, "Let there be light," and there was light. And God saw that the light was good.

GENESIS 1: 3−4

HAD JOHN DEE BEEN born a hundred years later, he might have resembled Isaac Newton, the brilliant scientist who ground the mirror to make the first reflecting telescope, who dissected light to create modern optics, and who propounded the laws of physics, including gravity, which explained the then-known forces that allowed the universe to function. Like Dee, Newton was fascinated with alchemy and the occult, but because he was born later, with the Scientific Revolution well under way, he changed the way we view the world instead of trying to talk to angels.

Newton famously pointed out that his work stood "on the shoulders of Giants." The scientific tradition—intimately tied to mirrors, astronomy, and the study of light—stretches back to ancient times, often in tandem with magic, and the men on whose shoulders Newton stood were a brilliant, quirky, independent lot who hatched their ideas while contemplating rainbows, looking at their own distorted faces in oddly shaped mirrors, or watching sunlight filter through the dust motes in their prison cells. We'll meet them in this chapter (including Dee in scientific garb) and the next, culminating with Newton.

The science of mirrors begins with the study of light. From the earliest times, humans worshipped the sun, our principal light source, but they also tried to understand what light was and how their eyes could see. They soon realized that light could bounce and bend—that is, it could be reflected or refracted. The study of optics was born, and its gradual development provided an important impetus for scientific advances leading up to Newton and beyond. For ancient and medieval scientists, the study of light and its behavior was the most fundamental of the sciences—and the one promising to illuminate the deepest secrets.

What is light? Even though it allows us to see, it is itself invisible, traversing space without a trace, unless it bumps into something like dust, which allows us to see that it travels in straight lines. It isn't readily apparent that it has a finite speed—maybe it simply instantaneously *is*—or that "it" is an *it* at all. In later chapters, we will consider more modern ideas about light. Here, suffice it to say that we still really don't understand it, though we know a great deal about how it behaves. We should remain humble, however, in reviewing theories of vision and optics that turned out to be wrong. The marvel is that humans have tried, with some success, to figure it all out.

Plato's Mirror Worlds

The first ancient opticians were priest-shamans in Mexico, China, Egypt, and elsewhere. The Olmecs and Chinese used highly polished concave mirrors (of obsidian and bronze respectively) to light sacrificial fires, and the Egyptians constructed the great temple of Karnak so that the sun shone down its long corridor only at sunset of the summer solstice. The Great Pyramid, originally covered with white marble, was a huge mirrored surface, and New Kingdom obelisks were capped with polished electrum, a gold-silver alloy on which sunlight flashed dramatically just before sunrise.

Although the priests were excellent optical technicians, they were not interested in scientific progress for its own sake. Rather, they used their secret knowledge to mystify and enthrall. The modern scientific enterprise begins with the individualistic, argumentative, ever-curious Greeks.* "Egypt and Phoenicia love money," Plato remarked. "The

---

*In his classic book *Science and Civilization in China*, Joseph Needham argued that passages in the *Mo Ching*, "truncated and fragmentary though they are," in-special characteristic of our part of the world is the love of knowledge."

Many early Greek scientists, including Pythagoras and Plato, traveled to Egypt to learn mathematics and optics. When Socrates was executed in 399 B.C.E., Plato, then around thirty years old, left Greece in disgust. After a decade of travel, he settled back in Athens and founded his Academy, where he lectured on science and politics.

According to Plato, our eyes, the first of the sense organs installed by the gods, contained "so much of fire as would not burn, but gave a gentle light." Plato believed that we see because our eyes send out visual rays, which combine with sunlight to produce vision. Eventually, this somehow streams back to the soul (brain and heart), causing us to see.

Even though we can still send forth this gentle visual fire at night, it falls upon an unlike element and so is extinguished. "And now there is no longer any difficulty in understanding the creation of images in mirrors and all smooth and bright surfaces," Plato confidently affirmed. "For from the communion of the internal and external fires, and again from the union of them and their numerous transformations when they meet in the mirror, all of these appearances of necessity arise." Plato thought the "fire" from the eye somehow fused on the bright mirror surface. He also tried to explain why right and left are reversed when we look into a mirror (a puzzle that has plagued philosophers and optical theorists ever since). "Right appears left and left right, because the visual rays come into contact with the rays emitted by the object in a manner contrary to the usual mode of meeting."

Plato's extramission theory, in which visual rays proceed *from* the eye, was disputed by Democritus (ca. 460 B.C.E.–ca. 370 B.C.E.) and his followers, who believed that all visible objects constantly shed a thin skin of tiny atoms, maintaining an ever-shrinking outline until they are reflected in the mirror of the eye.

---

dicate that the Mohists understood a good deal about light and mirrors in the fourth century B.C.E., the same time that the Greeks were looking into the same subject. Mohists knew that light rays were linear, and they studied plane, convex, and concave mirrors. But the Chinese study of physics never flourished.

Aristotle's Rainbow Visions

Plato apparently passed on an appreciation of optics to his student Aristotle, but the pupil disagreed with his master about the nature of light, vision, and reflection, among other things. Perhaps because of such conflicts, Aristotle left Athens upon Plato's death rather than taking over the Academy. In 342 B.C.E., King Philip of Macedonia hired Aristotle as the tutor of his thirteen-year-old son, Alexander. Seven years later, when Philip died, Aristotle returned to Athens, where he formed the Lyceum as a rival school to the Academy. There, he lectured while walking restlessly up and down, thus securing the nickname "the Peripatetics" for himself and his followers. Unlike Plato, Aristotle embraced scientific observation and produced a huge body of work explaining how everything worked. Although much of what he wrote was grounded in real-life observations, other "facts" he conveyed were folkloric hearsay. Nonetheless, his work dominated much of Western intellectual thought, either directly or indirectly, until the seventeenth century.

Aristotle mocked Plato's extramission theory of vision. "If vision were the result of light issuing from the eye as from a lantern, why should the eye not have had the power of seeing even in the dark?" he asked. Aristotle argued instead that light traveled to the eye and that light itself wasn't simply a *thing* but an *activity* of sorts—light interacted with air and water to actualize its potential transparency, just as it produced red or blue when it hit such colored objects. He insisted that light itself had no body. Unlike matter, it was not made up of Democritus's atoms but was simply an interaction in an invisible medium, just as sound was. Light came into the eye and thence to the brain and heart. "Soldiers wounded in battle by a sword slash on the temple," he observed, sometimes went blind, even though their eyes remained intact. In these injuries the eye was "cut off from its connection with the soul."

That is why Aristotle denied that eyesight resulted from "mere mirroring" in the pupil, as Democritus had asserted. Aristotle wrote condescendingly that "in his [Democritus's] time there was no scientific knowledge . . . of the formation of images and the phenomena of reflection. It is strange, too, that it never occurred to him to ask why, if his theory be true, the eye alone sees, while none of the other things in which images are reflected do so."

Aristotle hypothesized that colors arose from the interaction of light with "translucent" objects and that the color varied depending on the strength of the light or a person's visual acuity. Rainbows fascinated him, as they did succeeding generations of scientists. He wrote that rainbows always occurred with the observer between the sun and the rainbow and that there could be two parallel bows, perhaps even more. "In the inner [primary] rainbow the first and largest [outside] band is red," he noted, and the colors were reversed in the secondary outer bow. He explained rainbows as reflections from tiny mirrors—droplets of water in the sky. "In some [larger] mirrors the forms of things are reflected, in others [very small] only their colors," he asserted. Tiny droplet mirrors were too small to reflect the entire sun. "But since something must be reflected in them," only color appeared.

Aristotle observed that rainbows sometimes also appeared much nearer, in the spray from oars, for instance, or when "a man sprinkles fine drops in a room turned to the sun." In speaking of the rainbow and reflection, however, he appeared to contradict his theory that light coming *into* the eye causes vision. Here, he wrote that "sight is reflected from all smooth surfaces," as though a visual beam came out of the eye toward the mirror. Curiously, the philosopher said that air could act as a smooth reflective surface. "Air must be condensed if it is to act as a mirror, though it often gives a reflection even uncondensed when the sight is weak." As an example, he cited the case of a man with "faint and indistinct" eyesight who always saw a reflection of himself in the air as he walked. In this odd example, Aristotle clearly thought vision issued from the eye and that its quality determined the quality of reflection.

Another anecdote is even more startling, and it, too, contradicts Aristotle's assertion that light goes into the eye rather than flowing from it. "If a woman chances during her menstrual period to look into a highly polished mirror," Aristotle wrote, "the surface of it will grow cloudy with a blood-colored haze."

Perhaps because of this waffling, it was a version of Plato's extramission theory—that vision resulted from rays emitted by the eye—that was to dominate scientific thought for more than a thousand years.

Hellenistic Geometry and Burning Mirrors

Although Athens remained an intellectual haven after Aristotle's death in 322 B.C.E., the real action soon moved elsewhere. Alexander the Great turned to world conquest rather than a career in science, but at least Aristotle had instilled a proper reverence for learning. Alexander founded the city of Alexandria in Egypt, where his general, Ptolemy, who had also studied under Aristotle, built the new city's huge library and museum (a combination temple-school) and made Alexandria the center of learning of the Hellenistic world.

In Alexandria, around 300 B.C.E., Euclid synthesized everything then known about mathematics and geometry in his great work, *Elements of Geometry*. We know almost nothing of Euclid other than his reply to a potential student who asked him, "What shall I get by learning these things?" The irritated teacher reputedly told a slave, "Give him three-pence, since he must make gain out of what he learns." Euclid also wrote *The Optics*, in which he espoused a variant of Plato's theory that visual rays issued from the eye. Euclid wasn't really interested in the physiology of vision, however, but in its mathematics. He argued that rays travel in straight lines from the observer's eye, forming a cone with the vertex at the eye and the base on the visible object. He also assumed that the angles of incidence and reflection of a ray on a mirror are equal. It appears that by Euclid's time this fundamental law of physics—that light bounces off a flat mirror at the same angle at which it hits it—was well known from practical observation, though it was incorrectly envisioned as visual rays going the opposite direction (see Figure 3.1).

Mirror knowledge may have been put to use in a big way during Euclid's lifetime. Ptolemy initiated construction of the huge Lighthouse of Alexandria, one of the Seven Wonders of the Ancient World, on the island of Pharos. At night, a great fire atop the lighthouse provided a beacon to sailors. Legend has it that a large curved metal mirror projected the firelight by night and sunlight by day and that the mirror could also be used to magnify views of far-off Constantinople. This early telescopic use seems highly improbable, and sunlight reflected by a concave mirror would only have incinerated whatever lay at the focus. A large mirror may have been used at night to make a directed beacon of the firelight, however.

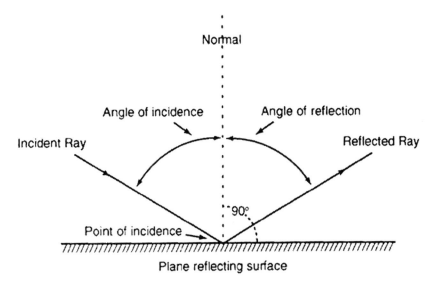

FIGURE 3.1 The law of reflection.

The greatest Greek scientist of all, Archimedes, was born in the city of Syracuse on the island of Sicily in 287 B.C.E. From his father, an astronomer, Archimedes learned a love of math and the heavens, but he also displayed a more practical bent. In his youth, he traveled to Egypt, where he invented an ingenious water-screw mechanism that revolutionized irrigation methods. Back in Syracuse, King Hieron continually pressed Archimedes to apply his intellect to practical affairs; the scientist preferred abstract theory.

The quintessential absent-minded scientist, Archimedes would become so absorbed in thought that he frequently forgot to eat as he drew geometrical figures in the ashes of a fire or on the oil anointing his body. He described his original discoveries in geometry, arithmetic, mechanics, hydrostatics, and astronomy with elegant clarity. We know from subsequent references that Archimedes wrote a book on mirrors in which he dealt with reflection and refraction, but it has now been lost.

Although Archimedes regarded his practical applications as mere "diversions of geometry at play," he also wrote that "certain things first became clear to me by a mechanical method, although they had to be demonstrated by geometry afterwards." He probably played with metal mirrors of various shapes and curves to determine their mathematical characteristics, which may have led to one of the last practical applications of his life.

When Archimedes was in his seventies, the Roman fleet, under the direction of General Marcus Claudius Marcellus, attacked Syracuse. To defend his city, Archimedes designed catapults that heaved huge stones to varying distances, as well as gigantic cranes that could grab nearby boats, lift them into the air, and then drop them. The frustrated Marcellus marveled at the old man, who, "sitting at ease by the sea, plays pitch and toss with our ships." The Roman sailors were so terrified of Archimedes' inventions that whenever they saw a piece of rope or wood projecting over a wall, they would cry out "There it is again!" and flee.

Historians who wrote relatively near the time of Archimedes repeat these stories, but the story about mirrors arose much later. Archimedes allegedly arranged mirrors—either flat or concave—to set fire to Roman ships more than a bowshot away. This legend inspired many attempts to imitate it and thus led to much new experimentation with mirrors, as we shall see. Although Archimedes did write about paraboloid surfaces (the shape of a mirror that directs sunlight to a sharp focus), it is unlikely that he used just one paraboloid mirror to concentrate the sun, since it would have had to be gigantic.*

It's more likely that Archimedes used multiple flat or slightly concave mirrors in the form of reflective shields. Given the urgency of the situation, and the faith that King Hieron placed in his adviser, it is possible that such mirror-shields were made just for the occasion. With minimal training, the soldiers could have lined up on the walls of Syracuse and, when the sun was behind the fleet, directed each separate reflection onto a designated spot on a ship, with devastating results.

Archimedes' ingenious devices held off the Romans for three years. In 212 B.C.E., however, the Romans managed to scale an unguarded tower and take the city. Marcellus gave orders to capture Archimedes alive. According to one story, when a soldier found him, the seventy-five-year-old scientist was absorbed in drawing geometrical figures in

---

*A parabolic mirror focuses not to a point but to a plane, and the focal plane is larger, and the light more diffuse, the longer the focal length. When such a mirror reflects the sun, a small replica of the sun appears at the focus. With a longer focal length, that replica is larger and concentrates less heat.

the dust, oblivious to all else. When the soldier ordered Archimedes to come with him, the irritated old man said, "Stand away from my diagram, fellow," and the enraged soldier killed him.

The Magic of Conic Sections

Modern mirror technology owes a great debt to the Greek obsession with geometry, particularly the rather odd study of cones, begun around 350 B.C.E. by Menaechmus, a contemporary of Plato. Picture an upside-down ice-cream cone whose pointed top makes a right angle (90 degrees). By taking a slice of the cone parallel to its side slope, Menaechmus produced a curve later named a parabola.

He discovered that he could generate two other interesting conic section curves, producing an ellipse by slicing off the top of the cone at an angle, and a hyperbola by cutting a perpendicular slice.* Hundreds of years later, these curves, too, would turn out to have important applications for telescope mirrors. Though Menaechmus generated his conic sections a bit differently, all three can be shown clearly by cuts through a cone with a common axis laid point to point (see Figure 3.2).

Once they discovered conic sections, the Greeks, including Euclid and Archimedes, studied them assiduously. Apollonius of Perga (ca. 262 B.C.E.–ca. 190 B.C.E.), born in Greek Ionia (in modern-day Turkey), was the first to name the conic sections parabolas, ellipses, and hyperbolas. As a young man, Apollonius moved to Alexandria, where he studied with followers of Euclid. Known as the "Great Geometer," Apollonius loved his work, referring to the "prettiest of these theorems" with pride. He proved, among other things, that an ellipse (oval) has two foci, that the sum of the focal distances to any point on the ellipse remains the same, and that a straight line (such as a light ray) from one focus will "bounce" from any point on the ellipse to the other focal point.

Oddly, Apollonius didn't mention the focus of a parabola. That was left to a contemporary, Diocles, who lived in rural Greek Arcadia. Dur-

---

*For a hyperbola, it is actually important to picture two cones point to point, so that the perpendicular slice produces two curves, each with its own focal point.

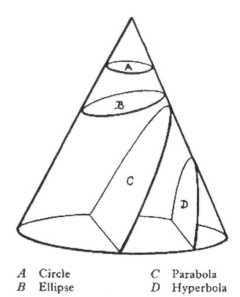

FIGURE 3.2
Cutting a cone produces curves
important for reflecting light.
An angled cut through the top of
the cone (B) produces an ellipse.
A cut parallel to the side of the
cone (C) makes a parabola.
A cut perpendicular to the   *A* Circle   *C* Parabola
base (D) makes a hyperbola.   *B* Ellipse   *D* Hyperbola

ing the Hellenistic period, mathematicians often pursued their work in isolated areas, communicating with one another in letters and through travel. Diocles explained in his book, *On Burning Mirrors*: "When Zenodorus the astronomer came down to Arcadia and was introduced to us, he asked us how to find a mirror surface such that when it is placed facing the sun, the rays reflected from it meet a point and thus cause burning." In response, Diocles proved that a paraboloid mirror (a reflective metal surface in the shape generated by spinning a parabola on its axis) concentrated parallel beams of light at one focal plane.

Of course, burning mirrors were already in wide use, but most of them were spherical—that is, they were reflective concave sections of a ball. Diocles proved that in spherical mirrors light beams parallel to the axis are reflected close to one another but that they don't meet precisely on one plane, resulting in what we now call spherical aberration. Thus, a sphere is not the most efficient shape for a burning mirror (see Figures 3.3 and 3.4).

The Last Flash of Greek Brilliance

Hero of Alexandria was born about fifteen years after Jesus. In encyclopedic fashion, Hero wrote practical guides on mathematics, physics,

Parabaloid Mirror

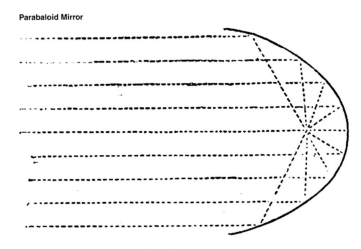

FIGURES 3.3 and 3.4 Parallel light beams coming straight into a paraboloid mirror are perfectly focused, unlike a spherical mirror, which produces spherical aberration.

The focal length of a sphere is half its radius of curvature, but a sphere does not focus light to a point.

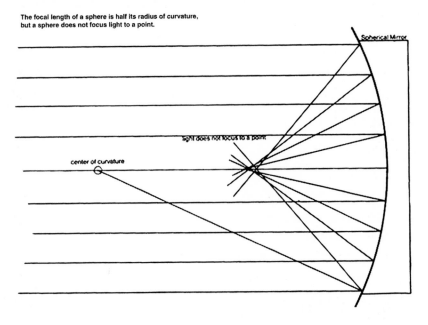

pneumatics, mechanics, and optics, including detailed instructions for making remarkable devices such as singing birds, water clocks, coin-operated vending machines, steam-powered engines, and war machines. He also wrote *Catoptrica*, a book entirely devoted to mirrors. "Catoptrics," he wrote, "is clearly a science worthy of study and at the same time produces spectacles which excite wonder in the observer."

With the aid of mirrors, he wrote, it was possible "to see our own backs, and to see ourselves inverted, standing on our heads, with three eyes, and two noses, and features distorted, as in intense grief." He also showed how to make what he called a *polytheoron* mirror that displayed "many images." Hero included a mathematical proof that angles of incidence and reflection are equal. He also understood that this law held for convex mirrors: the angles are equal in relation to the straight line tangent to the curve at the point of reflection.* Hero noted that two plane mirrors at right angles to one another would reverse left and right, thus showing an observer the way they looked to other people.

Shortly after Hero's death (around 75 C.E.), another major figure in optics, Claudius Ptolemy (not Alexander's General Ptolemy), was born and lived his life somewhere near Alexandria. Ptolemy is known primarily as the astronomer who created an elaborate model of the universe with nested spheres and special wheels-within-wheels to account for the seemingly odd behavior of the planets. He also wrote an *Optics* in five books, dealing with mirrors in two of them. Although the first book is missing, the remaining work is comprehensive and oppressively detailed. Ptolemy conducted numerous experiments with highly polished strips of iron bent into spherical convex and concave shapes. Although he relied heavily on Euclid and Hero, he went beyond them in his attempt to explain complex visual phenomena. He was interested in mirrors primarily as examples of optical illusions—in them, we see objects in places where they really do not exist. In a plane mirror, the image always appears to lie *behind* the mirror by the same distance that the real object actually lies in *front* of the mirror (see Figure 3.5).

Similarly, people looking into convex spherical mirrors always see images behind the mirror, but they are somewhat distorted, smaller, and appear to be farther away. Ptolemy treated concave spherical mirrors last because they are the most complicated. An object placed between the mirror surface and the mirror's focal plane appears right side up and enlarged, with its image behind the mirror. But when placed outside the focal plane, the reflected object flips upside down and shrinks

---

*A *tangent* is a straight line touching a curve only at one point. Picture a ruler balanced on a bowling ball.

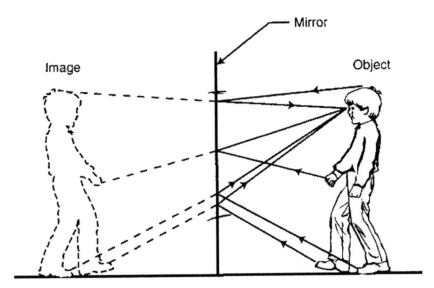

FIGURE 3.5 The illusory image in a plane mirror appears to be behind the reflecting surface. Note that the mirror must be half the boy's height for him to see his whole body.

the farther back it is placed. This image appears in the air in *front* of the mirror.

As the Roman Empire declined, so, too, did Hellenistic science. Theon of Alexandria, whose dates we can approximate because he observed a solar eclipse in 365, revised Euclid's *Optics* and may have written a book on mirrors (incorrectly attributed to Euclid) that owes a great deal to Hero's work. Theon was aided by his daughter, Hypatia, who was also a scientist and who wrote a commentary on Apollonius's *Conics*. In 415, she was murdered by a Christian mob that was threatened by her "pagan" learning, and none of her work survives.

Hypatia's death marked the beginning of Alexandria's decline as a center of learning, as many other scholars hastily departed. A century later, Anthemius of Tralles (the architect of the reconstructed Sancta Sophia Cathedral, with its 100-foot-diameter dome), attempted to answer the question: "How shall we cause burning by means of the sun's rays in a given position, which is not less distant than the range of bowshot?" As he pointed out, the required distance was too great to allow a single mirror—unless impracticably large—to do such damage. "But since Archimedes cannot be deprived of the credit of the unanimous tradition which said that he burnt the enemy fleet with the rays of the sun," he continued, "it is reasonable to suppose that the problem can be solved."

Anthemius proceeded to illustrate mathematically how to build an approximation of a parabola, using flat mirrors along the tangents of the paraboloid curve. Through experimentation, he apparently concluded that at least twenty-four mirrors were required to produce combustion. "To avoid giving trouble by enlisting the help of many persons," however, he suggested a contraption that could be operated by one man. This consisted of a central hexagonal (six-sided) mirror, with slightly smaller mirrors attached to each side by leather strips or ball-and-socket joints, followed by other smaller mirrors attached outward in concentric circles. By folding the mirrors inward and experimenting, "combustion will occur at the given position."

It is possible that Anthemius put his experiment to real use. According to a twelfth-century historian, Anthemius's teacher Proclus (not to be confused with the more famous Proclus Diadochus) used burning mirrors to destroy an enemy fleet besieging Constantinople harbor in 515, nineteen years before Anthemius's death. If this legend is true, it seems likely that Anthemius helped his teacher construct the burning mirrors.

In his book on mirrors, Anthemius also solved another intriguing problem: how "to cause a ray of the sun to fall in a given position, without moving away, at any given hour or season." He accomplished this (at least hypothetically) with an array of small flat mirrors tangent to a portion of an ellipse. From Apollonius, he knew that the sum of the focal distances to any point on the ellipse was constant. From this, he deduced an ingenious method to draw an ellipse. Hammer two nails into a piece of wood at the two desired focal points. Place a loop of string around the nails. Now trace a line with a pencil inside the string, keeping it taut. The result will be an ellipse.

Anthemius also knew that a ray passing through one focus of an ellipse would always bounce from the inner surface and pass through the other focus. To solve his problem, Anthemius located a "slit or door" through which the sunlight would enter. Then, taking into account where the sun fell at the winter and summer solstices, and all places in between, he drew an ellipse with the slit (sun-source) at one focal plane, with a series of small mirrors tangent to the ellipse, which always reflected the sunlight to the other focus. To avoid having so many little mirrors, he suggested making "a curved reflector with the required properties."

Anthemius evidently used his scientific expertise to harass Zeno the orator, his self-important next-door neighbor. Perhaps using a concave mirror, Anthemius sent blinding light through Zeno's windows while somehow making a thunderous noise to terrify him. He also simulated an earthquake by piping pressurized steam under the orator's floorboards. When Zeno discovered who was responsible for these practical jokes, he hauled Anthemius before Emperor Justinian, seeking justice. The emperor laughed him off, observing that he couldn't combat the combined power of Zeus, the God of Thunder and Lightning, and Poseidon, the Maker of Earthquakes.

The Arabian Candle

The Christian dogma that condemned magicians and scryers also discouraged scientific inquiry in the West during the early medieval period, about 500 C.E.–1000 C.E.* Fortunately, Arab scientists valued the inquisitive Greek spirit, saving and translating many ancient manuscripts while advancing the study of mirrors and optics. The first great philosopher of the Islamic world was Alkindi.

Born in the late eighth century, Abu Yusuf Ya'qub ibn Ishaq al-Kindi pursued a scholarly career in Baghdad under three caliphs, from 813 to 847, before falling out of favor for the last few years of his life; as in Greece, the precarious life of an intellectual relied on the whims of those in power. For Alkindi, optics was vitally important. In a sense, he thought light held the universe together. "Everything in this world," he wrote, "whether it be substance or accident, produces rays in its own manner like a star. . . . Everything that has actual existence in the world of the elements emits rays in every direction, which fill the whole world."

---

*It would be a mistake to condemn all Christianity as antiscientific or anti-intellectual during the early medieval period. "The church was one of the major patrons—perhaps *the* major patron—of scientific learning," David Lindberg stresses in *The Beginnings of Western Science*. Nonetheless, the monasteries were dedicated primarily to spiritual pursuits, not science.

From this basis, Alkindi might have recognized that light rays entered the eye, but he continued to espouse the extromission theory, as had Euclid, Hero, Ptolemy, and Anthemius.

More than a hundred years later, however, Abu Ali al-Husain ibn Abdallah ibn Sina (980–1037), called Avicenna in Latin, finally restored the Aristotelian theory of intromission. In an unstable political era, Avicenna served various princes as a physician, adviser, and administrator and was imprisoned at times for supporting the wrong faction. In several different books—notably the *Kitab al-Najat* (Book of Deliverance) and the *Danishnama* (Book of Knowledge)—Avicenna systematically demolished the idea that visual rays shooting out of the eye could account for vision. Rather, he argued, "the eye is like a mirror, and the visible object is like the thing reflected in the mirror by the mediation of air or another transparent body." This image was then somehow perceived by the soul, or brain. "If a mirror should possess a soul," he asserted, "it would see the image that is formed on it."

But it remained for Ibu Ali al-Hasan ibn al-Hasan ibn al-Haytham—Alhacen in Latin—to integrate the anatomical, physical, and mathematical theories into a unified intromission theory of vision. Although Avicenna and Alhacen were contemporaries, neither appears to have read the other's work. Alhacen was born in 965 in Basra, Persia, where he served as a civil servant, studied science, and read the ancient Greeks. Apparently he was also an engineer, but when he failed to dam the Nile to prevent flooding, he feigned madness out of fear of Egyptian caliph al-Hakim, remaining in prison in Cairo until the caliph's assassination in 1021, whereupon Alhacen miraculously regained his sanity. Freed at the age of fifty-six, he produced more than 100 original works, including a seven-volume work on optics, *Kitab al-Manazir* (Book of Optics), as well as books on paraboloid and spherical burning mirrors.

Alhacen is a pivotal figure in the history of optics. He truly embraced the experimental method and synthesized the mathematical and physiological theories of vision. While he was imprisoned, sunlight streaming through the dust particles from a tiny window in his otherwise darkened cell may have made him ponder the nature of light. He had read Aristotle, Plato, Archimedes, Euclid, Hero, and Ptolemy. Now he began to improve on Ptolemy's experimental apparatus, seeking theories to fit the evidence of "the instruments," as he called his senses.

To study reflection, Alhacen made seven steel mirrors, "as even and polished as possible"—one flat, then three pairs of spherical, conical, and cylindrical mirrors (concave and convex). By placing the mirrors in a slot in a level wooden base, then surrounding them with a circular wooden barrier punctured by regular holes, Alhacen could control and observe the light that entered the chamber and hit whatever mirror was placed there, and he could precisely measure where it was reflected. He tried each experiment with different kinds of light—sunlight, candle-light, colored light admitted through a red screen, "accidental" light reflected by an opaque wall. No matter what produced the light, Alhacen found that it always traveled in a straight line, and it always bounced from the mirror with equal angles of incidence and reflection.

Alhacen understood that he was studying light itself—not a medium of transmission or visual rays emitted by the eye. He took the geometrical insights of Euclid and turned them around, sending the rays in the other direction, with light traveling down a cone from a perceived object into the eye. Although Alhacen mistakenly believed that the moon was a luminous body in its own right, he did understand that *every visible object that is not a direct light source is a kind of mirror*, because light bounces off of it. Otherwise, we wouldn't see it. Alhacen also understood that light is a form of heat energy.

Most famously, he posed what has come to be known as Alhacen's Problem, which involved reflection from concave and convex mirrors: "Given a light source and a spherical mirror, find the point on the mirror where the light will be reflected to the eye of an observer." Alhacen solved the problem geometrically through the intersection of a circle and a hyperbola. Mathematically, such a proof requires a remarkably complex fourth-degree equation (i.e., containing an unknown raised to the fourth power).

Alhacen was also one of the first Western scientists to describe a camera obscura, in which light flows into a darkened room or box through a small hole. An image of an object placed outside the hole is projected—inverted and reversed—onto the interior wall, a mysterious process that fascinated medieval writers and artists and that eventually contributed to an understanding of how the eye works.

Alhacen's achievements were astonishing, comparable to Archimedes in terms of practical genius, theoretical rigor, and mathematical sophistication. Arab optics and science went into decline soon after Alhacen's death at age seventy-six around 1040. Warfare among factions destroyed peace and patronage, and the triumph of conservative Islamic forces discouraged critical thought. A reawakened European spirit of curiosity and adventure took up the optical tradition.

Universal Light and the Rainbow Connection

In the twelfth century, European travelers and scholars realized that the Arabs had preserved and expanded the work of the ancient Greeks, and a rash of translations followed in the next 100 years. Aristotle, Plato, Apollonius, Euclid, Hero, Ptolemy, and others were translated into Latin, along with Alkindi, Avicenna, and Alhacen. As a result, scholars in England, France, Germany, and Italy began to look carefully into mirrors, seeking answers to the secret of light.

Adelard, born around 1075, was one of the first of these adventurer-scholars. From his birthplace in rural Bath—a southwestern British town founded by the Romans—Adelard ventured to France to study and teach, then traveled extensively throughout Europe and the Middle East, teaching himself Arabic along the way. Back in Bath, he translated scientific works from Arabic into Latin, including Euclid's *Elements*. He also wrote *Questiones Naturales*, an encyclopedia of natural philosophy presented as a lively intellectual dialogue between the well-traveled Adelard and his stay-at-home nephew. In it, Adelard explained self-observation in a mirror: a Platonic "visual spirit" zips from the eye to a mirror, is reflected back to the observer's face, thence to the mirror again, and finally back to the eye.

While Adelard's theory didn't advance the science of optics, his translations inspired many others, including Robert Grosseteste, a combative theologian who helped bring scientific inquiry back into the mainstream of Catholic thought. Born around 1168 in rural England, Grosseteste rose from a poverty-stricken background to excel in the study of law and medicine at Oxford University, where he became chancellor in 1221. He was appointed Bishop of Lincoln in 1235.

Grosseteste was a keen observer, writing on comets, thunder, falling leaves, rainbows, eclipses, and mirrors. He learned Greek in order to translate Aristotle and other ancient authors, and he created an extraordinary theology of light that made mirrors more than merely reflective surfaces. According to Grosseteste, light was "the first corporeal form." Taking the Book of Genesis as his starting point, he said that the universe began as one point of light in a formless void. "Multiplying itself and diffusing itself instantaneously in every direction," this light formed a perfect sphere, turning into the "firmament" at its outer limits, then reflecting back in on itself to create the nine heavenly spheres surrounding the earth. "And thus in a certain sense," Grosseteste concluded, "each thing contains all other things," and everything is ultimately made of light.

Grosseteste explained the law of reflection on mirrors in terms of rebounding life forces. Hence mathematics, with its study of "lines, angles, and figures," was of ultimate importance. Grosseteste explained that the intensity of heat and light was due to the concentration of rays, and he compared the propagation of light to that of sound. Thus, even though he was motivated by a desire to make science conform to biblical exegesis, Grosseteste sounded remarkably modern in some ways, foreshadowing Einstein, the Big Bang, and the wave theory of light.

He also appeared to understand how telescopes might be made, writing: "This part of optics, when well understood, shows us how we may make things a very long distance off appear as if placed very close . . . so that it may be possible for us to read the smallest letters at incredible distances." This might be accomplished because "the visual ray penetrating through several transparent media of different natures is refracted where they come together."

It is unlikely that Grosseteste actually made a telescope. His interest in refracted light did lead him to conduct experiments, such as concentrating sunlight in a water-filled urine flask, and he also made the first effort to explain rainbows solely by refraction. Ultimately, Grosseteste regarded mirrors and lenses as religious metaphors. "All created things are mirrors which reflect the Creator," he wrote. "Consider the smallest and most insignificant object in the universe, a speck of dust. . . . In its beauty of form, it is an image of the whole universe. . . . [Now] consider the human mind meditating on the speck of dust. It presents a mirror of the Trinity in the memory, intelligence, and uniting love within the human mind."

Grosseteste never read Alhacen's work and so continued to believe that vision was caused by rays leaving the eye. Grosseteste's disciple,

Roger Bacon, did read Alhacen and became convinced of the intromission theory, although Bacon also tried to embrace and synthesize all ancient theories, including extromission. This led him into some tortured logic. For instance, Bacon claimed that the "species," or form, of an object was reflected in a mirror and traveled to the eye. Yet when the sky was reflected in a still body of water, he asserted that the eye saw it by putting forth its own species to bounce up toward the heavens.

Bacon was born into a wealthy family around 1219. After studying at Oxford, he taught in Paris. Around 1247, Bacon began to devote himself to mirrors and optics, which he considered to be one of the four primary sciences. Bacon joined the Franciscan order around 1257. Sent from Paris back to England, where he suffered from ill health, Bacon chafed under strictures of his new order, which forbade him to publish without permission. He wrote *Perspectiva*, his primary work on optics, around 1263, followed by *De Speculis Comburentibus* (On Burning Mirrors), but both remained unpublished until his friend, Pope Clement IV, solicited his work in 1265. Bacon responded with his *Opus Majus* and other works in 1267 and 1268, sending them to Italy.

Though often regarded as the father of experimental science, Bacon could sound rational and credulous in the same paragraph. He asserted that life could be prolonged with a tonic of "gold, pearl, flower of seadew, spermaceti, aloes, bone of stag's heart, flesh of Tyrian snake and of Ethiopian dragon." He believed in the efficacy of magic, scrying, astrology, and alchemy. His troubles with the church probably stemmed not from his scientific contributions but from his insistence on these occult beliefs.

Yet Bacon was also a sophisticated optician and theorist, and he echoed Grosseteste in predicting the refracting telescope, so that "from an incredible distance we would be able to read the smallest letters and count particles of dust and sand." His work on burning mirrors, relying heavily on Alhazen, is a masterfully argued mathematical analysis of light propagation. Bacon understood that paraboloid mirrors focused parallel beams and that "the mirror contains nothing." Rather, it reflects light, which propagates itself through what Bacon termed the "multiplication of species," or rapid replication of its form in rays. Moreover, Bacon asserted that light has a finite speed, although "the time occupied by a luminous radiation may be so small as to be imperceptible to our senses, even when the distance traversed is very great." He knew that light reflects off uneven surfaces. "When the surface is rough, the parts being unsymmetrical scatter the radiations irregularly, and there is no image. With smooth surfaces . . . the radiation comes back to the eye uninjured."

Bacon took issue with Grosseteste's refraction theory of rainbows, asserting quite logically that they must involve reflection, since the sun is always behind the observer in a direct line with the top of the rainbow's arch. Bacon clearly spent a lot of time chasing rainbows in misty British dawns and twilights, which makes his description of *perspectiva* (optics) as "beautiful and delightful" more meaningful. We may imagine his cassock flapping as he ran sideways, forward, and backward, while always the rainbow remained in the same relative position. He correctly concluded that "each of a hundred men, facing [away from the sun], would see a different rainbow, to the center of which his own shadow would point."

Each observer would view his private rainbow through different drops of water. "Each drop of rain in the cloud is to be regarded as a spherical mirror; these being small and close together, the effect is that of a continuous image." That image, like the image in any mirror, didn't originate in the reflective surface. Watching a rainbow was tantamount to viewing millions of tiny reflected suns. Bacon was also the first person to observe that at sunrise or sunset a rainbow's crest reaches 42 degrees above the opposite horizon. Bacon could not satisfactorily explain rainbow colors, however, even though he produced a spectrum through a piece of hexagonal Irish crystal.

Through true piety as well as an awareness of his papal audience, Bacon stressed the Christian uses of optics. "Mirrors can be so constructed and so placed and arranged that one thing will appear to be as many as we choose." One soldier's image could be replicated, appearing as an army to ignorant infidels. Or "mirrors may be erected in elevated positions which may reveal the details of an enemy's camp." Such innovations would be "useful to friends and terrifying to enemies."

Yet Bacon's work seemed doomed to oblivion. Shortly after Bacon sent his writings to Italy, Pope Clement IV died without ever commenting on them. Gregory X, the new pope, was a Franciscan who had heard all about the troublesome friar. A few years later, Bacon was apparently thrown into prison for "certain suspected novelties"—perhaps referring to his belief that astrological conjunctions affected religion, perhaps because he was extraordinarily abrasive and contentious. He was released shortly before he died in 1292.* But Bacon had not labored in vain. His work probably influenced Witelo, a Polish cleric, and John Pecham, a younger British Franciscan brother. For the next 300 years, the works of Bacon, Witelo, and Pecham, backed by Alhazen, spread the gospel of mirrors and light.

A decade after Bacon's death, at the beginning of the fourteenth century, a German Dominican monk, Dietrich (known as Theodoric in Latin) of Freiberg, finally solved a major part of the rainbow puzzle. Having observed them in fountains, waterfalls, and dewdrops on spider webs, Dietrich concluded that "we will understand [the rainbow] when we have understood what happens in a single drop of rain or mist." So he created a giant artificial raindrop by observing the sun hitting a large water-filled glass globe. With his eye at a 42-degree angle to the sunbeam, he saw a red light streaming back to him. Lowering his head a little, he watched the light turn orange, then yellow, and on through the spectrum to violet. Not only that, he could see the light beam in the water. As it entered the globe, the light bent down a little before hitting the back. There, some of the light departed, but some of it was reflected internally, as if from a concave mirror. That was the beam that was again bent a little as it came down to Dietrich's eye. He had solved the riddle of the rainbow, concluding that both Grosseteste and Bacon had been right—light was both refracted and reflected.

Dietrich did not stop there. Lowering his head another few inches, he saw a fainter violet light coming from the globe, following by the other colors in reverse order, ending with red. He realized that he was seeing the equivalent of the secondary rainbow. This time, the sun hit the bottom part of the globe-raindrop, bending up, reflecting off the back, then ricocheting off the back again before bending down as it departed the globe at a 52-degree angle to the entering sunbeam. This backward route explained why the secondary rainbow's colors were reversed; it also made sense that it was fainter, since it lost more light with the two internal reflections (see Figure 3.6).

---

*Historian of science David Lindberg questions whether Bacon was actually imprisoned.

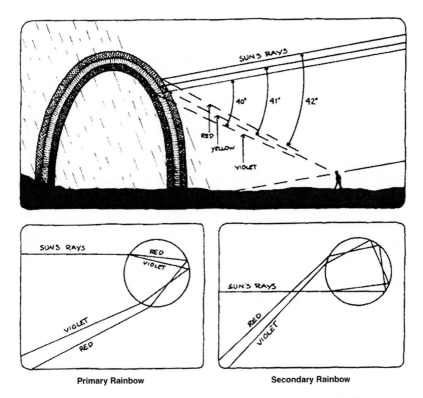

FIGURE 3.6   Dietrich of Freiberg figured out how the rainbow worked. In the bright primary rainbow, the sunbeam enters the top and reflects once from the back. In the fainter secondary, sunlight enters from the bottom and reflects twice.

But Dietrich of Freiberg still couldn't explain the rainbow's colors satisfactorily. That revelation would have to wait another three centuries.

Natural Magic Comes of Age

Optical theory languished in the fourteenth, fifteenth, and sixteenth centuries, but craftsmen and magicians continued to make practical advances, largely ignored by academicians, who looked condescendingly on lens-makers and mirror-gazers. Yet the "century of magic"—the 1500s, the age of Agrippa, Nostradamus, and John Dee—laid the groundwork for the Scientific Revolution.

Giambattista della Porta, born into a wealthy Neapolitan family in 1535, showed an early interest in medicine, astrology, optics, and esoteric facts. He and his friends met at his house, forming a society they called the Academy of Secrets, and the precocious twenty-three-year-old Porta published *Magia Naturalis* (Natural Magic) in 1558, though he later claimed to have written it at fifteen. Porta went on to write books on cryptography, memory enhancement, optics, physiognomy (the study of personality as revealed by facial appearance), horticulture, and palm reading. He also wrote popular plays.

In 1589, the now middle-aged scientist, renowned as a wizard whose horoscopes and prophecies had gotten him in trouble with the Inquisition, published a much-augmented second edition of *Natural Magic*, with twenty how-to chapters on subjects such as magnetism, farming, alchemy, fake jewels, cosmetics, perfumes, alcohol, fire, steel-tempering, cooking, hunting, invisible writing, engineering, practical jokes, and love potions. The seventeenth chapter offered a treatise on mirrors and lenses.

Like many such works, Porta's is a maddening mixture of credulity and folklore with shrewd observation and experiment. "Serpents have caused fennel to be very famous," he wrote, "for as soon as they taste of it, they become young again." But his work on optics seems to be largely based on actual experience. "I have often made sport of the most fair women," he observed, with his trick mirrors, which could distort the image to resemble an ass, dog, or sow or could color a face to give someone jaundice. He explained how to make an "Amphitheatrical" mirror, placing flat mirrors around a circle. "If you set a candle in the middle, it will seem so to multiply the images rebounding, that you shall not see so many stars in the sky, that you can never wonder enough at the order symmetry, and prospect."

Porta then moved on to concave mirrors, explaining how to find the focal plane at which a face would be inverted. He realized that burning paraboloid mirrors could also serve as searchlights: "Set a candle to the point of inversion, for the parallel beams will be reflected to the place desired." Even in the dark, then, "letters may be read, and things done conveniently, that require great light." What's more, you could set up a kind of time bomb so that when the sun came up it would hit the concave mirror and ignite gunpowder placed at its focus.

The Italian magician explained how a camera obscura worked, likening the small hole that admitted light to the pupil of the eye. He suggested the use of a lens to focus the scene, and he explained that this device provided a good way to observe a solar eclipse. By adding an appropriately placed concave mirror, he found that he could flip and reinvert the image so that "above the hole . . . you will see the images of the things which are outside so clearly and openly that you will never cease to be delighted and amazed."

Most intriguingly, Porta wrote "of Spectacles whereby one may see very far, beyond imagination," apparently using a combination of lenses, though his description is vague and confusing. Did he invent a refracting telescope? Had the ancients done it long ago, as Porta claimed when he wrote that General Ptolemy could see enemy ships 600 miles away?

It is possible that Porta made a rudimentary refracting telescope, as he later claimed. The first spectacles, using convex lenses to help the far-sighted elderly, were invented in Italy in the late thirteenth century. With the invention of the printing press and the subsequent increase in literacy, myopia increased, and by Porta's time, the lens-grinders were also making concave lenses for near-sighted people. In "The Invention of the Telescope," Albert van Helden argues that Porta and other Italians learned to combine relatively weak concave and convex lenses to produce mild magnification, but they thought of them only as another visual aid, not a way to see "beyond imagination," as Porta wished to claim.

Porta's imagination often outran his achievements. His instructions on how to make a paraboloid mirror, for instance, are laughable, and his assertion that he could make "a Parabolic Section that may burn to infinite distance" was preposterous. Porta hid behind mystification and the supposed need for secrecy lest his magic fall into the wrong hands. Nonetheless, his Academy of Secrets, shut down by the Inquisition, provided a model for future scientific societies, and he inspired others to look more closely at mirrors and lenses.

So did his contemporary, John Dee, who preached the gospel of light, which he called "the first of God's Creatures." Nothing else is "more important or more excellent than light," and without it, "the other forms could do nothing." Dee idolized Roger Bacon and wrote an entire book (now lost) with the grandiose title, *The Mirror of Unity, or Apology for the English Friar Roger Bacon; in which it is taught that he did nothing by the aid of demons but was a great philosopher and accomplished naturally . . . great works which the unlearned crowd usu-*

ally ascribes to demons. Dee also wrote five lost books on burning mirrors, and he praised Archimedes and Proclus for setting fire to ships with their mirrors. He pondered the mysterious reflective iridescence of the peacock's tail and dove's neck, and he once invited a six-year-old neighborhood girl to observe a solar eclipse projected through a pinhole into a homemade camera obscura.

The study of *Catoptrike* (mirrors), Dee concluded, "hath so many uses, both marvelous, and profitable," that it would take him too long to enumerate them. "The whole Frame of God's Creatures (which is the whole world) is to us, a bright glass [mirror]: from which, by reflection, reboundeth to our knowledge and perception, Beams, and Radiations: representing the Image of his Infinite goodness, Omnipotence, and wisdom." For Dee, as for Grosseteste, the universe is a gigantic light show, a mirror of God's creation, and the outermost heavenly sphere is "like a concave spherical mirror." Dee also realized that the eye receives light rather than emitting rays: "Our senses are not the causes of sensible rays flowing from things, but are witnesses of them."

"Strange things are done" with mirrors, Dee asserted. "As, to see in the Air, aloft, the lively Image of another man, either walking to and fro: or standing still. Likewise, to come into a house, and there to see the lively show of Gold, Silver or precious stones: and coming to take them in your hand, to find naught but Air."

Dee inspired a younger generation to delve into the scientific mysteries of mirrors. "If you were skilled in catoptrics," he asserted, "you would be able, by art, to imprint the rays of any star much more strongly . . . than nature itself does." What did Dee mean? Perhaps he foresaw the reflecting telescope. Perhaps he meant only that a scrying mirror could capture angelic messages. Thrilled to be living in a time when nature seemed to be on the verge of yielding up her secrets, Dee sought absurd short cuts, but his ultimate message was triumphant: "Let us embrace the gifts of God, and ways to wisdom, in this time of grace."

Among others, Dee inspired his friend Leonard Digges and Digges's son, Thomas. An expert in mathematics, optics, astronomy, surveying, and military engineering, Leonard died in 1559 at the age of thirty-nine. John Dee took the teenage Thomas Digges under his wing, tutoring him in math and astronomy. The younger Digges continued to revere the memory and genius of his father, and in 1571, when he was only twenty-five, he completed and published *Pantometria*, his father's work on geometry, mensuration, and surveying. In it, the younger Digges made the extraordinary claim that his father "hath by proportional Glasses duly situate in convenient angles, not only discovered things far off, read letters, numbered pieces of money with the very coin and superscription thereof . . . but also seven miles off declared what hath been done at that instant in private places."

The phrase *proportional Glasses* could have referred to either lenses or mirrors, but Digges continued in more detail: "By concave and convex mirrors of circular [spherical] and parabolic forms, or by pairs of them at due angles, and using the aid of transparent glasses which may break, or unite, the images produced by the reflection of the mirrors, there may be represented a whole region; also any part of it may be augmented so that a small object may be discerned as plainly as if it were close to the observer, though it may be as far distant as the eye can descry."

Did the elder Digges invent a reflecting telescope? Or was his son indulging in posthumous exaggeration? No one knows. As we approach the end of the sixteenth century, however, it is clear that someone is going to invent the telescope. Intrigued with possible spy applications, William Cecil, Lord Burghley, Queen Elizabeth's Lord Treasurer, commissioned William Bourne to write a report on optics in 1585 in which Bourne discussed concave mirrors but deferred to John Dee and Thomas Digges as ultimate authorities. He said that Digges's claims "may be accomplished very well, without any doubt of the matter: But . . . the greatest impediment is, that you can not behold, and see, but the smaller quantity at a time"—that is, you could magnify distant objects, but only with a limited field of vision.

Opticians had known how to grind lenses to make spectacles ever since they were invented in Italy in the late thirteenth century, and the technology of mirror production, as we will see in Chapter 5, was advancing steadily. Digges could have invented a telescope but kept it quiet. Among magicians, scientists, and craftsmen, there was a long tradition of secrecy in order to mystify, maintain a commercial advantage, or prevent one's methods from falling into the wrong hands. Dee himself warned his readers that some secrets were "barely credible to a few wise men" and that no "incautious person" should attempt to replicate them "to his own harm."

While the elder Digges used his hypothetical telescope as an early theodolite (a surveying instrument), his son would certainly have turned his to the heavens. He and Dee both wrote about the 1572 nova that pricked a bright hole in the Ptolemaic universe in which the "fixed stars" were unchanging, glued to a rotating sphere at a uniform distance from earth. They were also both Copernicans, though Digges was more outspoken in favor of a sun-centered universe. Digges was also the first astronomer to state that the universe extended forever. "This orb of stars," he wrote, "infinitely up extendeth . . . garnished with perpetual shining glorious lights innumerable, far excelling our sun both in quantity and quality." In this "wonderful & incomprehensible huge frame of God's work," Digges observed, we live on a "dark and obscure *Terrestrial Star*, where, wandering as strangers, we lead, in a short space of time, a life harassed by varied fortunes."

Unfortunately, Thomas Digges's varied fortunes forced him to abandon his ecstatic vision of the infinite heavens for politics and guns. He served in parliament, then sailed to the Netherlands as the master general of British forces there. Like his father, he indeed lived a "short space of time," dying in 1595. He was forty-nine.

A New Era

As the seventeenth century broke, the science of optics and mirrors had advanced to the edge of modernity. Men like Digges, Dee, and Porta understood that human vision took place because light entered the eye, something like a camera obscura. They understood the geometry of reflection and knew something about refraction as well, seeking to apply this knowledge to the miraculous rainbow. Mirrors and lenses could focus light either to burn or to magnify an image. They also made haphazard experiments to understand the workings of light and nature. In the new century, there would be no more "natural magic." Many people would continue to believe in the supernatural, but it would be divorced from the natural. Unraveling the secrets of nature would become the province of true experimentalists and theorists, culminating with Isaac Newton.

Chapter 4

THE RATIONAL MIRROR

But to determine more absolutely, what Light is, . . . is not so easy.

ISAAC NEWTON

THERE WAS NO CLEAR moment when Magic divided amoeba-like from Science, when the mirror gave back only the cold, hard light it received. Rather, there was a gradual movement that continues to this day, since magical thinking still invades our rational world, and mirrors haunt as well as reveal. But if a date must be named in which the true Scientific Revolution began, it would be 1609, the year John Dee died and both Thomas Harriot and Galileo Galilei first looked through telescopes.

Thomas Harriot, a young Dee associate, was by far the best British mathematician of his era. Soon after graduating from Oxford University in 1580, Harriot entered the service of Sir Walter Raleigh, who recognized that the math wonder could help train his navigators in making and interpreting sun and star sightings.

In 1585, Harriot helped found the doomed Roanoke colony in Virginia, where he learned the language of the native Algonquin tribe. "This people . . . void of all covetousness live cheerfully and at their hearts' ease," he observed. Harriot brought some experimental equipment with him, including "a perspective glass whereby was shown many strange sights, [and] burning glasses [which] were so strange unto them . . . that they thought they were rather the works of gods than of men." Harriot's "perspective glass" was probably a concave mirror, in which the Algonquin saw themselves distorted or upside down, while the "burning glass" was likely a convex lens.

Back in England, Harriot introduced tobacco, which he took as snuff. Harriot, Raleigh, and their friend Henry Percy, the ninth Earl of Northumberland (known as the "Wizard Earl"), considered themselves unconventional "freethinkers" who sought to plumb the mysteries of the world. In 1592, a sensational pamphleteer wrote "of Sir Walter Raleigh's School of Atheism . . . and of the Conjurer that is M[aster] thereof." Harriot was sure he was the "Conjurer," while John Dee thought it referred to himself.

As rumors swirled over the conjurer's identity, Harriot took up the serious study of optics. This was made easier in 1595 when the wealthy Percy admitted the scientist to the landed gentry by giving him a large piece of land. Harriot devised an elegant solution to Alhacen's Problem, mathematically determining the reflection point on a spherical mirror. In July 1601, he cracked the long-standing mystery of refraction, using trigonometry to prove that the sine of the angle of incidence is proportional to the sine of the angle of refraction. But he published none of his findings, perhaps afraid that he would be accused of wizardry. King James I locked Walter Raleigh and Henry Percy in the Tower of London—Raleigh would remain there until he was beheaded fifteen years later—and Harriot himself was imprisoned.

Harriot wrote a plaintive letter asking only to "study freely," and in a few months he was released. He then hired professional lens-grinder Christopher Tooke as his assistant in studying light's color dispersion as it passed through prisms of glass or crystal. Harriot realized that different colors were bent at different angles and computed the refractive indexes from the green to the red part of the spectrum. He also rediscovered Dietrich's secret of the rainbow, using a crystal sphere to prove that the rainbow requires both refraction and reflection.

In 1607, a comet (later named Halley's Comet) blazed across European skies. Harriot and other scientists rushed to observe it, trying to determine whether it lay outside the moon's orbit, a position that would contradict the Ptolemaic scheme of unchanging spheres. Unable to tell, Harriot was inspired to concentrate on his optical work, and within two years he and Tooke had made a telescope, using lenses held in a leather tube, that magnified normal vision by six times. On July 26, 1609, Harriot drew a picture of the new moon, the first astronomical drawing ever made with an instrument to extend human vision.

Although Harriot went on to make hundreds of observations, to discover sun spots, and to make other telescopes with magnification up to fifty times human vision, he didn't publish anything. "Do you not startle, to see every day some of your inventions taken from you?" his friend William Lower wrote in frustration in February 1610, begging him to publish.

Harriot never did. He died in 1621, his nose slowly eaten away by the cancer caused by his beloved tobacco.

Kepler's Visions

In October 1606, Thomas Harriot received a letter from Prague, where Johannes Kepler had heard of the unpublished British optician. Could Harriot tell him about his views on refraction and the rainbow? Harriot sent Kepler a table of light refraction through water, wine, vinegar, oil, and turpentine, but he didn't reveal the law of sines. The key to the rainbow, he wrote, lay in refraction and reflection inside a water droplet. Though the two intellectual giants exchanged a few more letters, nothing much came of it.

This is a shame, since Kepler was one of the greatest minds of the age. That mind was trapped inside a frail body, subject to fevers and stomach ailments, attended by myopic vision. Kepler was born premature in Weil der Stadt, Germany, in 1571 and survived an abusive childhood. At the University of Tübingen, where he learned Copernican astronomy, Kepler was commended for his "superior and magnificent mind."

He would lead a hard, even tortured life: Chronically underpaid, moving from town to town, hounded by the Counter-Reformation for his devout Lutheranism, at one point he had to drop his scientific work as he struggled to defend his seventy-year-old mother against a charge of witchcraft. In 1630, while trying yet again to find a home for his family, Kepler died of an acute fever at the age of fifty-eight.

In 1600, Kepler joined the irascible, brilliant astronomer Tycho Brahe in Prague as his assistant, and when Brahe died the following year, Kepler took over as the Imperial Mathematician for Rudolph II.

Kepler threw himself into his lifelong task of compiling Brahe's observations into useful tables, but he was also determined to understand how the planets really moved. Eventually, he realized that the planets' orbits were gigantic ellipses with the sun at one focus. Thus, Kepler liberated the planets from their fixed positions on mythical spheres, sending them careening through space, speeding up as they approached the sun, retarding as they swung farther away. He hypothesized that a turning, magnetic sun spun them on their way.

Before coming to this conclusion, however, Kepler revolutionized the study of optics. He realized that the behavior of light itself, and its perception by the human eye, was crucial to astronomical observations. He had to take account of the refraction of light by the earth's atmosphere to know precisely where a planet was actually located, and he also wanted to understand the miracle of vision. With dogged thoroughness, he studied Witelo and Alhazen. Kepler once wrote that his soul "seeks its way through tough brambles and is entangled in them," but he took a perverse delight in difficult tasks. "To walk over rugged paths uphill, through thickets, is a feast and a pleasure to me."

The book resulting from his studies—modestly titled *Ad Vitellionem Paralipomena* (Supplement to Witelo) but always referred to by Kepler as "my *Optics*"—was published in 1604. Following Grosseteste and Dee, Kepler considered light "the most excellent thing in the whole corporeal world . . . and the chain linking the corporeal and spiritual world." Like Dee, Kepler was a profound believer in astrology, mystical mathematical harmonies, and a living universe of which we are but a microcosm. Kepler likened the sun to an animal's heart, wherein beat a soul alight. He believed that light traveled instantaneously with infinite speed, and he defined reflection as light's "rebound in the direction opposite to that whence it approached." He also regarded light as a form of heat: "For light alone is always and everywhere accompanied by some heat, according to the measure of its brightness."

Kepler's third chapter ("The Foundations of Catroptrics and Place of the Image") was on mirrors. With great glee, he corrected Ptolemy's misconception that the image always appeared to lay on a line dropped perpendicularly from the viewed object to the mirror. Although that was true for flat mirrors, it wasn't for convex or concave. Kepler proceeded to give "*the true cause of the place of the image*, ignorance of which is a disgraceful stain in a most beautiful science." The image was an optical illusion that, "as regards place, [is] torn away from its object." But Kepler did not always get things right himself, claiming that "in convex mirrors the image appears to be both smaller and nearer," when in fact such images appear to be farther away.

Kepler's real triumph, however, was to follow Giambattista della Porta's hint that vision worked similarly to a camera obscura. Felix Platter, a German professor of medicine, had recently published an anatomy book, which Kepler read, suggesting that the retina, at the back of the eye, was the crucially sensitive instrument that recorded visual impressions. Until then, virtually every optical writer had followed Galen (ca. 129–ca. 199 C.E.) in considering the ocular lens called the crystalline humor as the screen upon which pictures were formed. Porta himself had thought the same thing. Acknowledging his debt to the "most ingenious Porta," Kepler nonetheless corrected him, surmising that the pupil acted like the pinhole of a camera obscura and that the admitted light was focused by the crystalline humor to form an inverted image on the retina. He introduced the term *focus* to describe the convergent point or plane of refracted or reflected light.

Wisely, Kepler stopped there. In his mirror chapter, Kepler had already pointed out that the brain's perception of an image was "torn away from its object." Clearly, the brain was capable of mentally flipping that inverted image so that it perceived it as upright—but how it did that wasn't Kepler's affair.

Despite his difficult life, Kepler never relinquished hope. He once described the view from a high mountain, looking down on scenery of "incredible brightness," delighting in the greens of the meadows and fields, the red of newly plowed soil. A twisting river gleamed like a jeweled serpent. "Overflowing into pools and turbid, [it] easily overcame the dimming brightness of the earth with its exceeding splendor." How? The glory could not result from "simple reflection," since the sun was not at the appropriate angle. He realized with joy that he was viewing "the brightness of the air by day . . . bounced back to me on the mountain from the smooth surface of the water." God's entire creation vibrated with light, and from the mountaintop it was obvious that the earth itself was a gentle mirror.

"A Vast Crowd of Stars"

On July 19, 1609, an underpaid, frustrated, ambitious, middle-aged Italian professor of mathematics at the University of Padua left to visit friends for a week in nearby Venice. While there, he heard rumors of newly invented Dutch spectacles "by means of which distant objects might be seen as distinctly as if they were nearby." Galileo Galilei rushed back to Padua to work out what combination of lenses might work.

The refracting telescope Galileo was trying to emulate had been invented by Hans Lippershey, an obscure Dutch spectacle-maker, in October 1608. Two children playing with lenses in his shop had noticed that the weathervane of a nearby church looked a lot bigger if they held two lenses up in a certain position. Putting the lenses in a tube, Lippershey promptly applied for an exclusive patent. But word got out, spreading across Europe. A salesman was hawking one of the Dutch instruments in Padua while Galileo was in Venice.

Returning to Padua, Galileo worked late into the night testing different lens combinations. "I solved it," he wrote succinctly, "and on the following day I constructed the instrument." He placed a plano-convex lens at the far end of a lead tube and a plano-concave lens near the eye, producing a noninverted image. Six days later, he took an improved model featuring his own hand-ground lenses to Venice to demonstrate how people could discern approaching ships that were still far out to sea. For the city of canals, the telescope was a miracle that could help trade and warn of approaching enemies.

Galileo soon abandoned Padua for his native Florence, where he had wangled a post as Chief Mathematician and Philosopher to Cosimo de Medici, the Duke of Tuscany. But the ever-curious scientist wasn't content just to look for ships with his telescope. "Forsaking terrestrial observations, I turned to celestial ones," he wrote in *Sidereus Nuncius* (The Starry Messenger), published in March 1610. "It is a very beautiful thing, and most gratifying to the sight, to behold the body of the moon," even though his telescope revealed that the moon was not a perfect mirrored sphere but "rough and uneven, covered everywhere, just like the earth's surface, with huge prominences, deep valleys, and chasms."

The observer also deduced that the dim light illuminating the shadowed portion of the moon came from sunlight reflected off of the earth—earthshine. Kepler had realized the same thing. Our entire world was itself a celestial mirror, part of the luminous, grand waltz of the universe. "The earth must [not] be excluded from the dancing whirl of stars," Galileo asserted. Rather, the earth was "a wandering body surpassing the moon in splendor, and not the sink of all dull refuse of the universe."

Galileo then turned his telescope on the Milky Way. "The galaxy is, in fact, nothing but a congeries of innumerable stars grouped together in clusters. Upon whatever part of it the telescope is directed, a vast crowd of stars is immediately presented to view." Finally, Galileo revealed his most startling discovery. On January 7, 1610, while looking at Jupiter, he noticed three new stars lined up with it. The next night, they had moved. After many nights' observations, he concluded that Jupiter itself had four moons, thus proving that the Ptolemaic universe was dead. "All the disputes which have vexed philosophers through so many ages have been resolved," the ebullient Galileo declared, "and we are at last freed from wordy debates about it."

Not exactly. Although Kepler warmly supported Galileo's work, others strove to save Ptolemy and Aristotle from this heretic by asserting that the moon's mountains were encased in an invisible crystal sphere so that it only *looked* as if it were imperfect. Critics mounted similar attacks on sunspots, the unsightly blemishes on the solar complexion that Galileo discerned. The Roman Catholic Church, which initially had been supportive, even enthusiastic, gradually turned against the loudmouthed scientist. In 1633, at the age of sixty-nine, Galileo was hauled before the Inquisition, forced to recant his Copernican beliefs, and placed under house arrest for the rest of his life. He slowly went blind but still managed to write his masterwork on mechanics. Just short of his seventy-eighth birthday, he died at his villa in Florence, having transformed the universe.

Despite Galileo's fate, it would be misleading to portray all contemporary religious figures as reactionaries. One of his most ardent admirers was Bonaventura Cavalieri, a Jesuit who studied under Benedictine monk Benedetto Castelli, himself a former student of Galileo. Having mastered Euclid, Archimedes, and Apollonius, Cavalieri became one of

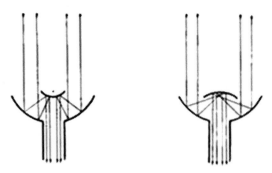

FIGURE 4.1 Marin Mersenne's two
proposed reflecting telescopes.

the era's leading mathematicians. In 1632, the year before Galileo was tried by the Inquisition, Cavalieri wrote *Lo Specchio Ustorio* (The Burning Mirror), in which he discussed whether Archimedes could have burned the ships at Syracuse, and in which he theorized that a reflecting telescope could be built. Twenty years later, another Roman Jesuit, Niccolo Zucchi, claimed that he had proposed the use of a concave mirror in a telescope, with a lens as an eyepiece, back in 1616.

Because it was well known that a concave mirror produced an enlarged image, putting such a mirror into a telescope was an obvious step. Galileo tinkered with the idea, too. But none of them could overcome a major problem: How could you see anything in the mirror if your head was blocking the view?

In 1636, the mathematician Marin Mersenne—a monk of the Minim order, whose members humbly called themselves the least of the religious—proposed an ingenious solution. In two designs for a reflecting telescope, Mersenne proposed drilling a hole in a large primary paraboloid mirror and placing another, smaller paraboloid mirror in front of it to reflect the light back through the hole to the eye. True, the smaller mirror would block some of the incoming light, but most of it would be redirected to the eyepiece at the end of the telescope, where people were used to viewing the heavens with refractors (see Figure 4.1).

In one of his designs, a small secondary concave paraboloid set beyond the focal plane would intercept the light and reflect it through the hole in parallel rays. In the other, a secondary convex mirror, set before the primary focus, reflected the rays back through the hole, though

Mersenne apparently didn't realize that a paraboloid shape wouldn't quite work here—it required a hyperboloid to bring the rays to a focus.

For years, no one followed up on Mersenne's innovative designs, in large measure because the art of mirror-making couldn't yet produce workable models, but also because René Descartes, one of Mersenne's most influential correspondents, dismissed the mirror arrangements and tried to dissuade his friend from pursuing his silly ideas.

Descartes' Clockwork Universe

Enormously self-confident, brilliant, and as often wrong as he was prescient, René Descartes was almost single-handedly responsible for plunging European intellectual culture into a mechanistic universe in which magic was firmly dissociated from science. Mirrors and lenses in this world helped scientists understand light as simple cause and effect.

Born in 1596 as his mother died, Descartes inherited enough money so that he never had to work. Sent to a fine French Jesuit school, the sickly, spoiled Descartes was allowed to stay in bed all morning, where he read, meditated, and learned to doubt everything other than mathematics, "because of the certainty of its demonstrations." He was only fourteen when Galileo's *Starry Messenger* changed the universe, and Descartes became the ultimate representative of a new generation dismissive of ancient doctrines and eager to define its own truths.

On November 10, 1619, the twenty-three-year-old "remained all day alone in a heated room," where he had "complete leisure to review my own ideas." He concluded that "the simple reasoning that a man of good sense can naturally make about things which he experiences" was infinitely better than tradition.

But Descartes considered that he was too young and inexperienced to arrive yet at his own system. "During all the following nine years, I wandered here and there in the world, trying to be a spectator rather than an actor." In 1628, he moved to the Netherlands, where he remained for the next two decades. By 1633, he had written his masterwork, *Le Monde, ou Traité de la Lumièr* (The World, or a Treatise on Light), in which he proposed his own cosmology of how the universe had evolved from primeval chaos. He posited a God who "did nothing except lend His ordinary support to nature, and left it to act according to the laws which he established."

Those laws had a great deal to do with the behavior of light—"what sort of light would be found in the sun and the stars, and how from there it would traverse the immense spaces of the heavens in an instant, and how it is reflected from the planets and comets toward the earth." But just as he was prepared to publish, word arrived that the elderly Galileo had been hauled before the Inquisition. Descartes dared not go to press with his Copernican cosmology. His book remained unpublished in his lifetime.

Four years later, Descartes published a more modest text incorporating four short works, the *Discourse on Method, Optics, Geometry*, and *Metrology*. In the *Discourse*, he revealed the logical methodology used in the accompanying scientific works: "The first [principle] was never to accept anything as true that I did not know evidently to be such . . . so clearly and so distinctly that I had no occasion to place it in doubt." From these a priori truths he could deduce, as in a mathematical proof, everything else. In effect, he was attempting to become a second Aristotle, a great systematizer—and like that of his Greek forebear, Descartes' system was fatally flawed. "There can be [no truth] so remote that we cannot eventually come upon it or so hidden that we cannot discover it," he declared. Despite his self-confidence, Descartes' self-evident truths, as well as his deductions, weren't always true.

Yet they weren't always wrong. Among other things, he invented analytical geometry. He explained the laws of mirror reflection, revealed the sine law of refraction, provided a working model of the eye, analyzed the way hyperboloid and ellipsoidal lenses focused light, suggested improvements in refracting telescopes, and explained how rainbows were formed.

It is unclear how much Descartes owed to previous researchers, since he seldom gave any credit. He repeated the rainbow experiments of Dietrich of Freiberg, expounded Kepler's theory of reversed images on the retina as his own, and failed to acknowledge that both Thomas Harriot and Willibrord Snel, a Dutch scientist, had already discovered the sine law of refraction. Yet it is possible that Descartes rediscovered everything independently. Although his deductive approach could mislead him, he also realized the value of experiments, noting, that "they become more necessary in proportion as our knowledge advances."

He dissected human cadavers, dogs, cats, rabbits, and fish to learn anatomy. In order to prove that human vision functions like a camera obscura, he plucked out the "eye of a newly deceased man" and carefully scraped away the back until only the thin retina remained. In a darkened room, he held the eyeball up to a hole, covering the retina with thin white paper to form a screen. Looking into daylight, he saw there, "with admiration and pleasure," the upside-down scene refracted through the eyeball. By gently squeezing the eyeball, he found that he could adjust the focus.

What Descartes couldn't ascertain by experiment, however, he made up. The picture formed on the retina was carried by "animal spirits, which are like a very subtle wind or air," into the brain to the pineal gland, where the soul, or "common sense," reconstructed it. "Sometimes the picture can pass from there through the arteries of a pregnant woman," Descartes continued, "right to some specific member of the infant which she carries in her womb, and there forms birthmarks."

More plausibly, Descartes defined light as "a certain movement or action, very rapid and very lively." He couldn't explain its "true nature," so he approached it by way of three somewhat contradictory analogies. First, it travels instantaneously. Just as a blind man's cane immediately feels what it touches, light from an object hits the eye. Thus, Descartes dismissed "those small images flitting through air" that Democritus and Bacon hypothesized. Next, he compared the transmission of light to half-pressed grapes in a wine vat in order to explain a "very subtle and very fluid material, extending without interruption from the stars and planets to us," through which light traveled. Finally, he asked his readers to think of light particles as little tennis balls. His book featured an illustration of a midget tennis player hitting tiny balls onto a mirror, where they bounced in a graphic display of the law of reflection.

Descartes used his tennis balls to explain refraction as well, although he somehow concluded that light, unlike tennis balls, bends in water and glass because it travels *faster* than in air. He explained colors rather ingeniously—they are caused by giving a different spin to light particles, just as a tennis player can slice a ball or give it topspin. When studying rainbows, he sent sunlight through a prism to produce a spectrum. Baffled, he pictured "small balls rolling in the pores of earthly bodies . . . in various ways according to the various cases which determine them." The rapidly spinning particles are red; the slower are yellow, green, and blue. "I do not believe it possible," Descartes concluded, "to doubt that the matter is as I have just explained it."

In 1649, Descartes finally abandoned his happy seclusion to become the Royal Philosopher for Queen Christina of Sweden, a twenty-three-year-old bundle of energy who sometimes spent ten hours in the saddle and who demanded the same brisk commitment from her retinue. With a lifelong habit of lolling abed until midmorning, Descartes suddenly found himself teaching classes at 5 A.M. Within a few months, he developed pneumonia and, after extensive bloodletting, died in Stockholm in February 1650, just short of his fifty-fourth birthday.

Descartes left behind a mixed legacy. "Neither through the promises of an alchemist, nor the predictions of an astrologer, nor the impostures of a magician" could truth be found. Cartesians, as his followers called themselves, relied only on what was undoubtedly true. Taken to their extreme, Descartes' ideas led to a terrifying nihilism in which nothing was certain. The senses could be deceived, as dreams, optical illusions, and trick mirrors showed, and so the world itself, and the body, might be illusory. There might be no God. "What am I?" Descartes asked in desperation. Unable to tolerate the ambiguity, he asserted that he knew he existed because he could think, famously stating, "I think, therefore I am."

He managed to "prove" that God existed, too, and that He had placed an immortal soul into man's mortal body, inserting it in the brain's pineal gland. This soul, this ability to think, was all-important to Descartes, and it distinguished humans from other animals. We could and should, therefore, "make ourselves the masters and possessors . . . of nature."

Essentially, then, Descartes split the mind from the body as well as science from religion. Cartesian dualism made the universe rather bleak, a clockwork mechanism over which a distant God presumably presided but that could all be explained rationally and manipulated without recourse to Him. As for mirrors, you could bounce tennis balls off them, but they were no longer magical. Neither, Descartes ruled, could Archimedes have used them to set fire to the ships at Syracuse, and they were useless in telescopes. For a long time, the influence of Descartes discouraged much mirror research.

Athanasius Kircher's Ecstatic Journey

One unabashed Archimedes fan wasn't discouraged, though, and he chose not to live within Descartes' sterile new universe. For the flamboyant Athanasius Kircher, the world was marvelously alive, light was "the exuberance of God's great goodness and truth," and mirrors were glorious means to reflect that truth. One of his books, *The Ecstatic Heavenly Journey*, featured Kircher romping through the starry universe with Cosmiel, his angelic guide, in a "fictitious rapture," but Kircher's real journey on earth had more than its share of ecstasy and terror.

Kircher, the last of nine children, was born in Fulda, Germany, in 1602. He survived a childhood swim through a churning mill wheel and near-fatal scrapes while traveling around Protestant Europe during the Thirty Years' War. By 1633, Kircher, a Jesuit priest, was teaching mathematics, philosophy, and oriental languages at Avignon, in his spare time studying Egyptian hieroglyphics, setting off fireworks, making little mechanical curiosities, playing with magnets, and observing sunspots with a telescope. He designed a planetarium, using mirrors to direct sunlight and moonlight into a tower at the Jesuit college. He arrived in Rome as a professor of mathematics in 1635, just two years after Galileo's trial there.

Following the tense trial, the mood in Rome was still somber. The pontiff and his cardinals were ready for a little diversion, and Kircher gave it to them, assembling a museum of curiosities to match that of Rudolph II, including a stuffed crocodile, skeletons, geodes, ostrich eggs, telescopes, microscopes, and oddly shaped mirrors.

One of his illusions with a hidden mirror showed water pouring upward into a "watery heaven." Another featured a "catoptric theater" in which greedy viewers reached out to grab gold pieces, only to find their hand grasping air. Kircher put a cat into this mirror box, where it attempted to entice or claw its illusory fellow felines until it yowled in "indignation, rage, jealousy, love and desire," as one observer noted.

Kircher gave public talks and demonstrations, featuring one of the first magic lantern shows in which he projected an image of a soul in

Purgatory, with the candlelight source providing a realistic flicker. Kircher sent up dragon-shaped hot-air balloons with "Flee the Wrath of God" painted on their bellies and put on a Baroque light show, flashing beams over hills and down into valleys with oddly shaped mirrors.

For the next four decades, Kircher entertained Rome and the world, churning out more than forty lavishly illustrated books on an astonishing range of topics, including magnetism, optics, geology, astronomy, music, archaeology, theology, medicine, philology, and natural history. He was fond of quoting Plato: "Nothing is more beautiful than to know everything," and he certainly tried, learning two dozen languages. He left Rome only once, going on a trip to Sicily and Malta. When Mount Etna erupted during his visit, the ever-adventurous Kircher climbed down into the crater to make a sketch of the lava pouring out. In Sicily, he was eager to see the harbor of Syracuse and pleased to discover that Roman ships could have come within thirty paces of shore.

Back in Rome, Kircher set up an experiment, reflecting sunlight onto a target more than 100 feet away, first with one, then two, three, four, and five flat mirrors. With four, the heat was just bearable; with five, his assistant couldn't stand in it comfortably. Like Anthemius before him, Kircher concluded that Archimedes could have burned ships with enough properly directed flat mirrors. In 1646, he wrote *Ars Magna Luci et Umbrae* (The Great Art of Light and Shadow), in which he discussed this experiment, described his magic lantern, detailed the workings of a camera obscura, and pondered the cause of the firefly's glow and other phosphorescence. He demonstrated how an object placed inside a reflective hollow cylinder, when viewed aslant, appeared to be floating in the air at the top of the cylinder. He created a lovely natural clock by floating a potted sunflower in a tub of water, with a stick in the flower pointing to the correct time as it followed the sun across the sky.

He dared to depict the sun not as a perfect crystalline sphere but as an erupting ball of fire, "a fiery, rough, and uneven body." After devoting an entire chapter to the color of angels, Kircher burst forth with a Neoplatonic ode to light: "What else is [light] in Heaven if not the abundance of life among the Angels, and . . . the laughter of the Heavens?"

In his long life, the enthusiastic Kircher got a lot wrong. Insects are not spontaneously generated in animal dung, all languages did not descend from Hebrew, cavernous lakes do not lie under great mountain ranges, and there are no dragons, mermaids, or griffins. But he also pointed the way to modern bacteriology, hinted as broadly as he could (under the circumstances) that the planets moved around the sun, and, mostly, encouraged experimentation and openness to life.

The 600-Foot Telescope and Monstrous Insects

Athanasius Kircher's telescope contained no mirrors; neither did anyone else's for most of the seventeenth century. Refracting telescopes ruled. In 1610, Johannes Kepler had suggested an improvement over the Galilean device. Why not use two convex lenses, he said, rather than a convex and a concave? By allowing light to focus and cross within the telescope, in front of the second lens, viewers could gain greater magnification and a larger field. Although this meant that the view would be upside down, that made little difference to sky-watchers, and Keplerian telescopes eventually became popular.

But there were two problems. The easiest lenses to grind were spherical, because rubbing two hard surfaces together at random—for a long time with water and grit between them—naturally produces a concave spherical curve on the top piece and a convex spherical curve on the bottom. Light refracted through spherical lenses does not focus precisely, producing spherical aberration. Descartes was right in calling for hyperbolic and other aspheric shapes, but they weren't so easy to make, and no one knew how to test them.

The second problem was even more intractable and inexplicable. Not only did the stars look a bit fuzzy; they were surrounded by a strange halo of colors, and the stronger the lens—the more curved and the shorter the focal length—the worse was this effect, called chromatic aberration. The solution was longer telescopes, with flatter lenses, ground just enough to bend the light to a very long focus. That way, both types of aberration, though present, weren't too disruptive, and the image size increased, albeit with a slight decline in image brightness.

Two Dutch brothers, the sons of an intellectual diplomat who corresponded with Mersenne and befriended Descartes, pioneered longer telescopes, spurred on by their curiosity about Saturn. Galileo and Kircher had both called attention to the "arms" of Saturn, fuzzy attachments that came and went mysteriously. In an effort to resolve the matter, Christiaan and Constantijn Huygens, then twenty-five and twenty-seven, respectively, devised an improved method of grinding and polishing lenses.* Christiaan was primarily a theoretician, his brother an expert craftsman. In March 1655, the brothers hoisted a 12-foot-long refracting telescope with a 2-inch aperture toward Saturn, and they discovered Titan, the brightest of Saturn's moons.

Toward the end of the year, they used a 23-foot telescope that revealed a dark line on Saturn. In January 1656, they erected an enormous 123-foot telescope, holding it up with a big pole and adjusting it with ropes and pulleys, but they still couldn't see any appendages to Saturn. In October, however, something like a thin plate opened out, and by 1657 the fascinated Huygens boys could clearly see the miraculous halo that has seduced and inspired astronomers ever since—"a ring, thin, plane, nowhere attached," as Christiaan Huygens wrote in his 1659 book, *Systema Saturnium*. They hadn't been able to see it before because the rings had presented a razor edge to earthly observers. The Huygens siblings also observed stars lighting up the gorgeous Orion Nebula.

A wealthy Polish brewer named Johannes Hevelius read Huygens's account, which included a description of his ever-lengthening telescopes. Hevelius had a home observatory in Danzig that he had named Sternenburg (Star City), and in 1647 he had published *Selenographia*, the first complete lunar atlas. Now he ordered telescopes constructed of 60 and 70 feet and, finally, a 150-foot monster composed of wooden slats connected by round black rings, suspended from a 90-foot-high mast and operated by a group of assistants hauling on various guy ropes. Its objective lens had a diameter of 8 inches.

The telescope never worked very well, due to wind, wood warp, and rope stretch. Keeping the lenses aligned and aimed in the right direction was incredibly difficult, which may explain why Hevelius reverted to

---

*By this time, Christiaan Huygens had already exchanged letters on scientific subjects with Marin Mersenne, but they didn't concern Mersenne's ideas about reflecting telescopes, and there is no evidence that Huygens ever considered using mirrors rather than lenses.

the naked-eye observations of Tycho Brahe for positional determinations. On one of his engraved title pages, Hevelius inscribed: "I prefer the unaided eye."

Christiaan Huygens responded by eliminating the tube. He mounted his object-glass in a short iron tube mounted with a ball-and-socket joint to a tall pole, then held the eyepiece in his hand. By moving around and manipulating the higher lens with a rope, he could make relatively good observations, though stray light and atmospheric turbulence interfered. As the century progressed, craftsmen in Italy, France, and Holland ground lenses with ever-longer focal lengths. Constantijn Huygens made one with a 210-foot focus. Not to be outdone, French scientist Adrien Auzout made lenses of 300- and 600-foot focal lengths and speculated that with magnifications of 1,000 he might see animals on the moon.

In this spirit of scientific ferment, with new worlds opening up through exciting technological breakthroughs and experiments, learned societies and national observatories sprang up. Instead of working in isolation, scientists traded ideas, and the pace of discovery picked up, along with ego clashes and arguments over who could claim important insights. In England, the Royal Society evolved from informal weekly meetings in Cromwellian Oxford to a formal London charter in 1662 under King Charles II, recently restored to the throne. Three years later, Adrien Auzout convinced Louis XIV to construct L'Observatoire Royal, where the first director, Italian-born Jean-Dominique Cassini, peering through midlength refracting telescopes (17 feet and 34 feet), discovered four more moons of Saturn, as well as a gap in Saturn's ring that came to be known as Cassini's Division. In 1666, the Académie Royale des Sciences commenced in Paris, and founding member Christiaan Huygens moved there.

Along with new discoveries in the heavens came wondrous revelations in the microscopic world. In 1665, Robert Hooke, one of the founding members of the Royal Society, published *Micrographia*, featuring his startling drawings of gigantic insects, forest-sized mold, and air spaces in a slice of cork that Hooke called "cells," since they reminded him of the small rooms in monasteries. Unfortunately, Hooke's compound microscope (using two or three lenses) suffered badly from chromatic aberration, worsened by every additional lens. That is why Antonie van Leeuwenhoek, a Dutch draper, could see "animalcules"—

protozoa, bacteria, and spermatozoa—better with his tiny single-lensed microscopes, as Hooke had to admit. Still, he hated van Leeuwenhoek's tiny devices, which he found "offensive to my eye." Many things and people offended Hooke throughout his life.

Born on the Isle of Wight in 1635 as the son of a minister, the sickly Hooke suffered awful headaches when he studied theology, but he was a natural mechanical genius. At Oxford, Hooke served as Robert Boyle's assistant, making an effective air pump to produce a vacuum. In 1662, upon the founding of the Royal Society, he was appointed Curator of Experiments, providing hundreds of demonstrations and ingenious mechanisms at subsequent meetings. A brilliant experimentalist, Hooke was an indifferent theoretician, and his confrontational manner alienated many people, including poor old Hevelius, whom Hooke took to task for his naked-eye observations.

Hooke's pugnacity may have stemmed in part from his physical appearance. "He is but of midling stature," wrote one acquaintance, "something crooked." He apparently suffered from scoliosis. London diarist Samuel Pepys observed that he "is the most and promises the least of any man in the world that ever I saw."

Hooke spewed forth ideas and experiments involving blood transfusion, mechanics, cartography, skin grafting, optics, botany, geology, clocks, engines, telescopes, and microscopes. In *Micrographia*, he espoused a vague theory of light, defining it as "pulses of motion." He also wrote about the colorful reflections of soap bubbles, thin pieces of mica, and air between sheets of glass, though his attempt to account for the colors was not convincing. To shorten the absurd length of refracting telescopes, Hooke suggested bouncing the long beam required with two or three flat mirrors within a shorter scope, but he never built a working model.

Meanwhile, in Scotland, James Gregory resurrected Mersenne's idea of a reflecting telescope with a hole in the primary paraboloid mirror. Gregory published his *Optica Promota* in 1663. "Moved by a certain youthful ardor," he wrote, "I have girded myself with these optical speculations, chief among which is the demonstration of the telescope." After detailing fifty-nine theorems on the reflection and refraction of light, the twenty-five-year-old proposed a small secondary concave ellipsoid mirror that would re-reflect the light to the ellipse's second focal

FIGURE 4.2   Gregorian telescope.

plane in the middle of the hole in the primary mirror and thence to an eyepiece (see Figure 4.2).

Gregory commissioned a London optician to make the mirrors, but the results were abysmal. He hoped to find Italian craftsmen who could make the requisite mirrors and in 1664 ventured to Rome and Padua. Giving up on his telescope, he studied mathematics at the University of Padua, eventually returning to Scotland to teach math. There, he turned a bird's feather into the first diffraction grating. "Let in the sun's rays by a small hole to a darkened house, and at the hole place a feather ... and it shall direct to a white wall or paper opposite to it a number of small circles and ovals ... whereof one [in the center] is somewhat white and all the rest severally colored." But why?

Research into the nature of light advanced quickly in the 1660s, along with efforts to improve telescopes and microscopes. Father Francesco Maria Grimaldi, a Jesuit professor of math at the University of Bologna, died in 1663, two years before the publication of his book, *A Physical and Mathematical Thesis on Light, Colors, the Rainbow, and Other Related Topics.* Even though light was so common, he wrote, "to explain its nature is a most difficult task." Grimaldi proved that light could be not only reflected and refracted but also *diffracted*, a word he coined to explain what happened when he put a small opaque object in a stream of light coming through a pinhole. Because light travels in straight lines, he expected a clean shadow line along a mathematically predictable path. Instead, the shadow was larger and more diffuse than it should be, and part of it was colored. "There are bands of colored light such that the center of each is pure white, whereas at the edges there is color, always blue on the edge nearer the shadow ... and red on the farther side."

Grimaldi wasn't sure what was going on, but he realized that color wasn't a quality inherent to a particular object. It was some kind of special motion of light. "Light is a kind of fluid that moves very fast and sometimes passes through a transparent body in the form of a wave." Colors involved some sort of special light wave. Thus, when you look at a bluebird, it looks blue not because of the inherent blueness of the feather but because the light reaching your eye is somehow made to say "blue" to the brain. What's more, a pigeon's or peacock's iridescent feathers change that message somehow, so that the reflected light is sometimes one color, sometimes another.

Isaac Newton Makes a Telescope

Grimaldi's book appeared in Italy the same year that Isaac Newton, twenty-two, received his bachelor's degree from Cambridge University. By that time, the young Newton had already reached some of the same conclusions about the nature of color as Grimaldi. Born prematurely on Christmas Day in 1642, several months after his illiterate father's death, the baby Isaac was not expected to live. He did, only to be abandoned by his mother three years later when she remarried the much older Reverend Barnabas Smith, who didn't want her brat around.

Raised by his maternal grandmother on the nearby family farm in rural Woolsthorpe, the lonely, introverted Isaac entertained himself with homemade toys and dark revenge fantasies. He constructed a tiny mouse-powered mill, sundials, and fiery kites that terrified the neighbors. When he was about ten, Isaac threatened "to burn them [his mother and stepfather] and the house over them," as he later confessed to his journal. He found himself "wishing [for] death and hoping it to some." The following year, the hated Reverend Smith died, and his widowed mother moved back in with him, along with his three younger half-siblings. Young Isaac was soon sent to a private boarding school in Grantham, where he excelled academically but remained aloof and friendless. In 1661, he went to Cambridge University.

Newton hated his first roommate. Noticing Isaac walking, "solitary and dejected," fellow student John Wickins, who didn't like his roommate either, suggested that they move in together, and the arrangement stuck for most of the next two decades. Wickins must have been a tolerant companion, since Newton's obsessive nature, weird experiments, and odd hours made him less than the ideal companion. During the summer of 1663, Newton strolled through the nearby Stourbridge Fair, where he bought a book on astrology. In reading it, he couldn't understand a heavenly conjunction without more math, which eventually led him to Euclid's *Elements*. Then he moved on to Kepler, Galileo, and Descartes, among others.

Reading Descartes, Newton found a congenial spirit who, like him, was fascinated by light. Newton once observed that for him truth was "the offspring of silence and unbroken meditation," as it was for Descartes. More clearly than the French philosopher, however, Newton relied on experiments to provide appropriate data for his solitary ruminations. In his quest to understand optics, Newton stared repeatedly into a mirror at the sun's reflection in order to induce odd after-images. Then he "turned my eyes into a dark corner of my chamber & winked to observe the impression made & the circles of colors which encompassed it & how they decayed by degrees & at last vanished." He nearly blinded himself.

Still puzzling over odd visual phenomena, Newton carefully inserted the blade of a small knife into the corner of his eye. "I took a bodkin and put it between my eye and the bone as near to the backside of my eye as I could: & pressing my eye with the end of it (so as to make the curvature in my eye) there appeared several white, dark and colored circles." In another experiment, he gazed through a feather at the setting sun, as James Gregory did, and found that it made "glorious colors."

At the August 1664 Stourbridge Fair, Newton picked up a cheap glass prism, no doubt inspired by Descartes' experiment in which he split sunlight into a spectrum of colors. About this time, too, Newton began to stay up nights to observe the sky. On December 10, he saw a comet, and a week later, another appeared at 4:30 A.M. He continued to watch it every night through Christmas and his twenty-second birthday, until it finally vanished a month later. Newton desperately wanted a telescope but, unable to afford it, decided to make one, and he set about grinding his own lenses. He wasn't satisfied with the blurry results, surrounded by confusing light halos. To avoid spherical aberration, he tried to make nonspherical lenses—hyperboloids or ellipsoids as Descartes had suggested—but he couldn't figure out how.

As a diversion, Newton played with his prism. He darkened his room and made a small hole in his window shutters, allowing a sunbeam to pierce the gloom. Placing the prism near the entrance hole, he found it a "very pleasing divertisement, to view the vivid and intense colors produced thereby" on the far wall. But then he noticed something odd. "I became surprised to see them in an *oblong* form; which, according to the received laws of Refraction, I expected should have been circular." The circular ball of white light had been stretched out into a rainbow of color five times longer than he expected.

Newton put one of his convex lenses into the rainbowed beam coming from the prism, and as it reconcentrated the beams, they turned white again at the focus before spreading out again in colors. Could it be that sunlight was actually an aggregation of colored lights, each of which was refracted at a different angle? Then Newton bought another prism. He turned it at 180 degrees to the first prism, and the elongated rainbow once again reunited into a circular white beam.

"I began to suspect," Newton recalled later, "whether the Rays, after their trajection through the Prism, did not move in curve lines." Undoubtedly influenced by Descartes' tennis-ball analogy, he "remembered that I had often seen a Tennis ball, struck with an oblique Racket, describe such a curve line." But no—it was easy enough to put some chalk dust in the air and to see that the light traveled in straight lines.

Then Newton performed what he came to call the *Experimentum Crucis*, the "crucial experiment." Just beyond the first prism, he placed a board with a small hole drilled in it. Twelve feet beyond that, he set up another board, also with a hole drilled in it. Finally, he set the second prism beyond the second board in a position to intercept just one color of the light beam and project it onto the wall. By turning the first prism in the beam of sunlight slowly about its axis, Newton directed first one color, then another, through the hole in the second board. When a single colored beam refracted through the second prism, it did not elongate but formed a roughly circular colored light on the wall. But that wasn't all. As he progressed from red through orange, yellow, green, blue, indigo, and violet, the little balls of light gradually climbed the wall, so that together they would have formed the elongated rainbow he had first seen.

From these experiments, Newton concluded, "*Light consists of Rays differently refrangible,* which, without any respect to a difference in their incidence [the angle at which they hit the prism], were, according to their degrees of refrangibility, transmitted towards divers parts of the wall." Once he had absorbed this stunning revelation, Newton stopped trying to grind aspheric lenses. "When I understood this, I left off my aforesaid Glass works; for I saw, that the perfection of Telescopes was hitherto limited, not so much for want of glasses truly figured, [but] because that Light itself is a *Heterogeneous mixture of differently refrangible Rays.*" This accounted for chromatic aberration, and Newton discovered that its effects far outweighed those of spherical aberration. "Nay, I wondered, that seeing the difference of refrangibility was so great, as I found it, Telescopes should arrive [even] to that perfection they are now at."

Having given up on the improvement of refracting telescopes, Newton recalled James Gregory's plans for a reflecting telescope, which he had read about in the 1663 *Optima Promota.* Unlike lenses, mirrors did not separate different colored lights—they all bounced together. He concluded that through reflecting telescopes, "Optic instruments might be brought to any degree of perfection imaginable, provided a *Reflecting* substance could be found, which would polish as finely as Glass, and *reflect* as much light, as glass *transmits*, and the art of communicating to it a *Parabolic* figure be also attained." In theory, Newton was right, but he realized that he faced "very great difficulties," because "every irregularity in a reflecting superficies [surface] makes the rays stray 5 or 6 times more out of their due course, than the like irregularities in a refracting one." In other words, mirrors demand far more precision than lenses.

"Amidst these thoughts," Newton recalled, "I was forced from *Cambridge* by the Intervening Plague."* The university was closed in July 1665 by an outbreak of bubonic plague.

---

*The chronology of Newton's optical experiments is confusing. Elsewhere, Newton wrote that he began trying to grind aspheric lenses and bought a prism at the beginning of 1666, but here he says that the plague interrupted this work, and that occurred in the middle of 1665. Newton was apparently one year off. He later told someone that he bought his first prism at the 1665 Stourbridge Fair, when it had to have been 1664, because the fair was canceled in 1665 and 1666 due to the plague. Thus, it seems likely that Newton's crucial experiment took place early in 1665, not 1666.

Back at his mother's home in Woolsthorpe, Newton became absorbed in higher mathematics. He also thought deeply about the mysterious force of gravity and, allegedly inspired by the fall of an apple, began the thought process that was to lead to the publication of the *Principia Mathematica* twenty years later. "I was in the prime of my age for invention & minded Mathematics & Philosophy more than at any time since," he remembered wistfully in his old age.

Not until 1667, when he returned to Cambridge, did he follow through on his plans to make a reflecting telescope. He cast his own alloy—three parts copper to one part tin, with a touch of arsenic—and, "having thought on a tender way of polishing, proper for metal," he ground it against a convex copper tool, then polished it with pitch. This produced a concave mirror 1 1/3 inches in diameter. To try to turn the spherical shape into a parabola, he ground the middle "with all my strength for a good while together," but the mirror was basically spherical, with a turned-down edge. Mounting it at the end of a 6-inch tube, Newton avoided Gregory's solution—to drill a hole in the primary and make a difficult ellipsoidal secondary—by mounting a small secondary flat mirror in the center of the tube at a 45-degree angle, thus reflecting the light to an eyepiece on the side (see Figure 4.3).

Late in life, when asked where he had the telescope constructed, Newton said he made it himself. Where did he get the tools? Newton laughed—a rare event in his somber life—and said he had made them, too. "If I had stayed for other people to make my tools and things for me, I would have never made anything of it." With his new telescope, completed in 1668, Newton could see Jupiter's four moons as well as the "horned" phase of Venus. He made an improved model in the fall of 1671.

By that time Newton had been appointed the second Lucasian Professor of Mathematics, after the first holder of that office, Isaac Barrow, stepped aside in his favor, explaining that Newton was "of an extraordinary genius & proficiency." Aside from Barrow, few knew of Newton's work or genius. Asocial and introverted to the point of paranoia, Newton had published nothing in his own name, only reluctantly agreeing to one anonymous mathematical paper. "For I see not what there is desirable in public esteem," he wrote. "It would perhaps increase my acquaintance, the thing which I chiefly study to decline."

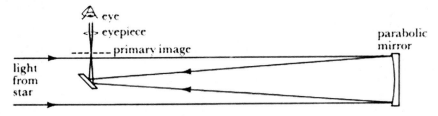

F I G U R E 4.3   Newtonian telescope.

Newton fulfilled his professorial duties by giving lectures and filing them at the university library. In January 1670, he gave his first class in a droning Latin, describing his experiments and reflecting telescope. The few students who showed up yawned. No one attended his second lecture. "So few went to hear him, & fewer that understood him, that oftimes . . . for want of hearers, [he] read to the walls," his assistant recalled.

Unperturbed, Newton continued his research. He seldom left his room, often forgetting to eat. He rarely went to bed before 3 A.M. and often stayed up until dawn. He took no exercise other than to wander in the garden, deep in thought, only to rush back up the stairs to scribble furiously at his desk, without taking the time to sit down.

Near the end of 1671, Newton reluctantly allowed Isaac Barrow to take his telescope to a London meeting of the Royal Society, where it caused a sensation. Christopher Wren and other members took it to Charles II for a personal demonstration. A few days later, just after New Year's, Henry Oldenburg, the secretary of the Royal Society, wrote Newton that his telescope had been "examined here by some of the most eminent in optical science and practice, and applauded by them." He was anxious to "secure this invention from the usurpation of foreigners"—ironically, Oldenburg was a transplanted German—and wanted to send a description to Christiaan Huygens to establish priority.

The Oddest Detection

Newton wrote back to say that he "might have let it [the reflecting telescope] still remain in private as it has already done some years" had not the Royal Society paid attention. He approved of the letter to Huygens, asking Oldenburg to stress that his telescope (unlike Huygens's refractor) "represents things distinct & free from colors." A few days later, Oldenburg wrote to tell Newton that he had been voted a member of the Royal Society. Finally, Newton felt comfortable enough to reveal that the reflecting telescope was simply a logical by-product of his optical research, which involved "the oddest if not the most considerable detection which has hitherto been made in the operations of Nature." He followed up with a long letter explaining his prism experiments and his discovery that white light was a mixture of differently refractable colored lights. It was published in the February 19, 1672, issue of the *Philosophical Transactions* of the Royal Society.

"Colors are not *Qualifications of Light*," he wrote, "derived from Refractions or Reflections of natural Bodies (as 'tis generally believed), but *Original* and *connate properties*, which in divers Rays are divers. Some Rays are disposed to exhibit a red color and no other; some a yellow and no other," and so on. He noted that "the most surprising, and wonderful composition was that of *Whiteness*," which was always the compounded result of all other colored rays. "Light is a confused aggregate of Rays indued with all sorts of Colors." This finally explained why the rainbow colors appear as they do, due to their different indices of refraction. It also explained why Grimaldi had been right that color was not inherent in objects. And it explained why refracting telescopes could not easily be cured of chromatic aberration—hence, the need for him to make the reflecting telescope that had attracted all this attention in the first place.

In conclusion, Newton wrote that "it can be no longer disputed, whether there be colors in the dark"—there were not—"nor whether they be the qualities of the objects we see"—no again—"nor perhaps, whether Light be a Body." Yes, he thought, though he qualified it with "perhaps." Here Newton slid into conjecture. Because colors were the *qualities* of light rays, how could the rays themselves simply be qualities? Light must be a substance. Then he backed off. "But to determine more absolutely, what Light is, after what manner refracted, and by what modes or actions it produceth in our minds the Phantasms of Colors, is not so easy. And I shall not mingle conjectures with certainties."

Newton's exposition was a model of rigorous scientific logic, presented in a clear, compelling style. He wasn't just throwing off hypotheses—he was proving a new theory. Consequently, he was shocked when Robert Hooke criticized him. "As to his hypothesis of solving the phenomenon of colors," Hooke wrote, "I confess I cannot yet see any undeniable argument to convince me of the certainty thereof." Indeed, Newton's experiments, as well as Hooke's own, "do seem to me to prove that light is nothing but a pulse or motion propagated through an homogeneous, uniform and transparent medium."

Hooke had zeroed in on Newton's assertion that light was a body, that it consisted of tiny particles of some sort. For Hooke, light was a kind of wave action. In response, Newton admitted that he wasn't sure that light consisted of particles. Indeed, Newton had said as much in his original paper, warning about mixing conjecture with certainty. The important point, for Newton, was that light—whatever it was—*did* consist of variously colored strands that refracted to varying degrees.

Newton came to despise Hooke, for whom the feeling was mutual, even though they were similar in many ways. Both suffered traumatic childhoods and lost their fathers young, and as youths both constructed ingenious mechanisms. Hooke was a hypochondriac and insomniac; so was Newton. Their similarities may have been part of the problem. At any rate, the randy, gregarious Hooke regarded the cloistered, sexless Newton as an arrogant stewed prune who overrated himself.

Only months after Newton's telescope created a sensation in London, a French physics professor named Guillaume Cassegrain came forward with an alternative design for a reflecting telescope. Like Gregory's, it featured a paraboloid primary mirror with a hole in the center. But Cassegrain's called for a convex hyperboloid secondary that would intercept the reflected light before it focused, sending it back through the hole in the primary to focus at an eyepiece. It was an elegant design, shorter than the Gregorian telescope, with the added advantage that grinding errors would tend to cancel out one another between the concave and convex mirrors (see Figure 4.4).

Cassegrain's supporters asserted priority over Newton, since the Frenchman had supposedly invented it several weeks before Newton's paper appeared. The irritated Newton, who had actually made his telescope years before, dismissed the Cassegrain design as profoundly flawed. "The advantages of this design are none," he incorrectly asserted, and the disadvantages "great and unavoidable." Finally, New-

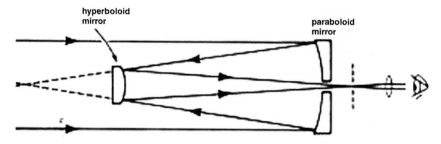

FIGURE 4.4 Cassegrain telescope.

ton emphasized that at least *he* had produced a working model. "I could wish, therefore, Mr. Cassegrain had tried his design before he divulged it. . . . Such projects are of little moment till they be put into practice." No opticians were capable of making the new French design, however, and no one took up the challenge.

Trouble from abroad kept coming. Christiaan Huygens, initially enthusiastic about Newton's color theory, which he called "highly ingenious," gradually turned against it.* In 1673, he wrote that Newton "hath not taught us, what it is wherein consists the nature and difference of Colors, but only this accident . . . of their *different Refrangibility*." He suggested that "it will be much more easy to find an *Hypothesis* by Motion"—that is, he wanted Newton to consider a wave theory of light, one that Huygens detailed later in his 1690 book, *Treatise on Light*.

Newton responded: "To examine how colors may be thus explained Hypothetically is besides my purpose." He would "leave it to others to explicate by Mechanical Hypotheses," but he added dismissively that it should be "no very difficult matter."

Brooding over the issue, Newton eventually did come up with his own "Hypothesis Explaining the Properties of Light," which he deliv-

---

*Newton's proof that different-colored lights refracted differently devastated Huygens, who had written an unpublished *Dioptrics* to reveal his theory of how to construct a telescope without spherical aberration. Once he realized that chromatic aberration was worse, he abandoned his book, scribbling that his previous conclusions were "useless." Ironically, Huygens invented a compound eyepiece for telescopes that was indeed achromatic, although he didn't recognize it. The Huygens eyepiece is still used on many telescopes.

ered to the Royal Society in December 1675, demonstrating that he was every bit as capable as Descartes of grand theories with little empirical support. Newton proposed the existence of an all-pervading "ether"— similar to air but "far rarer, subtler, and more strongly elastic"— through which all forces in the universe acted. "Perhaps may the Sun imbibe this Spirit copiously to conserve his Shining, & keep the Planets from receding further from him. And they that will, may also suppose, that this Spirit affords or carries with it [the] material Principle of Light." This hypothetical ether, Newton explained, was the medium that transmitted gravity, cohesion, electricity, and animal sensation, as well as light.

Newton defined light as "something or other capable of exciting vibrations in the ether," which appeared to leave the door open to waves. But then he spoke of "corpuscles" (particles) of light. As these particles of varying sizes passed through the ether, their speed and direction was altered by its density, which pervaded all things. As for glass and water, Newton oddly followed Descartes, asserting that the ether's lesser density there speeded rather than retarded light particles, turning them variously depending on the size of the particle, which accounted for colors. On a mirror surface, the ether's density was so great that the particles all bounced off at a similar angle.

In this paper, Newton also tried to account for the varying colors in soap bubbles, mica, and thin pieces of glass, phenomena first described by Hooke in *Micrographia*. Newton used a barely convex lens and a flat piece of glass. "I pressed them slowly together, to make the colors successively emerge in the middle of the circles, and then slowly lifted the upper glass from the lower, to make them successively vanish again." Then Newton sent light of a single color through, noting that red light made bigger circles and that it was "very pleasant to see them [the circles] gradually swell or contract accordingly as the Color of the Light was changed."

With only a compass, Newton measured the diameter of each ring to within 1/100th of an inch and the air space between the pieces of glass, down to 1/78,000th of an inch. He found a periodicity for each color, with rings appearing at predictable multiples of space between the two pieces of glass, with dark rings between them. Hooke fumed as the world called the phenomenon "Newton's Rings."

Newton struggled to explain this bizarre occurrence. He concluded that "the Air between the Glasses, according to its various thickness, is disposed in some places to reflect, and in others to transmit the Light of any one Color." He called these "Fits of easy Reflection and Transmission." He declined to speculate "whether it consists in a circulating or a vibrating motion of the Ray, or of the Medium, or something else." In other words, he had no idea what was going on, but if it was something like an epileptic seizure, at least it occurred regularly.

And there Newton left his optical studies for many years, proceeding to the *Principia*, which was finished in 1687. In 1690, Huygens published his *Treatise on Light*, in which he attacked Newton's theories. "I am astonished," he wrote, that Newton (to whom he referred without naming him) could have offered, "as assured and demonstrative, reasonings which were far from conclusive." Light could not possibly consist of particles, since it traveled much too fast, and the particles from different sources would bump into one another on the way to the eye. Instead, Huygens said that light traveled in waves, just as sound did.

But light traveled far more quickly than sound. Huygens cited the ingenious conclusions of Ole Römer, a Danish astronomer who in 1676 had approximated the speed of light by observing the varying rates at which Jupiter's moons appeared to move, depending on how far away the earth was. "Hence the velocity of Light is more than six hundred thousand times greater than that of Sound," Huygens wrote. "This, however, is quite another thing from being instantaneous." To explain this incredible speed, Huygens adopted Newton's concept of ether, but rather than it being "rare," Huygens's ether was so dense with hard little invisible particles that the motion of light could be communicated nearly instantaneously, just as one billiard ball, striking a group of contiguous balls, instantly sent balls rolling from the outside of the pack. With this wave theory, Huygens explained every aspect of light, including its apparent travel in straight rays, reflection from a mirrored surface, and refraction.

Newton's Queries

In response to Huygens, Newton began to compile his masterwork on optics, but he waited to publish it until 1704, when both Huygens and

Hooke were safely dead. In *Opticks*, Newton summarized all his previous work, experiments, and theories. He observed that light at a sufficiently oblique angle to glass or water is totally reflected rather than refracted and that the same thing is true once light gets inside glass—it is totally reflected if it hits the glass at a small enough angle. In Book Three, he covered Grimaldi's diffraction, which Newton called "inflection." He confirmed that, even in single-colored light, what he called "fringes" were produced surrounding the shadows of thin objects—Newton tried hair, a knife edge, threads, pins, straw—held in a narrow light beam.

He then launched into a series of "Queries," in which he felt free to hypothesize without proof, suggesting that "a farther search . . . be made by others" on these matters. Among the first Queries, he suggested that the diffraction phenomenon might be explained by light bending back and forth "with a motion like that of an Eel."

In subsequent editions of *Opticks*, Newton added more Queries as he pondered the mysteries of light and its reflection, refraction, and diffraction. For him, as for Grosseteste and Dee, light might be God's glue to hold the universe together, along with gravity and other mysterious forces. "Do not Bodies and Light act mutually upon one another?" he asked. "That is to say, Bodies upon Light in emitting, reflecting, refracting and inflecting it, and Light upon Bodies for heating them, and putting their parts into a vibrating motion wherein heat consists?" Later, he stated it more boldly: "Are not gross Bodies and Light convertible into one another?"

In his description of light's effects, Newton veered close to a wave theory. "Do not several sorts of Rays make Vibrations of several bignesses, which according to their bignesses excite Sensations of several Colors, much after that the Vibrations of the Air . . . excite Sensations of several Sounds?" The most refractable rays excited the shortest vibrations, producing violet, and the least refractable produced the largest vibration, creating a "Sensation of deep red" upon the retina.

Elsewhere, however, Newton clearly espoused the particle theory of light. "Are not the Rays of Light very small Bodies emitted from shining Substances?" Then he attacked the wave theory. "Are not all Hypotheses erroneous, in which Light is supposed to consist in Pression or Motion, propagated through a fluid Medium?" If it were, light would bend around corners into shadows (more than diffraction), the waves would bump confusingly into one another, and the necessarily thick ether would slow the planets in their orbits. Instead, Newton thought that particles of light set up waves in objects when they hit them, which accounted for visual perception and heat.

Newton still believed in an exceedingly subtle, thin ether, but he frankly admitted that "I do not know what this Ether is." Somehow, it facilitated action at a distance, such as gravity, electricity, and magnetism. He hypothesized that "there may be more attractive Powers than these," some that cover such small distances "as hitherto escape Observation." Newton emphasized that he did not consider these forces as mysterious qualities in the old-fashioned sense. "To tell us that every Species of Things is endowed with an occult specific Quality by which it acts and produces manifest Effects, is to tell us nothing."

Newton sounds completely modern and rational, reacting against critics who accused him of occult mumbo-jumbo with his theory of gravity, in which every object, no matter how far away, exerted some mysterious force on every other object. Yet Newton did have his credulous side. He spent a huge amount of time on useless alchemical experiments, attempting to transmute mercury. Some have suggested that when Newton suffered an acute period of irrationality, late in 1693, it was due in part to mercury poisoning.

Even when he was in his right mind, however, Newton was profoundly religious. In a majestic passage in the Queries, reminiscent of God's speech from the whirlwind in the Book of Job, Newton explains that the "first Cause" was certainly "not mechanical":

Whence arises all that Order and Beauty which we see in the World? To what end are Comets . . . and what hinders the fixed Stars from falling upon one another? How came the Bodies of Animals to be contrived with so much Art, and for what ends were their several Parts? Was the Eye contrived without Skill in Optics?

The Seashore of the Universe

Isaac Newton, the paranoid loner, eventually became Sir Isaac Newton, the infallible sage. In 1696, he left Cambridge to take over the British

Mint, where he reformed the currency. In 1704, he was elected president of the Royal Society, ruling it with an iron hand for nearly a quarter-century. The vindictive, defensive aspects of Newton's personality now found ready outlet. For years, he engaged in a nasty battle with Gottfried Wilhelm Leibniz over who could claim priority in inventing calculus.

Yet under the tyrannical bluster lurked the sad, lonely, imaginative little boy who wanted to understand everything, who yearned for love. "He had such a meekness and sweetness of temper," reported a companion of his old age, "that a melancholy story would often draw tears from him, and he was exceedingly shocked at any cruelty to man or beast."

One of Newton's biographers comments that there was "precious little of the child in the man," but that observation applies only to his crusty exterior. Newton spent much of his adult life playing with light—bouncing it off mirrors and refracting it into rainbows, blowing soap bubbles to study the writhing bands of color on their filmy surfaces, collecting feathers through which to gaze into the sun, making telescopes to view comets. "I do not know what I may appear to the world," Newton told a visitor shortly before his death at eighty-four, "but to myself I seem to have been only like a boy, playing on the seashore, and diverting myself, in now and then finding a smoother pebble or prettier shell than ordinary, whilst the great ocean of truth lay all undiscovered before me."

Like this chimpanzee, early hominids probably learned to recognize their reflections in still waters.

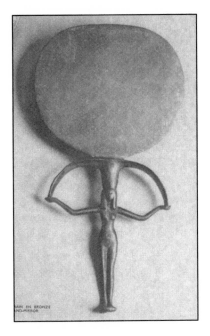

Hathor, the goddess of love, fertility, beauty, and dance, holds a bronze Egyptian sun-mirror.

The backs of Etruscan mirrors were illustrated with elegant line engravings, often with sexual themes. The inscription identifies the central lovers here as Mexio and Fasia.

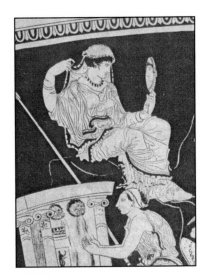

The goddess Hera admires herself in a mirror on this Greek vase.

Chinese magic mirrors, which reflect the design on the mirror back, mystified opticians until they realized the polishing technique caused imperceptible irregularities on the mirror surface.

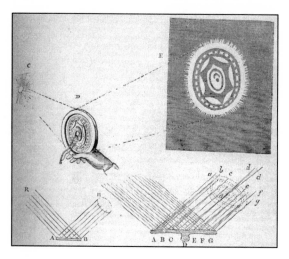

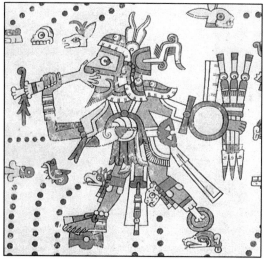

The Aztec god Tezcatlipoca, whose name means "Smoking Mirror," had a scrying mirror instead of one foot. In this drawing he is eating the palm of a sacrificial prisoner, considered a great delicacy.

Elizabethan scientist-occultist John Dee was fascinated by mirror illusions and optics. In seeking "the best knowledge that man might attain unto in the world," he fell under the influence of scryer Edward Kelley.

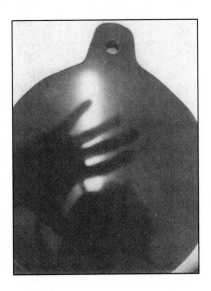

In this Aztec obsidian mirror, John Dee believed his scryer could see and speak to angels. The author's hand is reflected in this picture taken at the British Museum.

Legend has it that Archimedes used mirrors to set fire to Roman ships during the siege of Syracuse. Probably untrue, the story nonetheless inspired useful experiments for centuries.

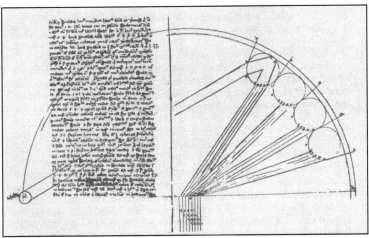

Around 1300, Dietrich of Freiberg solved part of the rainbow puzzle by observing sunlight in a large water-filled glass globe. His drawing here shows a sunbeam refracting as it enters a raindrop, reflecting from the back, and refracting again as it leaves.

Concave spherical mirrors can create startling optical illusions—in this case, it appears you can shake your upside-down virtual hand.

Convex spherical mirrors, like this small polished ball, distort images, as in this self-portrait of the author, who doesn't usually look so menacing.

In 1620 a British visitor marveled at the landscapes Johannes Kepler drew with the aid of his camera obscura, "me thoughts masterly done . . . surely no Painter can do them so precisely."

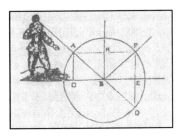

In his *Dioptrique (Optics)*, Descartes illustrated the law of reflection with a midget tennis player hitting tiny balls of light onto a mirror.

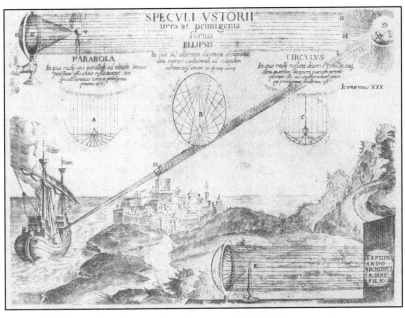

This illustration from Athanasius Kircher's *Ars Magna Lucis et Umbrae* (The Great Art of Light and Shadow) shows reflective parabolas, ellipses, and spheres, with an early candle searchlight. But Kircher believed that Archimedes used a mirror to burn the ships, not a lens as shown here.

17th-century Polish brewer Johannes Hevelius built long refracting telescopes, but they suffered from buffeting winds and misalignment.

The ultimate loner, Isaac Newton made the first functional reflecting telescope, grinding and polishing his own metal mirror. "If I had stayed for other people to make my tools and things for me, I would have never made anything of it."

When he split a beam of sunlight with a prism, Newton was surprised that the resulting colors were not circular. He eventually concluded that white light was a "Hetergeneous mixture of differently refrangible Rays."

The *Romance of the Rose*, also called *The Mirror of Lovers*, was an international literary sensation that praised the "marvelous powers" of mirrors and sex. This 14th century French ivory mirror case shows the climactic storming of the Castle of Love.

Dante's *Paradiso* is filled with images of light, glass, and mirrors. In the mystical finale, Dante is allowed to gaze directly at the "Living Light."

Elizabethan poetry and plays glitter with looking-glass references. In Shakespeare's *Richard II*, the dethroned king complains that "a brittle glory shineth in this face," then smashes the mirror.

In this 1403 illustration, a nun uses a hand mirror to paint a self-portrait. But the illustration itself is awkward. Note how the square floor tiles lack perspective.

In Jan van Eyck's 1434 painting, Giovanni Arnolfini apparently weds a woman who appears to be extremely pregnant. Between them, a convex mirror reflects the scene. Van Eyck may have achieved such realism by using a convex mirror as an aid.

A close-up of the convex mirror shows the couple's back, the distorted window and bed on either side, and two tiny figures in the doorway, one of whom must be van Eyck.

Albrecht Dürer painted this self-portrait in 1484 at age 13. He may have used two mirrors to see himself from the side.

In this self-portrait at age 21, Dürer stares directly at himself in the mirror with worried intensity. He was the third of eighteen children, of whom only three survived to adulthood.

In this 1509 painting by Hans Baldung, the beautiful young woman admiring herself in the convex mirror is oblivious to death holding an hourglass over her head. The vanity-death theme ran through European art in a time of high mortality rates and increasing mirror usage.

Satirical artists mocked women (though men too were concerned with appearance) for spending too much time in front of mirrors, as this anonymous 1493 engraving demonstrates with anal humor.

Artists did not satirize all mirror usage. In this 1515 painting, Bellini passes no judgment on the lovely young woman using *two* mirrors to examine herself.

Parmigianino's 1524 self-portrait deliberately emphasized the distortions produced by the convex mirror.

This 1568 German print of *Der Spiegler* (The Mirror-Maker) shows the craftsman cutting circular convex mirrors with special scissors while talking to customers.

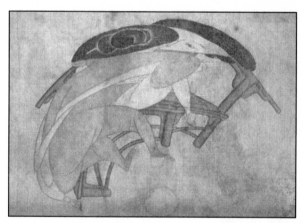

Around 1600 the Chinese invented cylindrical anamorphic art, in which distorted paintings can be "solved" by reflection in a tubular mirror. Many of the Chinese (and subsequent European) images showed erotic scenes such as this.

This anamorphic painting followed the 1660 beheading of King Charles I, which gives gruesome meaning to the skull (not seen here) over which the cylindrical mirror is placed.

In the late 17th century the French broke the Murano mirror monopoly in one of the world's first examples of industrial espionage. They also invented a new casting method (shown here) to make larger glass plates.

King Louis XIV, the Sun King, opened the Hall of Mirrors at Versailles in 1682. A contemporary reporter called it a "palace of joy . . . a dazzling mass of riches and lights, duplicated a thousand times over in just as many mirrors."

Mirrors + humans = sex has been true throughout history, as voyeurs enjoy observing themselves. In this 1810 Rowlandson drawing entitled *The Curious Wanton*, a woman examines her genitals.

This early 19th century caricature mocks the male dandy, with Russian Oil, Curling Fluid, and rouge on his dressing table. Note that mirror size is growing.

William Hershel, a German musician who became obsessed with astronomy after he moved to England, pursued ever-larger metal mirrors for more "light grasp." He was sponsored by King George III.

Herschel's 40-foot telescope had a mirror four feet in diameter. One visitor called it "a mighty bewilderment of slanted masts, spars and ladders and ropes, from the midst of which a vast tube . . . lifted its mighty muzzle defiantly towards the sky." The unwieldy telescope never worked very well.

As an old man, John Herschel (William's son) posed for a photograph — a new medium he helped invent. "In the midst of so much darkness, we ought to open our eyes as wide as possible to any glimpse of light," he advised shortly before his death.

At Birr Castle in Ireland, the Leviathan of Parsonstown held a four-ton metal mirror six feet across.

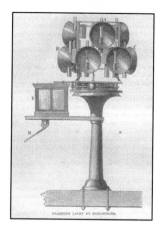

Traditional lighthouses used concave metal mirrors to project the light, but they were inefficient. These rotating mirrors flashed a distinctive series of lights.

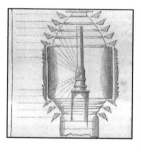

The Fresnel lens, shown here to the right on display at the 1851 Crystal Palace exhibit, used a series of prisms to send out parallel light beams, as the drawing to its left indicates.

(Right) Alexander Graham Bell invented the photophone in 1880. As Bell spoke, a small, thin mirror vibrated, causing minute fluctuations in the light reflected from it.

(Left) Across the room, a paraboloid dish-mirror concentrated the light waves onto a selenium cell and a telephone receiver. "I have heard a ray of the sun laugh and cough and sing!" Bell exulted.

In 1895 Wilhelm Conrad Röntgen accidentally discovered high-energy x-rays and took this picture in which the subject's skeleton and the keys in his pockets appeared. X-rays are so powerful that they plow into regular mirrors instead of reflecting from them.

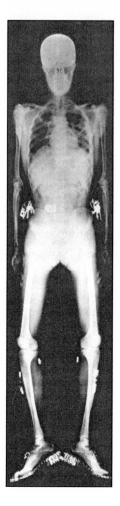

With a diamond stylus and an accurate screw mechanism, Henry Rowland's machine could incise 43,314 lines to the inch to make accurate reflective gratings for spectroscopes. Rowland wanted his ashes interred near his machine.

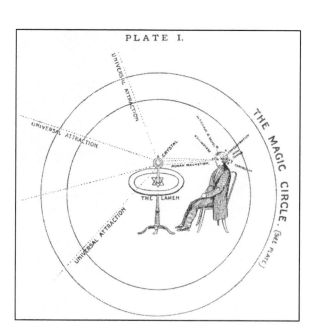

Scrying and magic mirrors enjoyed a revival in the late 19th century, as this illustration from John Melville's *Crystal Gazing and Clairvoyance* (1896) shows. "The surface of the mirror or crystal becomes charged," Melville wrote, "the Brain being as it were switched on to the Universe."

The Pepper's Ghost illusion used a large sheet of glass (invisible to the audience) as a beam splitter, so that the actor out of sight in the pit appeared as a translucent ghost on stage. This 1871 illustration isn't quite accurate, since the actor had to lie on an inclined platform in order to appear upright on stage.

The Sphinx illusion, in which a disembodied head in a box opened its eyes, turned from side to side, and spoke. The audience was fooled by the mirrors between the table legs, though this 1865 illustration is unrealistic, if you think about it.

The "Mystic Maze" consisted of three large mirrors facing each other in an equilateral triangle. Through infinite reflection, a few people appeared to be a huge crowd.

Mathematics don Charles
Dodgson was inspired to write
Through the Looking-Glass
by young Alice Raikes' answer to
the mirror reversal puzzle. Here
the fictional Alice climbs through
the mirror to a world where
"things go the other way."

Certainly the glass *was* beginning to melt away.

Dodgson's fascination with mirrors is
apparent in his photograph of his younger
sister Margaret.

George Malcolm Stratton literally sought to
broaden his vision with various mirror
contraptions. In homage, modern optics professor
Nicholas Wade put together this upside-
down/rightside-up composite picture of Stratton.

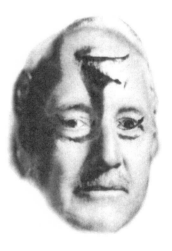

Chapter 5

LOOKING-GLASS LITERATURE

Glass is more gentle, graceful, and noble than any metal, and its use is more delightful, polite, and sightly than any other material at this day known to the world.

ANTONIO NERI, *L'ARTE VETRARIA*, 1612

ISAAC NEWTON BELIEVED that mirrors could, in theory, bring telescopes to "any degree of perfection imaginable," but his toy-sized telescope's metal mirror tarnished quickly, and every time he rubbed it, he changed its shape. Also, "more Light was lost by Reflection in the Metal, than by Refraction in the Glass."

Newton suggested using glass mirrors for future telescopes, since glass was easier to polish. He wanted an optician to grind a piece of glass of uniform thickness, with a perfectly concave front and convex back. The back should be "quick-silvered" with highly reflective mercury, then protected with paint or varnish. Unfortunately, he couldn't find a sufficiently skilled craftsman in London. Newton didn't address the problem of light refracting through the front glass surface as it traveled in and out of the mirror, and the metallic tin amalgam reflected well only when applied to the rear surface. Newton's idea died with him.

In 1664, when Newton bought his first prism at the Stourbridge Fair, the world's best mirrors were made on the island of Murano, 3 miles north of Venice, from amazingly large sheets of clear glass with a highly reflective back coating. The secret of their manufacture was known only to the Italian workers. By 1704, when Newton suggested a glass-mirrored telescope, the monopoly had been shattered through one of the world's first acts of industrial espionage. That event, described in Chapter 6, eventually transformed the mirror from a relatively rare item to an everyday household ornament. The fine glass mirror also changed the world of literature, art, and architecture and fundamentally altered the way people viewed themselves and their world.

Congealed Air

Glass is nearly as magical as light itself. Made primarily from sand, it can be virtually invisible, as clear as water or air. Because its chaotic molecules are not held together in rigid crystalline form, glass is a kind of solid liquid. Yet it is hard enough to be molded, sanded, blown, ground, polished, melted, colored, and twisted. In the right light conditions, it can act as a good mirror on its own, and with proper reflective coatings, it can make an extraordinarily good mirror—so good that it seems to disappear.

In the first century, Pliny the Elder recounted a creation myth. In Phoenicia, where the sluggish Belus River flowed into the sea, it deposited muddy sand that was eventually washed clean by the waves. "A ship belonging to traders in soda once called there, so the story goes," wrote Pliny, "and they spread out along the shore to make a meal. There were no stones to support their cooking-pots, so they placed lumps of soda from their ship under them. When these became hot and fused with the sand on the beach, streams of an unknown translucent liquid flowed, and this was the origin of glass."

The story sounds plausible, since the Phoenicians were great traders, and natron, a natural soda, was much in demand as an embalming agent. By itself, pure sand (silicon dioxide) makes perfectly good glass, but only if heated to an extraordinary temperature. An alkaline substance such as soda (sodium oxide or sodium carbonate) acts as a fluxing agent, allowing lower melting temperatures, and an alkaline earth such as lime (calcium oxide, chalk) makes the glass less viscous and protects it from water damage. Perhaps bits of seashell in the beach sand provided the chalk in Pliny's anecdote.

Yet man-made glass probably predated these traders. It may have been discovered originally as an intriguing by-product in a Sumerian pottery kiln or bronze smelter. By 2500 B.C.E., the Egyptians were making gorgeous glass beads. Perhaps the first craftsmen learned from nature, which creates imperfect glass. When lightning strikes a beach or desert, it fuses the sand into fulgurites, thin glass tubes. A meteor smashing into rock creates scattered tektites, dark-colored little glass blobs. The most common type of natural glass is obsidian, the shiny, dark substance melted in volcanic hearts and used by early man for tools, weapons, and mirrors.

As we have seen in Chapter 1, the Romans blew glass spheres and poured molten lead into them to produce cheap convex mirrors. Very few large glass reflectors were made, since the glass produced by the Romans was generally discolored, wavy, and full of bubbles and lines. Mirrors made of silver and bronze predominated for another millennium, and the quality of European glass declined along with the Roman Empire.

The art of the convex glass mirror was apparently kept alive (barely) following the fall of the Roman Empire, though the archaeological evidence is thin.* In the Near East, convex mirrors from Roman through Islamic times have been found in graves. A few tiny mirrors or shards are found in Viking gravesites of the ninth through eleventh centuries. There is no question, however, that such convex mirrors were made again in northern Europe, beginning in the twelfth century. Throughout the Middle Ages, they appear in art and literature. While most of the thin, fragile mirrors have disappeared (and many glass shards were undoubtedly recycled), quite a few mirror holders made of wood, bone, ivory, and metal survive from the period.

As they did with optics, the Islamic countries learned from and improved upon European practices, creating gorgeously engraved and enameled glass objects in Baghdad and elsewhere from the eighth century on. In Islamic Spain, craftsmen produced glass mirrors by the eleventh century.

Meanwhile, in Europe, soaring Gothic cathedrals required brilliantly colored stained glass, much of it produced by glasshouses connected to monasteries. In the abundant forests of northern Europe—notably in

---

*Such convex mirrors are still made in Gujarat, India, where the cheap, thin convex mirror pieces are sewn into fancy, glittering clothing.

the area between the Rhine and Meuse, in Lorraine, and near Nürnberg in Bavaria—craftsmen produced *Waldglas* (forest glass), using local timber as fuel for the furnace and wood ash for soda in the mix. "The men stand continually half-naked . . . near very hot furnaces and keep their eyes fixed on the fire and molten glass," one visitor marveled. Local mothers invoked the terrifying soot-blackened workers to threaten misbehaving children.

Such sweaty laborers may have seemed half-savage, but they produced some very fine glass as well as thick-bodied mugs. Their mirrors improved in quality and got larger over time. They blew thin-walled spheres about 2 feet in diameter—the size of a large beach ball—then, while the glass was still hot, poured molten lead in and swirled it to a thin coating. Then the cooled ball was broken and smaller pieces cut with scissors. These *Ochsenaugen* (oxen-eyes) became common in Europe in the late medieval period. Most were only 2 or 3 inches in diameter, though some were as large as 8 inches across. The quality varied considerably. Some convex mirrors, made with thick, striated glass and poorly applied lead, were called *Schattengesicht* (shadow-face). Others, made with thin glass that reduced the greenish tinge, and well-coated with lead, yielded good images, albeit somewhat distorted by the convexity.

At some point, certainly by the early fifteenth century, glassmakers in Germany, France, and Italy learned to blow relatively large cylinders of glass, then open the ends and slit them down the side to produce sheets of glass as large as 30-by-45 inches, which could then be coated (though imperfectly) with a hot lead-antimony layer. Florentine artisans apparently learned to apply unheated lead or tin to the glass, thereby avoiding much danger and breakage. Carnival song lyrics of the Florentine Mirror-makers' Guild bragged: "Our trade demands experience and adroitness in managing the blown glass forms. . . . But worth more than this or anything else is the secret of the substance which one places behind the glass with great skill."

The Mirror Island of Murano

Florence may have produced good mirrors, but it was in Venice that glass was perfected and where the modern mirror industry was born.

From the eleventh century on, Venice held a monopoly on commercial shipping to and from the East. Its craftsmen probably learned the art of glassmaking from Germans or Islamic exporters. Venetian importers sold luxuries from abroad, including cotton, silk, oranges, figs, rugs, gems, drugs, pepper, incense, perfume, swords of Damascus steel, porcelain, glassware, and mirrors. In the main Square of San Marcos, Protestants mingled freely with Catholics, Greek Orthodox, Muslims, and Jews. All were subservient to the mighty gold ducat, and the laws ensuring economic stability were strict. Forgers had one hand cut off. Trade secrets were strictly kept, with official detectives snooping vigilantly.

A Venetian glassmakers' guild formed in the early 1200s. In 1291, city authorities forced the dangerous furnaces to move to the nearby island of Murano. There, accidental fires wouldn't burn down the city—and, perhaps as important, the glassmakers could be better protected from spies and prevented from leaving. Although they earned top wages and were allowed to marry the daughters of nobles, the glass experts were prisoners. Flight warranted the death penalty. And so the glass families produced generations with venerated names—Barbini, Beroviero, Briati, Bertolini, La Motta, del Gallo—who learned the art of mirror-making from childhood and who knew every alley and inlet on their tiny island.*

Around 1450, the Muranese glassmaker Angelo Beroviero, using ash from sea plants rich in potassium oxide and magnesium, created an extraordinarily clear type of glass, which he christened *cristallo*, since it resembled the clearest rock crystal. In 1507, Andrea and Domenico d'Anzolo del Gallo petitioned the Venetian Council of Ten, seeking a kind of patent for a new foiling method, which involved pounding tin to a uniformly thin layer, then rubbing mercury over it, forming a shiny amalgam. Covering it with paper, a workman lowered a sheet of glass over it with one hand while carefully pulling the paper out with the other. Weights atop the glass assured a bubble-free reflective surface that clung tightly to the glass and was then covered with a protective

---

*German glass historian Ingeborg Krueger thinks that early Italian glass-mirror expertise is a myth. She cites a 1215 contract to deliver to Genoa "glass of the very best quality for making mrrors" from German glass houses, and points out that in 1317 three Venetians hired a German master to help them make mirrors, but the enterprise failed when the master departed.

varnish. The Venetian authorities granted the del Gallo brothers a twenty-year monopoly.

A two-century rage for the exquisite, expensive Murano mirrors ensued. At the beginning of the sixteenth century, a Venetian mirror in an elaborate silver frame was valued at 8,000 pounds, nearly three times the contemporary price of a painting by Raphael. The French monarch François I, a lover of luxury and Italian art, helped create the mirror craze by ordering a Murano mirror decorated with gold and precious stones in 1532, followed by thirteen more the next year and eleven additional mirrors in 1538.

Upon François's death in 1547, his son, King Henri II, kept buying Venetian mirrors for his Italian wife, Catherine de Medici. After Henri's death in 1559, she had a *cabinet de miroirs* (mirror chamber) installed to honor his memory. There, above the fireplace, a portrait of the departed monarch was reflected in 119 Venetian mirrors set into the chamber's paneling.

Sacred and Secular Specula

Over the course of six centuries, ever-finer and larger glass mirrors gradually supplanted those of metal, with unexpected cultural and social consequences. As we have seen, humans had always regarded mirrors with awe. Saint Augustine, among other early Christian theologians, saw the perfect mirror as a metaphor for divine wisdom, so it was natural that mirrors provided sacred book titles such as *Speculum Ecclesiae* (The Ecclesiastic Mirror, ca. 1100).

Hugh of Saint Victor (1100–1141) explained in his commentary on the Rule of Saint Augustine: "[This book] is rightly called a mirror; for we can see in it as in a mirror in what state we are, whether beautiful or deformed, just or unjust." Hugh's contemporary, Peter Lombard, observed that "the soul is a mirror in which in some way we know God," followed a few years later by Alanus de Insulis, who elaborated a more complex mirror metaphor: "O Man! See yourself in this three-fold mirror . . . the mirror of the Holy Scriptures, the mirror of nature, and the mirror of creatures." Only the scriptural reflection was a happy one, of course. The false mirror of the flesh was concave, inverting reality. In the thirteenth century,

Saint Bonaventura, a Franciscan mystic, wrote of the *speculum inferius*, the mirror of creation, and the *speculum superius*, the mirror of God.

Thomas Aquinas, a contemporary of Bonaventura, offered one of the first etymological links between mirrors as *specula* and the modern meaning of *speculation*: "To see something by means of a mirror is to see a cause in its effect wherein its likeness is reflected," Aquinas wrote. "From this we see that 'speculation' leads back to meditation."

Around 1260, Vincent of Beauvais, a French Dominican, published his encyclopedic compendium of medieval knowledge, the *Speculum Maius* (Large Mirror), which included three parts—the *Speculum Naturale*, covering theology, psychology, physiology, cosmography, physics, botany, and other sciences; the *Speculum Doctrinale*, on subjects such as logic, rhetoric, poetry, and astronomy; and *Speculum Historiale*, a sweeping history. After Vincent, innumerable mirror titles flowed from clerical pens.*

Beginning in the twelfth century, a revived worship of the Virgin Mary took the mirror as her symbol, derived from the *speculum sine macula*, or "spotless mirror," of the second-century *Book of Wisdom*. The *Speculum Virginium*, a twelfth-century "Mirror of Virgins," urged virtuous women to emulate the Virgin Mary in maintaining their sexual purity. "Maidens look into mirrors to see whether there is any increase or decrease of their adornment, but Scripture is a mirror from which they may learn how they can please the eternal spouse." Small mirrors began to appear in church interiors as symbols of purity.

Gradually, however, mirrors invaded secular literature, a shift exemplified by the lovers' mirrors of *Roman de la Rose* (Romance of the Rose), by Guillaime de Lorris and Jean de Meun. De Lorris, who wrote in the early thirteenth century, created an allegory of courtly love. The lover, gazing into the eyes of his beloved, sees them as two crystalline mirrors in a garden fountain. Yet her mirror-eye is also dangerous. "Out of this mirror a new madness comes upon men: Here hearts are changed; intelligence and moderation have no business here, where there is only the simple will to love."

---

*Chinese and Japanese mirror titles independently paralleled European developments at nearly the same time. In China, the *Tzu-chih T'ung-chien* (Comprehensive Mirror for Aid in Government) appeared in 1085, followed by the Japanese *Ôkagami* (The Great Mirror), an exemplary biography, around 1100.

De Lorris died before finishing. Forty years later, Jean de Meun picked up the story, naming his lengthy conclusion *Le Mirouer aus Amoureus* (The Mirror of Lovers) and taking mirror imagery to a new erotic level. He also used his literary mirror as a harsh satiric spotlight on all classes of society, attacking magistrates, soldiers, nobles, and monks for their foibles.

In a lengthy digression, de Meun put an impressive speech about mirrors into the mouth of Nature. Heavily influenced by the renewed interest in optics and mirrors sparked by Alhacen, Robert Grosseteste, and Roger Bacon, de Meun's Nature praises "the causes and the strength of the mirrors that have such marvelous powers" to reflect and alter images. Nature laments, "If Mars and Venus, who were captured in the bed where they were lying together, had looked at themselves in such a mirror before they got up on the bed," they would have seen the fine net Vulcan had spread to capture them.

By implication, the lovers would also have enjoyed the sight of their own lovemaking in the mirror. Jean de Meun's poem—a best-selling title for several centuries, translated into many languages—was more than anything else a celebration of sex, ending with a near-pornographic climax in which the lover finally satisfies the now-inflamed virgin.

By the time Jean de Meun wrote in the late thirteenth century, wealthy women primped and preened with "these head ornaments, these coifs with golden bands, these decorated head-laces, [looking into] ivory mirrors." He referred to small cases of exquisitely carved ivory, holding either glass or metal mirrors and usually depicting the Crucifixion or scenes from the life of the Virgin Mary. By the fourteenth century, such ivory mirror cases frequently depicted the Assault on the Castle of Love, in which knights attacked a fortified gateway while women threw flowers from the parapets, or the sequel in the Garden of Love, with a happy couple holding a heart between them. One late-fourteenth-century mirror case shows a couple with the woman holding a round mirror coyly behind her back, then holding it over her lover as he fondles her breast.

Although the secular mirror was firmly established in literature by 1300, a strong tradition of religious symbolism continued, particularly in Dante Alighieri's *Divine Comedy*. Dante was influenced by Jean de

Meun's work in many ways—including a scientific interest in mirrors and optics. *Paradiso*, which Dante finished shortly before his death in 1321, is filled with devotional mirrors.

Paradiso commences with a mirror experiment related to the moon. "Take, then, three mirrors," says Beatrice, Dante's virginal guide to Paradise, "and let two of them be set at equal distance from thyself, the third set farther off, between the others. Face toward them, and behind thy back install a lamp which so illuminates all three that thou canst see its image in them all. Thou wilt perceive that the more distant one, reflecting least in quantity of light, will be as brilliant as the other two." Just so, the light of God shines equally on all of His creation, regardless of the distance from Him.

Throughout *Paradiso*, imagery involving light, glass, and mirrors predominates. In the mystical finale, Dante is allowed to gaze directly at the "Living Light," a Trinity of self-mirroring glory. In contrast to the scientific experiment described at the beginning of the poem, in which the light source was hidden behind the observer, the three mirrors are now gloriously self-illuminating. "Of that Exalted Light, I saw three rings of one dimension, yet of triple hue. One seemed to be reflected by the next, as Iris is by Iris; and the third seemed fire, shed forth equally by both."

Following Dante's devotional mirrors, the most popular mirror title of the fourteenth century, *Speculum Humanae Salvationis* (Mirror of Human Salvation), was also religious. Nonetheless, secular mirror titles increased dramatically after 1300, gracing medical, astrological, and alchemical texts, and the attitude toward the once-sacred mirror became ambivalent. Poet Jean Molinet (1425–1507) called his midnight mirror "an impossible and contrary monster," imagining that in the Garden of Eden a glorious mirror perfectly reflected the image of God before Adam and Eve ate the forbidden fruit, whereupon, when they looked in the mirror, it cracked in two. In Sebastian Brandt's 1494 *Das Narren Schiff* (The Ship of Fools), morons use mirrors—a vain young man changes his clothes before one, a conceited old man believes himself wise in his, a lady contemplates her beauty in her mirror while the Devil sets fire to her bench. François Rabelais placed 9,332 mirrors in as many bedrooms in *Gargantua* (1525).

The Elizabethan Looking Glass

By 1500, more than 350 European books had mirror titles of one sort or another. With the invention of printing in the middle of the fifteenth century, the number of such titles exploded, particularly between 1550 and 1650. In *The Mutable Glass*, the classic study of mirror titles, author Herbert Grabes calls the mirror "the central metaphor of a literary era," especially in Elizabethan England.

Glass mirrors had become so widespread that the words "mirror" and "glass" were interchangeable. Mirror metaphors, in titles and in literature, took on a rich layer of meanings, depending on the context. There were, of course, religious tracts, such as *A Christal Glasse of Christian Reformation*. Others, such as *A Myrroure for Magistrates* (1559), held up an exemplary mirror, exhorting politicians to behave well, as did the hard-hitting *A Trewe Mirrour or Glasse Wherein We Maye Beholde the Woful State of Thys Our Realme of Englande* (1556).

Some titles, such as the *Mirrour of Princely Deedes and Knighthood* (1578), were eulogies to heroic figures, whereas the *Christal Glasse for Christian Women* (1591) was a best-selling moral mirror, as was the titillating *A Glasse for Amorouse Maydens to Looke In* (1582). The *Mariners Mirror* (1588), *Myrrour for English Souldiers* (1595), and *A Mirror for Mathematiques* (1587) were practical how-to books, along with lengthy titles such as *This Is the Myrour or Glasse of Helthe Necessary and Nedeful for Euery Person to Loke In, That Wyll Kepe Their Body from the Syckenes of the Pestylence* (1531), or *A Mirrour of Loue, Which Such Light Doth Giue, That All Men May Learne, Howe to Loue and Liue* (1555). Similar titles, suitably modernized, still populate the self-help sections of modern bookstores.

In *The Steel Glass*, published in 1576, British satirical poet George Gascoigne poked fun at court life, contrasting it with country virtues. "What monsters muster here [in my mirror]," he wrote, "with angels' face, and harmful hellish hearts?" Such men used the newfangled "christall glasse" mirrors from Venice rather than the traditional mirrors of steel, which Gascoigne preferred.

Mirror titles reflected every conceivable point of view. *The Mirrour of Madnes, or a Paradoxe Maintyning Madnes to be Most Excellent* (1576) sat on the shelf beside *A Mirrhor Mete for All Mothers, Ma-*

trones, and Maidens (1579). *The Mirrour of Mirth and Pleasant Conceits* (1583) might provoke a smile, whereas reading the English title of a 1576 translation of Pope Innocent III's gloomy treatise was enough to depress anyone: *The Mirror of Mans Lyfe, Plainely Describing, What Weake Moulde We are Made of: What Miseries We are Subiect unto: Howe Uncertaine This Life Is: and What Shal Be Our Ende.* And then there were the polemicists whose cracked mirrors reflected their sour views, such as the 1594 anti-Catholic *A Mirrour of Popish Subtilties* or the 1587 diatribe against the theater, *The Mirrour of Monsters: Wherein Is Plainely Described the Manifold Vices, & Spotted Enormities, That are Caused by the Infectious Sight of Playes.*

In the popular plays themselves, mirrors and mirror metaphors often played an important role. William Shakespeare's works, composed between 1589 and 1613, glitter with witty mirror references, many of them central to his concern with identity, illusion, and reality. "Methinks you are my glass," says Dromio to his newly discovered twin brother at the end of *The Comedy of Errors.* "I see by you I am a sweet-fac'd youth."

In *The Two Gentlemen of Verona*, Julia, disguised as a sunburned man, describes herself to another character who doesn't realize the irony. "Since she did neglect her looking-glass, / And threw her sun-expelling mask away, / The air hath starv'd the roses in her cheeks, / And pinch'd the lily-tincture of her face, / That now she is become as black as I." Here, Shakespeare subtly mocked the manners of the times, which demanded that women have white complexions and rouged cheeks, which they frequently checked in their mirrors. "For there was never yet fair woman but she made mouths in a glass," the Fool tells King Lear.

In Sonnet LXII, Shakespeare toyed with the notion of male narcissism. "Sin of self-love possesseth all mine eye, / and all my soul, and all my every part. . . . Methinks no face so gracious is as mine." But then "my glass shows me myself indeed, / Beated and chopp'd with tann'd antiquity." In the final couplet, he explained that because his beloved *is* himself, and he adores her so totally, loving her is like loving himself.

By the late sixteenth century, mirrors were so common and had acquired such stereotypical symbolic weight that Shakespeare could pull off these clever reversals, knowing that his audiences would appreciate them. In courting a French princess, the rough-hewn King Henry V says that he cannot "gasp out my eloquence," but he hopes that she can love a fellow "that never looks in his glass for love of any thing he sees there. . . . I speak to thee plain soldier."

Sometimes the mirror is not just a metaphor but literally takes center stage. After he has been deposed, Richard II asks for a mirror. When he looks at himself, he is surprised to see that he looks the same as ever. "No deeper wrinkles yet? Hath sorrow struck / So many blows upon this face of mine, / And made no deeper wounds? O flatt'ring glass, / Thou dost beguile me! . . . Was this the face / That like the sun, did make beholders wink? . . . A brittle glory shineth in this face, / As brittle as the glory is the face"—with this, Richard dashes the mirror in pieces on the ground.

In other Shakespearean scenes, the mirror tells hard truths. Hamlet uses his mother's bedroom mirror to force her to face reality. "Sit you down, you shall not boudge; / You go not till I set you up a glass / Where you may see the [inmost] part of you." Such stark mirrors often reflect human mortality. "Death remembered should be like a mirror, / Who tells us life's but breath, to trust it error," Pericles asserts.

Ultimately, Shakespeare held the mirror up to nature, as Hamlet advises the itinerant players to do, and mostly that meant holding it up to *human* nature. In one of his most intriguing passages, in *Measure for Measure*, Shakespeare had one of his strong female characters spit:

> But man, proud man,
> Dress'd in a little brief authority,
> Most ignorant of what he's most assur'd
> (His glassy essence), like an angry ape
> Plays such fantastic tricks before high heaven
> As make the angels weep.

Man's "glassy essence" is the brave-looking but illusory and fragile figure he sees in his mirror, the one that struts and frets its hour upon the stage. He sees himself in an imperfect mirror, and he is most ignorant of that which he thinks he knows best—his own essential character. He has no more self-knowledge than the ape that, confronting himself in the mirror, angrily postures before the image of another posturing ape he sees there.

Many other British poets of the era, including Edmund Spenser, Philip Sidney, Samuel Daniel, John Davies, Michael Drayton, John Suckling, Henry Constable, Joseph Beaumont, George Chapman, Fulke Greville, and John Donne, used mirror metaphors to reflect on the human condition, especially in love poems, such as "Elegies of Love" by John Davies:

Within thine eyes (the Mirrors of my mind)
Mine eyes behold themselves, wherein they see
(As through a Glass) what in my Soul I find;
And so my Soul's right shape I see in thee.

Neither was the mirror limited to British literature. Italian poet Giambattista Marino reversed the metaphor, asserting that his beloved was so beautiful that her face itself served as a perfect mirror.

Miguel de Cervantes had Don Quixote compare his ideal lady to "a brilliant and polished crystal mirror that the slightest breath darkens and tarnishes. She should be treated like a relic, adored but not touched." Cervantes was, of course, mocking Petrarchan attitudes; Don Quixote's idealized love was in reality a prostitute.

Abraham Cowley twisted the theme to mock vanity: "Can that for true love pass, / When a fair woman courts her glass?" James Shirley cautioned beautiful women to use their mirrors to reflect their inner worth:

For not to make them proud
These glasses are allowed
But to compare
The inward beauty with the outward grace,
And make them fair in soul as well as face.

Because expensive Venetian mirrors were fragile, they sometimes became "that brittle Emblem of Corruption" for poets such as Joseph Beaumont. The same poet could turn around and utilize mirror metaphors in a completely different context, though. Thus, Beaumont also wrote of the "never-erring Glass . . . *Truth's Mirror.*"

Such a truthful mirror was the opposite of the "flattering glass," a type of mirror that supposedly made people look better than they did in reality. Perhaps these mirrors were slightly convex, to make people look thinner, but it is likely that they were simply poor mirrors that didn't reveal wrinkles or other blemishes. Queen Elizabeth, whose heavy white makeup usually hid her face, apparently used such a flattering glass most of her life. According to an early biographer, in her final days "she desired to see a true looking glass, which in twenty years she had not seen, but only such a one as was made of purpose to deceive her sight: which glass, being brought her, she fell presently into exclaiming against those which had so much commended her."

Spiritual Optics and Mirrors of Piss

After 1610, when Galileo's telescope revolutionized the world of science, authors quickly adapted the new technology to their mirror titles. John Vicars's *A Prospective Glasse to Looke into Heaven* (1618) was not, despite its reference to the telescope, a book on astronomy but rather a religious tract. In 1628, the *Prospectiue Glasse of Warre* offered a detailed how-to-battle book. The subjects didn't change much, other than to reflect the increasing influence of the Puritans. *A Prospective-glasse For Gamesters; or, a Short Treatise Against Gaming* was the title of a 1646 diatribe, and *Spiritual Optics: or a Glasse Discovering the Weaknesse and Imperfection of a Christians Knowledge of This Life* was a lively 1651 text.

Seventeenth-century literary mirrors often reflected strong sectarian views. As Oliver Cromwell took over and then Charles II mounted the throne, there were pro- and antimonarch books such as *A Looking Glass for Traytors . . . Who Contrived and Compassed the Death of His Late Sacred Majesty King Charles the First* (1660). Unflattering mirrors were held up to drunkards, mutineers, corn-hoarders, papists, Quakers, Jews, Anabaptists, married couples, women, corrupt lawyers, soap makers, the Irish, Parliament, fanatics, and New England colonists.

Sensationalistic titles appealed at once to prurient and puritanical sensibilities. *A Looking-Glasse for Young-Men and Maids* (1655) told the cautionary tale of two lovers who fell into a brewer's vat while "striving about a kiss." The most titillating item was *A Looking-Glass for Wanton Women by the Example and Expiation of Mary Higgs, Who Was Executed on Wednesday the 8th of July, 1677, for Committing the Odious Sin of Buggery, with her Dog, Who Was Hanged on a Tree the Same Day.*

By the end of the seventeenth century, the secularization of the mirror was clearly complete.* In England, a urine-filled chamber pot was humorously and euphemistically called a looking glass. Physician John Collop wrote doggerel to explain how he could make a medical diagnosis by examining human waste: "Hence looking-glasses, Chamberpots we call, / 'Cause in your piss we can discover all."

---

*Japanese mirror titles had similarly evolved toward the secular and erotic, with books such as the 1685 *Shikidô Ôkagami* (The Great Mirror of the Art of Love).

Chapter 6

A NEW WAY OF SEEING

The mirror—above all, the mirror is our teacher.

LEONARDO DA VINCI (1452–1519)

He who cannot see himself might as well not exist.

BALTHAZAR GRACIAN (1584–1658)

WHILE RENAISSANCE literature reflected the mirror's transformation from a largely sacred object to a common one, in Renaissance painting the mirror was an essential agent in transforming the medium itself. Before the fifteenth century, mirrors appeared infrequently in European artwork, usually in tapestries or illuminated manuscripts. The mirrors in such pictures are small, round, and convex.

This medieval art was flat and without perspective, but Giotto, an Italian artist of the early fourteenth century (and Dante's good friend), approached realistic three-dimensional figures. As an impish apprentice, Giotto once painted a fly on the nose of his master's work-in-progress so realistic that the older man fanned the bug with his hand to shoo it away.

How did Giotto "see" more clearly than other artists? He may have used some of the first flat Italian mirrors as an aid. In them, he saw three-dimensional reality represented on a conveniently framed flat surface. Giotto put himself into at least three of his frescos; he could have accomplished these self-portraits only with a mirror.

Giotto's work was merely early inspiration. The true artistic Renaissance arrived nearly a century later, when, almost miraculously, artists began to paint with near-photographic accuracy. "It remains a source of continual astonishment that so infinitely complex a genre should develop in so brief a space of time," observes art historian Norbert Schneider, "indeed within only a few decades." How did it happen? A 1403 illustration in a French edition of Boccaccio's *Lives of Famous Women* provides a clue. Marcia, a blonde nun wearing an elegant pink dress, sits at a desk. In her left hand, she holds a small, round mirror and, in her right, a paintbrush. She has nearly completed a realistic self-portrait.

But Marcia's painting itself is rather awkward. The square tiles on the floor, for instance, are all painted the same size, with no attempt at artistic perspective to create an illusion of depth—a skill the ancient Romans knew but that had been lost in the intervening centuries. It is perhaps not coincidental that the Romans also had many mirrors. In the first century, Pliny wrote about another female artist: "Iaia from Cyzicus, who never married, painted with a brush at Rome, and also drew . . . a self-portrait done with the aid of a mirror. No one produced a picture faster than she did."

A few years after Marcia put down her mirror, a short, wizened, bald genius named Filippo Brunelleschi would rediscover the art of perspective. Like Giotto, he lived in Florence. Also like Giotto, he used mirrors.

Brunelleschi's Perspective

Born in 1377, Brunelleschi was trained as a goldsmith and sculptor, though he also loved to tinker with gears, wheels, and weights, and he made a number of clocks, including one with an alarm bell. In the summer of 1400, he left Florence to avoid a terrible outbreak of the plague, which killed nearly one-fifth of the population, but he came back in 1401 to enter a competition for the design of bronze doors—intended to propitiate God and prevent future plagues—for the octagonal Baptistery of San Giovanni. When the judges chose two winners, Filippo Brunelleschi and Lorenzo Ghiberti, and asked them to collaborate, the infuriated Brunelleschi renounced sculpture forever. The twenty-four-year-old went to Rome, accompanied by his friend, the teenage sculptor Donatello. Together, they spent the next fifteen years, off and on, studying ancient Roman architecture and art while making a living by setting gemstones and making clocks.

On one of his visits to Florence, Brunelleschi conducted an experiment with a mirror that made him famous. He painted a picture of the octagonal Baptistery of San Giovanni—the one with Ghiberti's antiplague doors—that was to be viewed in an odd manner at a particular time of day from a specific location, a few feet inside the doorway of the nearby Santa Maria del Fiore, the huge cathedral then under construction. Brunelleschi drilled a small hole in his painting at the "vanishing point," where parallel lines appeared to meet at the central viewing point of the picture. While holding a rectangular mirror at arm's length, viewers were to peer through the hole so that they saw the front of the painting reflected in the mirror. The perspective was so realistic that the mirror reflection blended seamlessly with the real view. In place of the sky, Brunelleschi had substituted reflective silver, so that the real sky was reflected there.

To create his painting, Brunelleschi must have set up a flat mirror on an easel inside the doorway of the nearby cathedral so that, when he faced *away* from the Baptistery, he saw it reflected in the mirror. Perhaps he marked a crisscross grid on the mirror so that he could copy it perfectly onto a grid on the canvas he carefully painted. According to Georgio Vasari in his sixteenth-century biographical work *Lives of the Painters*, Brunelleschi used "intersecting lines" to create his perspective pictures.

In painting the picture shown in a mirror, however, Brunelleschi reversed right and left. Although the Baptistery is basically symmetrical, the reversal still would have been evident, and the shadows and light—let us say morning light—would have looked wrong if the picture itself had been held directly out at arm's length. Viewers looking through the hole in the back, however, and observing the picture in the mirror, saw it re-reversed, so that it appeared as in reality (see Figure 6.1).

Brunelleschi went on to design and supervise the construction of the dome for the new cathedral, which would be the world's largest dome, finally surpassing the Pantheon. In planning the octagonal dome, he probably viewed alternate designs on the already-constructed building, holding a mirror at arm's length and peering through the hole in the back of different pictures.

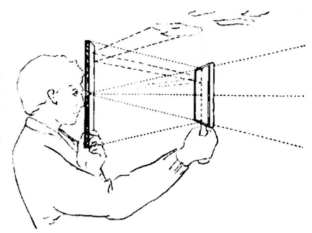

FIGURE 6.1
Brunelleschi asked the
observer to look
through the back of
his painting while
holding a mirror half
the size of the
painting to reflect it.

After Brunelleschi moved back to Florence permanently around 1417, he befriended a brilliant young painter named Tommaso, whose last name has been lost to history. Absent-minded, disheveled, and obsessed only with his art, this prodigy earned the nickname "Masaccio," the equivalent of "Sloppy Tom." Masaccio may have been sloppy about everything else, but he was meticulous about his painting. From Brunelleschi he learned the secret of perspective, applying it masterfully in *The Holy Trinity with the Virgin and St. John*, a 1425 fresco showing Christ on the cross, with God standing behind him in a vault that looks as if it recedes into real depth. Masaccio also painted a self-portrait "so skillfully with the help of a mirror that it seems to breathe," as Vasari observed. The artist died tragically young at the age of twenty-six in 1428.

Jan van Eyck Was Here

In the meantime, Brunelleschi's mirror perspective had reached the Netherlands, where there was an active trade with Italy and where wealthy Italians had settled among the Dutch.

There in the early 1400s, Jan van Eyck revolutionized Dutch art. The van Eyck family produced several talented artists, including Jan's older brother, Hubert, and sister, Margaret. Jan van Eyck served as the court painter for Duke Philip the Good of Burgundy, who occasionally sent him on secret diplomatic missions around Europe, where he undoubtedly learned from and taught other artists.

Van Eyck painted his most famous picture, *The Arnolfini Portrait*, in 1434. In exquisite detail and perfect perspective, it shows what has usually been taken to be a private wedding scene between Giovanni Arnolfini, a wealthy Italian businessman living in Bruges, and his unidentified bride. In a bedroom, the bridegroom, dressed in a rich fur cloak, holds his bride's hand in his left while holding up his right hand as if taking a vow. Between them, in the center of the picture, is an elegantly framed convex mirror.

In that mirror, you can see the entire room, including the back of the bride and groom, the canopy bed, the window—and two small figures, dressed in blue and red, standing in the doorway. One of them must be the artist, who signed his name in elegant script just above the mirror: *"Johannes de Eyck Fuit Hic 1434"* (Jan van Eyck Was Here 1434). Most critics have concluded that van Eyck and his companion served as witnesses to a priestless marriage ceremony. Many aspects of the picture have been interpreted as symbolic—the single lighted candle in the chandelier signifies marriage, the little dog faithfulness. The round mirror itself, whose frame contains scenes from the life of Christ, perhaps stands for the eye of God.

If so, God is observing a strange scene. The bride looks quite pregnant, her hand simultaneously holding up her green gown and resting on her protuberant stomach. Right next to the four-poster bed is a statuette of Saint Margaret, the patroness of pregnant women.*

There is another problem with the painting, however. By extending his *left* hand to his bride, Arnolfini would have signified a "morganatic" marriage between two parties of unequal social station, which seems unlikely. The answer may well lie in how Jan van Eyck created

---

*Art historians do not want the woman in the painting to be pregnant. Some assert that she represents only a symbolic wish for fertility, others that van Eyck and other contemporaries just painted women with big stomachs. It is difficult to believe, however, that van Eyck, who painted in such minute detail otherwise, would have misrepresented such a ballooning belly. As early as 1700, someone describing the painting for an inventory wrote: "The woman is a German and is pregnant." Perhaps this isn't a wedding scene at all, but a married couple celebrating an impending birth.

the portrait. Could he have observed the scene in a mirror in order to paint it? If so, because of the right-left mirror reversal, Arnolfini would have held out his right hand.

Critic Robert Hughes observes "the obsessed and simultaneous focus [van Eyck] imposed on every object, large or tiny, far or near." One method of noting such detail is to look into a convex mirror, which magically transforms the world into an exquisite, if somewhat distorted, fishbowl of a particular reality. Van Eyck was reputed to be an expert at geometry, and his patron, Philip the Good, called him a "paragon of science and art." Could it be that van Eyck used convex mirrors as an aid but that he corrected the distortions? In order to see the astonishing level of detail in his paintings, however, he would have required quite a large convex mirror with only a slight curve. It is possible that he used a smaller mirror from a distance to sketch out the entire scene, then moved in closer for detailed work, moving from place to place in the room.

Critic Elisabeth Dhanens, for one, thinks van Eyck did use a convex mirror as an aid. "It is as if Jan revealed his secret in a poetic fashion," she writes, by placing the mirror in the center of *The Arnolfini Portrait*. "When Jan succeeded in rendering architectonic space in a convincing fashion . . . perhaps he accomplished it by turning his back on the space he wanted to show and looking at it in a convex mirror." It may be no coincidence that in nearby Bruges the painters belonged to the same guild as the mirror-makers.*

Van Eyck saw the minute particulars of life—the intricate grain of wood, the subtleties of shadow—with an eagle's eye. Perhaps we see that eye in his *Man in a Red Turban*, which is probably a self-portrait. An intensely serious middle-aged man with compressed lips stares from a three-quarter profile directly at the viewer. Art historian H.W. Janson

---

*In his book *Secret Knowledge* (2001), David Hockney argues that van Eyck used a concave mirror in a darkened room opposite a small window to project various objects upside down onto a piece of paper tacked next to the window. Hockney has convincingly demonstrated that this method works, but he is incorrect in stating that van Eyck could have simply turned a convex mirror around to make it concave. Still, van Eyck had the resources and the knowledge, perhaps, to have a concave mirror especially made for him, and Hockney's arguments, bolstered by evidence found in the paintings by optical scientist Charles Falco, are fascinating.

believes that "the slight strain about the eyes seems to come from gazing into a mirror." By the time van Eyck died in 1441, he had mirrored
the world on canvas better than anyone previously. "In the work of Jan
van Eyck," Janson concludes, "the exploration of the reality made visible by light and color had reached a limit that was not to be surpassed
for another two centuries."

In 1435, Leon Battista Alberti published *On Painting*, in which he
gave specific directions for drawing in perspective. Instead of a mirror,
he suggested using a translucent veil to form a grid. Yet Alberti did
note: "A good judge for you to know is the mirror. I do not know why
painted things have so much grace in the mirror. It is marvelous how
every weakness in a painting is so manifestly deformed in a mirror."

Alberti was probably referring to flat Italian mirrors, but in northern
Europe the convex variety prevailed. Jan van Eyck's mirrors, which allowed a picture within a picture, influenced subsequent painters, particularly in the Netherlands. Four years after the *Arnolfini* portrait, the
Master of Flémalle (probably Robert Campin) painted Saint John the
Baptist with a convex mirror on the wall reflecting him, the room, and
the window. A 1449 picture by Petrus Christus, *St. Eligius and the Engaged Couple*, shows a jeweler with a young couple. The convex mirror
on his desk—useful for detecting thieves—reveals the neighboring
houses and another approaching couple.

There is indirect evidence that some fifteenth-century artists used convex mirrors to sketch landscapes and interiors and that they were not as
sophisticated as Jan van Eyck in correcting the resulting distortions. In
the church interiors of the Swiss artist Conrad Witz, for instance, and in
the outdoor scenes of the French painter Jean Fouquet, lines are bent as
if viewed in a convex mirror. As many of these scenes are miniatures,
they may have been copied directly from the reflecting surface. Witz
hinted as much in a charming pen-and-ink drawing in which the baby
Jesus is mesmerized by his reflection in a mirrorlike basin.

Leonardo's Mirrors

Mirrors of all varieties deeply influenced Leonardo da Vinci
(1452–1519). He used flat mirrors to critique his paintings rather than

to create them: "When you wish to see whether the general effect of your picture corresponds with that of the object represented after nature, take a mirror and set it so that it reflects the actual thing. . . . The mirror ought to be taken as a guide. . . . If you but know well how to compose your picture, it will also seem a natural thing seen in a great mirror." Thus, the mirror was "the master of the painter." In a query to himself, Leonardo wrote, "Why does a painting seem better in a mirror than outside it?"

Da Vinci wrote these words in his notebook backward, in mirror writing, and mirrors appear frequently throughout the notebooks. He recorded recipes for casting them out of metal, designed machinery for grinding and polishing them, told fables about them, and sought mirror-makers' secrets.

Leonardo pondered "why the mirror in its images of objects changes the right side to the left." He concluded that "the solar rays reduced . . . to a point by the concave mirror are redoubled in warmth and radiance." The moon, he correctly believed, reflects sunlight "after the manner of the mirror," although he also thought the stars were tiny mirrors and that the eye contained a mirror, because "if you look in the eye of anyone you will see your own image there." He thought that just as people can see themselves reflected in water from above, fish must view themselves when they look up. He was fascinated by the infinite reflections "seen with mirrors when placed one opposite another" or "a number of mirrors placed in a circle."

As an artist, Leonardo considered painters superior to sculptors because they could make a flat surface represent three dimensions. "The mind of the painter should be like a mirror," he advised, "which always takes the color of the thing it reflects, and which is filled by as many images as there are things placed before it." Above all, an artist must concentrate and "keep his mind as clear as the surface of a mirror." Elsewhere, Leonardo modified this metaphor, emphasizing that the true artist must have vision beyond vision: "The painter who draws by practice and judgment of the eye without the use of reason, is like the mirror that reproduces within itself all the objects which are set opposite to it without knowledge of the same." He realized that a mirror without an observer is just a shiny object. "No body is in itself defined in the mirror; but the eye on seeing it in this mirror puts boundaries to it."

Dürer's Evolving Self-Portraits

One of the artists whom Leonardo probably influenced was the German Albrecht Dürer (1471–1528). The son of a Nürnberg goldsmith of the same name, Dürer studied perspective in Italy, but he learned to consult mirrors much earlier. At the age of thirteen, Dürer painted the first of many self-portraits, "made out of a mirror," as he later wrote. It is a remarkable work for an adolescent, showing a solemn, longhaired boy with a prominent nose, sensitive mouth, high cheekbones, and pudgy cheeks. Dürer depicted himself looking off to the side, and one wonders how he did it. He probably used two mirrors in order to see himself from the side.

After learning to work gold with his father, Dürer apprenticed with a local artist, then spent four *Wanderjahre* (years of travel) in Germany and Switzerland. During his travels, he painted more self-portraits. At age twenty-one, head leaning on hand, his eyes stare from the page with a worried intensity. Two years later, in 1493, he drew another serious portrait as a study for a formal oil painting. In the framed work, he is a well-dressed, handsome young man, holding a symbol for "luck in love" in his hand. He apparently sent this picture home to impress Agnes Frey, his bride-to-be. In an arranged marriage, they wed when he returned to Nürnberg in 1494.

It was not a happy match. Dürer almost immediately took off for Italy, and in subsequent years the childless couple spent little time together. Professionally, Dürer prospered. He gained international fame through his woodcuts and engravings, which could be reproduced in the newly invented printed book. At twenty-nine, in 1500, the artist painted his most famous self-portrait. In it, his mature, gaunt, bearded face stares straight out, reminiscent of a van Eyck head of Christ.

One of Dürer's innovations was to capture the light in the human eye. In this self-portrait, he has looked searchingly into his own eyes and seen every detail, including the tiny reflection of a window on his cornea. This small touch, which Dürer put into all of his portraits, created an extraordinary illusion of depth, of eyes that truly look out from the painting. A few years later, Dürer drew a knee-to-head frontal portrait of himself, wearing nothing but a headband. He appears fit, self-confident, and uncircumcised.

By the time Dürer created his last self-portrait, in 1522, he was fifty-one. He had established himself as a painter of emperors, had written books on geometry, perspective, and the human form, and was loaded with honors. He was also suffering from malaria, which he had caught in the mosquito swamps of Zeeland. This picture shows a seated, naked, miserable Dürer, slumped over and holding self-flagellating instruments. His hair and beard are wild, and he looks apprehensively off to the side. Once again, he represented himself as a Christ-like figure, but this time he was the "Man of Sorrows." He died six years later, at fifty-seven.

The Death Mirror

Dürer's art is filled with images of death. In a plague-ridden age, the awareness of mortality was pervasive, and the ever-present mirror showed it, too. In 1520, Dürer drew *Allegory of Youth, Age, and Death*, a pen-and-ink sketch of a lovely young nude woman brushing her hair while admiring herself in a handheld convex mirror. Behind her, a skeletal death figure looms, holding the hourglass of time. In front of the woman, an old man sits and looks back over his shoulder at her.

Dürer's picture wasn't shocking or even original. He was copying a 1509 painting by Hans Baldung, a German artist obsessed with mortality even as a young man. Thirty years later, Baldung painted another sensuous nude looking at herself in a convex mirror, from which a skull stares back. In 1529, Lucas Furtnagel painted a middle-aged couple holding a convex mirror that shows their faces as skulls. "This is what we looked like," reads the inscription on the painting. "In the mirror, however, nothing appeared but that." To pound home the point, "Know Thyself" is inscribed on the mirror's frame.

The mirror theme of vanity and mortality spread to Italy, where an anonymous sixteenth-century artist engraved *Death Surprising a Woman*. Here, the nude woman stands with her back to a large, flat mirror, looking over her shoulder to admire her back, while a grinning skeleton holding an hourglass looms in the background.

The death mirror was linked to increasing concern for the mirror's secular use as an instrument of vanity. As mirrors became more common, they lost much of their mystery and religious aura. Previously, the Virgin Mary or Prudencia (Wisdom) usually held a mirror as a symbol of purity or self-knowledge. Now Superbia (Pride), Luxuria (Luxury or Lust), Accidia (Sloth), and Vanitas (Vanity) gazed all too fondly into mirrors. Sometimes, the only way to distinguish between Prudencia and Superbia was to look for accompanying symbols—a snake somewhat surprisingly indicated wisdom, whereas a peacock's spread tail meant vanity.

In the late fifteenth century, a woodcut (possibly by Dürer) illustrating *Der Ritter vom Turn* shows a young woman combing her hair before a mirror. Behind her, a horned demon exposes his rear end, which appears in the mirror, although she is blissfully unaware of it. Hieronymus Bosch, the hallucinatory Dutch moralist, depicted Superbia in a domestic scene in which we see a woman trying on a fashionable linen bonnet while looking into a convex mirror. A wolflike demon, which looks absurd in its headdress, is holding up the mirror.

Bosch's disciple, Pieter Bruegel, drew an even more elaborate Superbia in 1558. In the center of the crowded picture, a well-dressed woman looks at herself in a mirror while a peacock spreads his tail in full glory beside her. In the left foreground, a strange little armored monster with a peacock's tail looks at himself in a large, flat mirror while another creature contorts himself, bending over while spreading his buttocks to see his own anus in yet another mirror. Bruegel also drew a picture of Everyman in which a man looks into a mirror bearing the pessimistic inscription (obviously intended as an ironic counterpoint to the Furtnagel print) "Nobody knows himself."

Bruegel's bleak pronouncement marks a significant turning point in the shift from the mirror of faith to the secular glass. Rabelais observed that some jokers brought small distorting mirrors into the sanctuary "to annoy people and make them lose countenance at church." Women of the sixteenth century began to wear small mirrors attached to their waists by elegant chains. "Alas! What an age we live in to see such depravity as we see," the French moralist Jean des Caurres wrote in 1575, "that induces them even to bring into church these scandalous mirrors hanging about their waist! . . . It is true at present none but the ladies of the court wear them, but it will not be long before every citizen's daughter and every female servant will wear them."

Men, too, carried pocket mirrors, though they were more circum-spect about it. The male dandy, according to an anonymous 1617 British poet, "Never walks without his looking glass / In a tobacco-box or dial set, / That he may privately confer with it."

Mirror Games

Despite their disapproval of people's vanity, artists were seduced by mirrored beauty throughout the sixteenth and seventeenth centuries. In 1515, eight years after the del Gallo brothers applied for a patent on their new foiling method, two Venetian artists painted women examin-ing themselves without a trace of skulls or judgment. Giovanni Bellini's brunette stands next to a wall mirror while holding a smaller mirror in one hand and adjusting her hair with the other. By using the two mir-rors together, she can see the back of her head. Similarly, Titian's beauty holds her long hair in one hand while employing two mirrors.

Although women did (and do) use this strategy to see themselves, the Venetian artists were also responding to an argument between sculptors and painters. Sculpture was the foremost art, its proponents declared, because viewers could look at it from all angles. A Venetian named Giorgione responded by painting "a man in the nude with his back turned," wrote Vasari, "and, at his feet, a limpid stream of water bear-ing his reflection." His burnished armor to the right and a mirror to the left reflected both profiles. "This was a very fine and fanciful idea, and Giorgione used it to prove that painting requires more skill and effort and can show in one scene more aspects of nature than is the case with sculpture."

Giorgione died in 1510 in his thirties, and his painting has been lost, but eleven years later, Giovanni Savoldo—another Venetian painter—did something similar in *Portrait of a Man*, featuring a fashionable young soldier reflected by two mirrors, so that it appears that two other men are leaning back into alternate and unknowable realities.

Three years before the Savoldo effort, Raphael Santi demonstrated that painters could even *replace* sculptors and architects by painting the awkwardly shaped vaults of the Loggie of the Vatican. Using illusionist painting that created columns and curved ceilings where there were none, Raphael made the space appear much higher and more extensive than it was.

Raphael's leading assistant, Giulio Romano, designed another architectural illusion in 1534, creating a columned balustrade, convincing if viewed from the proper place on the floor. Romano built a three-dimensional model of the scene, then placed it above a mirror with a grid marked on it. When viewed from the appropriate distance and angle, it allowed the artist to copy the exact scene required for his illusion.

A century and a half later, the imposing Jesuit church of Saint Ignazio in Rome was nearing completion, but the Dominicans who lived next door complained that the planned dome would block the light for their library. The Jesuit painter and architect Andrea Pozzo therefore created a fake dome on a flat ceiling, completing it in 1685. It is still there, and when viewed from the proper spot, it looks uncannily real, with side windows and a vaulted dome capped by a bright skylight. Pozzo used math to calculate the angles, but he probably also built a model and used a mirror to design the illusion.

Like Brunelleschi before them, Raphael, Romano, and Pozzo used mirrors and perspective to create the illusion of reality. But some artists began to use mirrors to distort reality. In 1524, Parmigianino painted his round *Self-Portrait in a Convex Mirror*. The young artist looks blandly out of the picture from hooded eyes, but his hand (it appears to be his right but is in fact his reflected left), which lies near the mirror, looks like a giant flipper, and the window in the top of the picture is curved. All previous self-portraits had sought to hide the mirror; Parmigianino flaunted it, ostentatiously depicting and at the same time twisting reality.

His strange self-portrait fascinated viewers and artists, and it presaged anamorphic art—from the Greek *ana* (again) and *morphe* (form)—in which deliberately distorted images could be seen normally only from a particular angle or when reconstructed in a mirror. In 1533, Hans Holbein the Younger painted *The Ambassadors*, a typically meticulous Dutch painting with two erudite, well-dressed men—except for a strange black-and-white elongation toward the bottom that looks like a crumpled piece of paper. When viewed from the side, however, or in an obliquely slanted flat mirror, it jumps out as a skull, an intimation of mortality like the death mirror.

Such anamorphic distortions continued throughout the sixteenth century, including the 1538 work of Dürer student Erhard Schön called *Was Siehst Du?* (What Do You See?), in which a man defecates in the distorted picture. Another Schön work, *Aus, Du Alter Tor* (Out, You Old Fool), shows a lecherous old man dallying with a young woman, who surreptitiously steals his money and passes it to her young lover. The accompanying anamorphic panel, when properly viewed aslant or in a flat mirror, shows her fondling her lover's penis while he massages her breast.

Obscene or erotic art was the perfect subject for these hidden images, as the Chinese apparently realized when Jesuits introduced them to elongated distortions of more conventional themes late in the century. Sometime between 1573 and 1619, the Chinese, already familiar with convex and concave mirrors, took anamorphism to another level, creating distorted round paintings that could only be seen correctly in cylindrical mirrors placed in the center. One, for instance, depicted an absurdly elongated elephant. Many were unabashedly pornographic.

Simon Vouet, a twenty-one-year-old French artist, must have seen some of these Chinese pictures in Constantinople, where he spent a year with the French ambassador in 1611. Sultan Ahmed I collected Chinese objets d'art. Vouet went on to Italy, where he lived for fifteen years, becoming an established painter. He also played with anamorphic art, drawing a distorted elephant reconstructed in a cylindrical mirror. In 1627, Vouet was recalled by King Louis XIII to France, where he became a sought-after painter of the royalty and nobility, churning out larger-than-life portraits with hovering winged cherubs and cute animals. Yet Vouet also maintained an interest in anamorphism, helping mathematician Jean-Louis Vaulezard explain it to the public for the first time in *Perspective Cylindrique et Conique*, published in 1630. This manual showed how to create anamorphic pictures with the aid of a grid that could then be reconstructed in a cylindrical or conical mirror.

Eight years later the mathematician and theologian Jean-François Niceron—like his teacher, Marin Mersenne, a Minim monk and correspondent of Descartes—published *La Perspective Curieuse* (Curious Perspective), a more accessible and influential work in which Niceron explained how to construct mirror anamorphoses using cylinders, cones, and pyramids. The pictures spread in profound distortion—literally turned inside out—around the mirrored cones and pyramids and were to be viewed directly from above, so that the apex formed the center of the picture. Simon Vouet designed the frontispiece, an anamorphic portrait of Louis XIII.

Niceron was preparing a second edition of his book when he died in 1646 at thirty-three. His posthumous book, published in Latin as *Thaumaturgus Opticus* (Miraculous Optics), envisioned entire rooms devoted to anamorphic art, with huge mirrored columns reflecting otherwise unrecognizable floor mosaics. "It will be a new wonder," he wrote, "when, after seeing the shafts of these columns sparkling with light . . . and free of any image or painting, you see pictures . . . gradually surge up as you approach." He also imagined conical mirrors suspended from ceilings, so that viewers looking up would see reconstituted reflected art.

No one ever made such rooms, but anamorphic mirror art grew in popularity throughout the seventeenth and eighteenth centuries. At first, artists stuck to safe themes, such as portraits of royalty or crucifixion scenes. Eventually, painters played with the possibilities and implications of such hidden art. In 1660, a British anamorphic portrait of Charles I, who had been beheaded in 1649, featured a skull in the round circle where the cylindrical mirror was to be placed, and subsequent paintings formed skulls themselves in the mirror. The illusory, transitory nature of the world matched the mysteriously distorted reality made whole in the mirror. Other European artists rediscovered the erotic possibilities, sometimes mixing them with sacred themes, as in a late-seventeenth-century anamorphosis of Peter Paul Rubens's *Erection of the Cross*. Reflected in the other side of the mirror cylinder, a woman on her back throws her legs around a man who is obviously experiencing another type of erection.

Camera Obscuras and Other Aids

The ancient Chinese, Alhacen, and Leonardo da Vinci knew about the camera obscura, in which light admitted into a dark room through a small hole projects an upside-down image of the outside world. But the device was popularized only in the late sixteenth century by Giambattista della Porta in *Natural Magic*. By putting a convex lens into the opening, one could widen the opening to admit more light. Realizing that a flat mirror in the light path would correct the reversed image, Johannes Kepler invented a portable-tent camera obscura with a revolving mirror and lens built into a turret. In 1620, a British visitor marveled at landscapes Kepler had drawn with the aid of his device, "me thoughts masterly done . . . surely no Painter can do them so precisely."

In England, the Dutch scientist-engineer-magician Cornelius Drebbel wowed King James I and London society with his camera obscura. Dutch diplomat Constantijn Huygens—Christiaan Huygens's father, not his brother of the same name—wrote home in 1622: "It is impossible to express its beauty in words. The art of painting is dead, for this is life itself, or something higher, if we could find a word for it." Within a month, Huygens had changed his mind. Painting was not dead, but the new device should revolutionize it. "I am rather surprised at the folly of so many of our painters who are ignorant of . . . the aid of something both pleasurable and useful." Anyone who has seen a good camera obscura will understand his enthusiasm. The projected image looks like a painting, but the wind really blows the leaves, smoke rises, dogs run, people stroll.

Many artists did take advantage of the camera obscura. Samuel van Hoogstraten, a seventeenth-century Dutch artist who regarded anamorphic art as a mere diversion, was enthusiastic about the camera obscura: "I am sure that these reflections in the dark can give no small light to the vision of young artists." As artists were notoriously secretive about their methods, it is difficult to prove that they used camera obscuras, but the indirect evidence is plentiful. "The superb effects of spatial and atmospheric naturalism in Dutch painting [of the seventeenth century] exhibit a condensed precision which is worthy of the best camera obscura images," writes Martin Kemp in *The Science of Art*. In particular, Jan Vermeer's work showed a sensitivity to light, shade, and space. The mirrors in many of Vermeer's interiors indicate a fascination with different ways of reflecting reality.

The modern painter and photographer David Hockney believes the camera obscura accounted for Caravaggio's remarkably lifelike, evocative paintings of the early seventeenth century. "Notice the constant sense of assurance," Hockney says. "And with no drawings, no sketches!" Hockney also points out that Caravaggio, using a lens-only camera obscura, painted his 1595 *Bacchus* holding his goblet in his left hand (probably reversed), whereas Vermeer, some 60 years later, showed the clavichord player in *The Music Lesson* the right-way round, presumably with the aid of a camera obscura equipped with both mirror and lens.

Caravaggio was attacked by critics for being able to paint only in cellars (dark spaces) "with a single source of light"—the proper conditions for a camera obscura. He was certainly interested in mirror effects, in 1598 painting a round portrait of Medusa's snake-haired severed head as seen in Perseus's reflective shield. Two years later, Caravaggio painted an extraordinarily realistic portrait of Narcissus peering into a still pool. In the water, his dark, mysterious reflection gazes back up at him. Could the artist's model have posed in a leaky basement, lit by strong torches off to the side? A hundred years later, Canaletto used camera obscuras to paint his landscapes of Venice, explaining that the devices gave "a picture of inexpressible force and brightness; and as nothing is more delightful to behold, so nothing can be more useful to study."

Rembrandt van Rijn (1606–1669) probably used the camera obscura as an aid, too, although he relied on direct reflection from mirrors for his ninety self-portraits. When he was twenty-four, Rembrandt made faces in his mirror, capturing the dramatic emotions on paper. We see him angry, frightened, brooding, laughing, crazy, surprised. In more formal oil paintings, the artist playfully assumed different roles, as a gentleman, beggar, burgher, or oriental sultan. In his final self-portrait, just before he died, Rembrandt showed his weathered face lit by laughter, a reference to the Greek painter Zeuxis, who reputedly suffocated from an overdose of mirth. Always the trickster, Rembrandt left a final stoic, ironic, and ultimately hopeful message to the world—one should die laughing.

The list of artists who used mirrors—either directly or in camera obscuras—could go on indefinitely. Diego Velásquez (1599–1660), whose library contained works on catoprics and optics, owned a camera obscura and ten mirrors—an astonishing number, given their expense. In *The Toilet of Venus*, Velásquez painted a sensuous reclining nude. We see her back as she props herself on an elbow to look into a mirror held by a winged cherub. In his most famous picture, *Las Meninas* (The Maids of Honor), Velásquez portrayed himself painting a huge canvas while a self-possessed little Spanish princess is attended by servants. On the back wall, a large mirror reflects the king and queen, who are evidently posing for their picture but who are otherwise invisible. Their presence is implied, however, since most eyes are focused on where they would be standing.

At the same time, French artist Claude Lorrain (1600–1682) painted muted rural landscapes that achieved enormous popularity, along with the so-called Claude Glass, a slightly convex mirror made from black glass or with a dark reflective coating. Mounted in velvet-lined carrying cases, these artists' mirrors produced a wide-angled view on their small surfaces. Although Claude Lorrain may not have used the mirrors himself, their reflections produced moods similar to his paintings.

Mirrors thus granted humans new perspectives—indeed, just as the *speculum* spawned the modern notion of *speculation*, the study of *perspectiva* gave us the concept of different *perspectives* on our lives, all through the *miracle* of *mirrors*—another pair of words with the same root.* In 1654, the Jesuit philosopher Emmannuele Tesauro compared the human intellect to a perfect mirror, claiming that it was an appropriate metaphor because "it is a more curious and agreeable thing to look at several objects in perspective than to see the originals pass before the eye."

Breaking the Mirror Monopoly

Nowhere was the mirror's transformation more evident than in seventeenth-century France. In 1633, the queen attended a Parisian ball featuring six rooms lined with Venetian mirrors and tapestries, setting off a veritable mirror craze. An archbishop throwing a party for the Duchess of Longueville in 1651 bought fifty mirrors, elaborately framed in carved gilt wood adorned with fruits, flowers, and angels. The French poet Jean Loret described the "faces, grimaces and postures, / laughter, favors, charms, / the breasts, the hands, the arms / of the beautiful group of conspirators," all sneaking looks at themselves

---

*The Latin *specere* means "to look," so *perspective* means "to look at closely, or to look through." The Latin *mirare* means "to look at wonderingly," and *miraculum* is "an object of wonder."

and one another in the multiple reflections. Guests also wore mirrored jewelry as bracelets and necklaces, and small mirrors attached by silver chains dangled from their belts.

Intimate mirrored rooms called *cabinets de glaces* became fashionable, with the aristocracy vying to outshine one another. "In their enchanted chambers," one poet noted, "fabric no longer has a place. / On all the walls on all four sides / Encrusted mirrors show their face." Louis XIV, who became a boy-king at the age of five, grew up in this glittering world. Cardinal Mazarin, the prime minister who actually ran the regency, offered mirrors as raffle prizes, and Nicolas Fouqet, the corrupt finance minister, owned fabulously expensive mirrors framed in gold, silver, ivory, and tortoise shell.

Upon Cardinal Mazarin's death in 1661, the twenty-three-year-old Louis XIV assumed real power, reigning in the fractious, proud nobility and reforming the government. Replacing Fouquet, Louis charged Jean Baptiste Colbert, the ambitious son of a Reims clothier, with reforming the nation's finances. In 1664, Colbert was also named the superintendent of buildings, royal manufactures, commerce, and fine arts.

Determined to promote French industry as well as please his king, Colbert chafed at the exorbitant prices the French were paying for Venetian mirrors, so in 1664 he asked Pierre de Bonzi, who served as the French ambassador to Venice, to recruit Murano mirror-makers willing to relocate to Paris. Bonzi looked into it and wrote back that "whoever might suggest that they go to France would run the risk of being thrown into the sea." But for his king, Bonzi would try.

These were not the first attempts to lure Italian mirror-makers to France. In 1551, King Henri II offered Theseo Mutio, an artisan from Bologna, a ten-year monopoly on "glass, mirrors . . . and other sorts of Venetian style glass." In 1558, Theseo's brother Ludovic joined him in France, but they had trouble smuggling out tools. Although the Mutio brothers apparently produced some mirrors, their enterprise eventually faltered and died. Other Italian refugees followed, with greater or lesser success, but none were willing to divulge their secret process. Venice continued to hold a monopoly on the finest mirrors, and Colbert was now determined to break it.

The Venetian Council of Ten ruled the island of Murano with a watchful eye and iron hand. The law was clear: "If any worker or artist should transport his talents to a foreign country and if he does not obey the order to return, all of his closest relatives will be put in prison." If that didn't work, "an emissary will be sent to kill him and, after his death, his family will be set free."

Bonzi hired a junk merchant who roamed the island of Murano in search of potential masters in mirror-making. Three months later, he had found three disgruntled and somewhat unsavory workers (one had murdered a priest) who were willing to take the risk for a large amount of money and tax exemptions. In June 1665, Petro Rigo, Zuane Dandolo, and La Motta (who went only by his last name) arrived in Paris and began to build their ovens on the rue de Reuilly in the district of Fauboug Saint-Antoine. When the Venetians discovered their disappearance, they asked Sagredo, their ambassador to Paris, to offer the renegades suspended sentences if they would return. But Sagredo couldn't find them.

By the fall of 1665, twenty Murano fugitives had been hustled aboard moonlit gondolas by secret agents and ensconced in the new Paris factory. Colbert had appointed Nicolas Dunoyer, an Orléans tax collector, to oversee the operation. On February 22, 1666, Dunoyer sent the first flawless mirror to Colbert.

By this time, of course, Ambassador Sagredo knew where the Italian fugitives were. He blustered, promised, threatened. In response, Colbert increased the salary of Antonio della Rivetta, the senior master, and his three assistants, Morasse, Barbini, and Crivano, and all the mirror-makers received free room and board. Sagredo reminded the workers that they were putting their Muranese families and property at risk, but it was an empty threat, because too many artisans had left, too much was at stake, and the Venetians couldn't afford to antagonize them.

Sagredo was recalled, replaced by Ambassador Giustiniani, who proved more effective. Several homesick Italian underlings sought and received safe passage back to Italy. Colbert responded by writing to the wives back in Murano, offering them a good life with their husbands in Paris. The Venetian police intercepted the letters and sent fake replies in which the wives asked their mates to come and fetch them. The Murano mirror-makers weren't fooled, however, because the letters were far too literate. Besides, many beautiful, compliant Parisian women were intrigued by these exotic Italians, who anyway were in no hurry for their wives to join them. Nonetheless, Colbert did arrange for some spouses to flee to Paris.

In April 1666, King Louis XIV, along with his retinue, arrived at the rue de Reuilly to inspect the Royal Company of Glass and Mirrors. The king bestowed a handsome bonus on the workers. The Italians, puffed with their importance, became more difficult, hotheaded, and prone to brawls. Slow mercury poisoning also probably contributed to their erratic behavior. They refused to allow any Frenchmen to work alongside them during crucial moments in the secret process.

Dissension between La Motta and Rivetta split the immigrants into rival camps. They continued to make mirrors while arming themselves with matchlocks. When tempers flared, the guns came out, and La Motta and a companion were wounded amid the shattered glass. The factory closed down for a few days. A few weeks later, a worker who handled the critical splitting-open of the blown glass injured his leg, and no one else had the expertise to replace him. The foundries had to keep burning fuel, however, or the glass would be ruined.

The desperate Dunoyer suggested that the king give the mirror-makers free land and that an Italian, for every French apprentice he trained, would receive a bonus. Just when such enticements looked as if they might work, in January 1667 a Murano worker came down with a violent fever and died within a few days. Three weeks later, after complaining of horrible stomach pains, another worker died. In a time when poisons of lingering subtlety were so commonplace that special courts were set up to deal with them, tensions at the mirror factory escalated. Ambassador Giustiniani promised amnesty to workers who went home, and La Rivetta, Barbini, and Crivano departed for Murano in April.

By that time, French workers had observed enough that they could carry on, and in Tourlaville, near Cherbourg, a skilled artisan named Richard Lucas de Néhou was making excellent glass. Under pressure from Colbert, this rival glass company merged with the Paris firm, which thereby gained Néhou's expertise.

By 1680, the Italian ambassador lamented: "I have tears in my eyes when I see how these many factories, which by an admirable gift that Providence, nature and hard work had granted particularly to us, have been so easily transported and sustained by the unpunished spitefulness of a few of our fellow citizens."

Two years later, the public was admitted to the partially completed Hall of Mirrors at Versailles, which was lit by chandeliers and candelabras. A starry-eyed reporter called it a "palace of joy . . . a dazzling mass of riches and lights, duplicated a thousand times over in just as many mirrors, creating views more brilliant than fire and where a thousand things even more sparkling came into play." A poet celebrated the spectacle that magically turned night to day:

By the reflection of a large number of mirrors
That make beauty visible in various places
The fire of diamonds worn by the court
In the middle of the night give birth to a new day.

Louis XIV, the Sun King, had found the perfect reflection for his glory, and he hastened to decorate his coach and the apartments of his mistresses with equally resplendent mirrors. A few years later, merchants set up a corridor of mirrors on a Parisian bridge over which the king walked "in order to multiply his image." Louis gave French mirrors to foreign dignitaries such as the King of Poland, the Siamese ambassador and the Turkish sultan. The Siamese government subsequently purchased 400 mirrors from the Royal Company.

When it was finished in November 1684, the Hall of Mirrors featured seventeen huge composite mirrors (each made of eighteen squares) in window casements. Placed opposite real windows, they reflected the ordered gardens and grounds of Versailles in more than 300 panes of mirrored glass. The hall gave the illusion of being even larger than it was, with invisible, airy walls. Over the next century, it inspired imitative mirrored halls throughout Europe.

The mirror panels at Versailles were made from blown, slit, and flattened plate glass, limited to about 40-by-60 inches. In 1680, in an Orléans glass factory, the Italian glassmaker Bernard Perrot (Perroto) had invented a method for making larger mirrors by pouring molten glass onto a flat table. Louis Lucas de Néhou, nephew of the Tourlaville plant owner and now working at the rival Abraham Thévart glassworks in the Saint-Germain district of Paris, experimented with the new casting method and succeeded in making astonishingly large panes of glass, nearly 5 feet high. Few panes survived the complex manufacturing process, however, and Thévart was forbidden to sell glass as small as that made by the Royal Company.

In 1693, near bankruptcy, Thévart moved his factory to cheaper quarters northeast of Paris at Saint-Gobain, a derelict chateau surrounded by forests that could provide fuel for the glass ovens. Two years later, Louis XIV resolved the competition between his troubled glass company and Thévart's by forgiving all of their debts and creating a new company based in Saint-Gobain that encompassed both of them, giving the Manufacture Royale des Glaces de France (Royal Glass Company of France) a thirty-year privilege. Bernard Perrot, who had invented the casting process, was stripped of his business and tools.

In 1700, the factory cast a pane of glass nearly 9 feet tall and 3 feet wide. Two years later, a Swiss banking firm based in Geneva took a controlling interest in the company, bringing much-needed managerial skills to the chaotic enterprise.

The French mirror industry had finally come of age, and from then on, as the French produced large mirrors for less money, the Murano business fell into steep decline, though the Italians continued to produce beautiful, expensive mirrors in the old style. By 1765, exactly 100 years after the first three Italian mirror-makers arrived in Paris, only one glasshouse remained on Murano, and it was open just two days a week. Competing glassworks sprang up near French borders, recruiting restless workers willing to risk imprisonment for breaking their contract with the Royal Glass Company, which forbade them to travel more than a league from the factory. Smuggling glass and mirrors became a lucrative enterprise.

"The glass world was unique," one historian remarked, "a law unto itself. It had its own rules and customs, and a separate language, too, handed down not only from father to son but from master to apprentice. . . . Theirs was a closed community, with every man, woman and child knowing his place within the walls." Mirror-making at Saint-Gobain required a huge force of laborers, all of whom lived in the manufacturing area until 1775, when a separate compound for employees was finally built.

The casting process began with 200 workers to clean and sift the pure white sand and soda, which was then melted in heat-resistant white-clay crucibles for thirty-six hours. The ovens withstood the immense heat for less than a year, and the crucibles shattered in a matter of weeks. The master, dressed in protective clothing, kept the ovens fired and at a constant temperature. Then, in a dramatic, dangerous maneuver, a crucible was lifted on a huge fork, carted to the metal casting tables, and poured out. With a large iron roller, the workers spread the molten glass in an even sheet, the width determined by a frame around the table. Then it was reheated in the oven and gradually cooled in a three-day process called annealing, which prevented stresses in the glass that would otherwise cause it to self-destruct.

The roughcast sheets were then shipped from Saint-Gobain to rue de Reuilly in Paris for the final labor-intensive stages of polishing and buffing. On the journey, more than three-quarters of the sheets broke, which explains why they didn't polish in Saint-Gobain. In the polishing process, workers spread fine wet sand between two sheets of glass, then rubbed them back and forth for several days before the pieces were turned over and the process repeated for the other sides. The polishing continued with emery paper and fine iron oxide powder. "Six hundred men work at it daily," observed a factory visitor in 1698. "The noise is most unbearable."

The polished glass—still often flawed by marbling, clouds, veins, or bubbles—was then turned into a mirror by weighting it down on beaten tin and mercury in a process that would not change for another 150 years. Mirror-makers, like those who produced hats, suffered notoriously from toxic mercury fumes. While the Mad Hatter is well known, there were an equal number of mad mirror-makers. Meanwhile, wealthy young men admired themselves, as a French poet put it, "in four mirrors at once to see if their breeches are tight against their skin," unaware of the workers who suffered poisoning to produce their pristine looking glasses.

The Saint-Gobain Glass Works, as it was eventually renamed, survived periodic mismanagement, dangerous working conditions, and revolutions, and the company—today a diversified multinational corporation—still produces some of the world's finest mirrors. Half of all Parisian homes acquired a mirror during the last two decades of the seventeenth century, and in the eighteenth century mirrors—at least small, inexpensive ones—became commonplace. Company sales from 1725 to 1788 rose by 400 percent.*

Mirrors graced chairs, desks, beds, candelabra holders, chimneys, and overmantels. They lurked in grottos, alcoves, and galleries. Candle brackets were attached to the bottom of mirror frames to reflect more light into the room. The *cabinet de toilette*—the forerunner of the modern bathroom—acquired at least three mirrors for multiple views, allowing lovers, as Abbé de Torche wrote in 1668, an intimate view of themselves: "Although they were alone in this charming place, it seemed nevertheless, when their eyes fell on the mirrors, that an agreeable company surrounded them."

When Madame de Pompadour made mountainous hairstyles (called *fontage*) popular, dressing mirrors bulged upward with a dome to accommodate views of all that hair. Freestanding, full-length mirrors, aptly named *psychés*, allowed people to study themselves from shoes to cowlick. Ornate mirrors became a luxurious necessity for the aristocracy. "I had some wretched land which brought me nothing but wheat," a countess explained, "so I sold it and bought this fine mirror." Every lady of fashion carried a makeup box with mirror, brush, and powdered pads. One resourceful wife's wardrobe mirror hid a secret passage to her lover's house.

Giacomo Cassanova often led a new conquest into an octagonal room "hung, tiled, and covered with mirrors," whereby she would "fall in love with herself"—and hopefully with him. "I could spend my life contemplating myself," an innocent young heroine declared in an eighteenth-century French play in which she saw herself in a mirror for the first time. "How I am going to love myself now!"

British Self-Regard

The British glass and mirror industry came of age during the same period. As early as 1571, Venetian glassmakers came to London, but they

---

*During the eighteenth century, fur traders gave North American natives small glass mirrors in return for pelts. A 1745 observer wrote, "The Indians like [mirrors] very much and use them chiefly when they wish to paint themselves. The men constantly carry their looking glasses with them on all their journeys."

did not bring the secret of foiling mirrors with them. By 1615, England's limited forest resources forced the glassmakers to switch to coal. In 1673, George Villiers, the second Duke of Buckingham and a gentleman of the bedchamber to King Charles II, established a glassworks at Vauxhall, where, following Colbert's example, he lured some Murano mirror-makers. Villiers promptly sold Charles enough mirrors to panel the bedroom of Nell Gwyn, the king's mistress.

Three years later, the London glassmaker George Ravenscroft added lead oxide to his formula, producing clear, heavy glass that sparkled brilliantly when cut in facets because it refracted light at a greater angle than did regular glass. Prismatic chandeliers made from such lead glass became a popular export, and the tax-free Irish glass industry thrived in places such as Cork, Dublin, Belfast, and Waterford. The Villiers works quickly adopted lead glass, which remained workable longer than regular glass and allowed workers to blow much larger bubbles that could be slit to make plate glass as large as 82-by-48 inches. With the secret of Venetian foiling, these mirrors were impressive, although they were much more expensive than mirrors produced by the French casting method, so Parisian imports dominated. Still, the British were able to export looking glasses to India and the American colonies.

The eighteenth-century British loved mirrors nearly as much as their French contemporaries. Initially the large Queen Anne style—simply but elegantly framed—was popular, then rococo carved and gilt wall mirrors. Mirror designers signed their works: Thomas Chippendale, George Hepplewhite, Thomas Sheraton, Robert Adams. Vista mirrors, in which two parallel mirrors on facing walls produced infinite reflections, were popular in aristocratic homes. The practical British added a box with a hinged mirror—the forerunner to our medicine cabinets—to the *cabinet de toilette*.

In 1715, essayist Richard Steele visited a London mirror shop, where "[people] will certainly be well pleased, for they will have unavoidable Opportunities of seeing what they most like. . . . I mean their dear selves." The mirror, once a rare object that reflected sacred purity, had truly entered the modern era—individualistic, ironic, capitalistic, self-promoting, self-conscious, and vain.

In the late eighteenth century, the London resort Jenny's Whim featured distorting mirrors to amuse visitors, while the quack Dr. James

Graham created his Temple of Health and Hymen, where "garlands, mirrors, crystals, gilt and silver ornaments are scattered about," a contemporary observed, "so that from all parts they reflect a dazzling light." There, for fifty pounds a night, amorous couples could occupy a mirrored fertility shrine. "This Grand Celestial Bed," Graham wrote in 1781, "is 12 feet long and 9 feet wide, supported by forty pillars of brilliant glass. . . . The super-Celestial dome of the bed . . . is covered on the underside with brilliant panes of looking-glass."

Yet even as these secular mirrors reflected randy dandies and dilettantes, an expatriate German musician in a British resort town was polishing curved metallic mirrors that he would turn toward the heavens. His mirrors once again became sacred eyes, peering ever farther into the mysteries of God's universe.

GRASPING THE UNIVERSE

Seeing is in some respects an art which must be learnt. . . .
Many a night have I been practicing to see, and it would be
strange if one did not acquire a certain dexterity by such con-
stant practice.

WILLIAM HERSCHEL, 1782

Coelorum perrupit claustra [He broke through the barriers of
the skies]

WILLIAM HERSCHEL'S EPITAPH

ON AUGUST 11, 1781, William Herschel attempted to cast the biggest mirror ever made by man, a 36-inch-diameter monster requiring more than 500 pounds of copper-tin alloy, to be inserted in a tube that would reach 30 feet in the air. Another casting had failed when the mold—made out of an immense amount of horse manure—sprang a leak. Now the second casting was under way, the metal melted, when William Herschel noticed a small leak from the furnace, with bright, coppery tears dripping into the fire. The drops quickly became a flood, running out onto the flagstone floor of the basement room. As the molten metal hit the cool floor, the paving stones blew up. "Both my brothers [William and his younger brother Alexander] and the caster with his men were obliged to run out at opposite doors," sister Caroline wrote, "for the stone flooring . . . flew about in all directions, as high as the ceiling."

This was not the life Caroline had in mind when she left her home in Hanover, Germany, in August 1772 to come to the fancy British resort town of Bath to keep house for her adored composer brother and to sing in his choir. She was twenty-two; William was thirty-three. She knew of William's interest in the night sky, and she enjoyed identifying the constellations he pointed out to her on their journey through the Netherlands. Once they reached London, she might have guessed something was up when William dragged her to every optician's shop in the city.

Growing up in Germany, William had learned to play the oboe, violin, and piano, among other instruments, but he didn't enjoy his life in a military band, caught near battles and sleeping in damp ditches. In 1757, at the age of nineteen, Herschel fled to England, where he soon made a name for himself as a musician and composer, winding up at Bath.

During Caroline's first Bath winter—the height of the social season, when rich folk soaked in the hot springs—William Herschel worked fifteen-hour days as the musical director and organist at the famed Octagon Chapel and as a private music tutor. "I retired at night with the greatest avidity to unbend the mind," he recalled, by reading mathematical and astronomical books. Among others, he read James Ferguson's *Astronomy Explained Upon Sir Isaac Newton's Principles* (1757) and Robert Smith's *Complete System of Opticks* (1738). "When I read of the many charming discoveries that had been made by means of the telescope," Herschel said, "I wished to see the heavens and Planets with my own eyes thro' one of those instruments."

In May 1773, when the social season had ended and he had free time on his hands, Herschel bought lenses and tubes to make refracting telescopes. He made one with a 12-foot focal length, rigged a stand, and, after some trial and error, there were Jupiter and its moons. He was hooked, but he wanted to see more, peer farther into space. He made a 30-foot refractor but had trouble with the long tube, which he found "impossible to manage," so he rented a 2-foot-long Gregorian reflector. "This was so much more convenient than my long glasses that I soon resolved to try whether I could not make myself such another." Herschel set about making his own telescopes. Shortly thereafter, Caroline recalled, "to my sorrow I saw almost every room [in the house] turned into a workshop."

William continued to teach and compose music and conduct his choir, but every spare minute was devoted to grinding mirrors, and his music students sometimes found themselves involuntarily immersed in astronomy. "There it is at last!" he exclaimed in the middle of one lesson when the sky cleared. He dropped his violin and rushed to his telescope.

Telescopic Advances of the Eighteenth Century

By this time, the art of telescopic mirror-making had advanced considerably. In 1721, John Hadley, a self-taught inventor, ground a 6-inch mirror for a Newtonian telescope just over 6 feet long that he presented at the Royal Society. The Reverend James Pound wrote enthusiastically to the Royal Society, praising Hadley for reviving interest in reflecting telescopes. Pound, who had viewed Saturn and its satellites with Hadley's "curious Mechanism," compared it favorably to Huygens's 123-foot refractor, and he predicted the eventual triumph of reflectors: "It is to be hoped that [Hadley or others] will in a short Time find out a Method, either of preserving the concave Metal from tarnishing, or . . . of making a good concave *Speculum* of Glass quicksilver'd on the Back-part." Unfortunately, neither Hadley nor any contemporaries solved these problems.*

By the middle of the eighteenth century, several London opticians were making small reflecting telescopes for amateurs and wealthy dabblers. The finest mirror-maker by far was James Short. In 1732, at the age of twenty-two, Short abandoned his chosen career as a Church of Scotland minister because he had become obsessed with grinding telescope mirrors. At first he tried to follow Newton's advice and make glass mirrors back-coated with a tin-mercury amalgam, but he soon switched to speculum metal. He eventually moved to London, grinding nearly 1,400 mirrors over his long life. He made only the mirrors, subcontracting the manufacture of the brass telescope bodies. Mostly, he

---

*Hadley also used two mirrors when he invented the nautical octant, the predecessor of the modern sextant, so that sailors could take accurate sun sightings without blinding themselves. Hadley's invention contributed to an astronomical method of determining longitude at sea.

made mirrors for Gregorian telescopes, which were compact and produced upright images, suitable for royalty and nobility, who used them
to look out their castle windows.

Short's primary mirrors, up to 18 inches in diameter, were extraordinarily well figured, many of them approaching a real paraboloid shape,
and he produced enough small secondary concaves so that he could
"marry" them properly, as he put it, to the primaries, their minor faults
correcting one another. No other optician approached him, and he kept
his method a prized secret. Thus, James Short was considered a genius
whose work could not be duplicated.

When Herschel took up mirror-making in 1773, most telescopes
were refractors, in part because the problem of chromatic aberration—
caused by different-colored light bending differently—had been solved.
In 1729, the British gentleman-barrister Chester Moor Hall reasoned
that highly refractive lead glass—used for chandeliers and goblets since
its invention in 1676—might be useful in combination with regular
glass, since light bent a different amount in each of them. Might they
not, in a proper combination, correct one another's chromatic aberration? In an effort to keep his idea secret, Hall commissioned one London optician to make a concave lens of flint glass (the optical name for
lead glass), and a different craftsman to made a convex lens out of regular crown glass. Both opticians, however, subcontracted the work to
the same man, who deduced and revealed Hall's secret.

Still, no one paid much attention until another optician, John Dollond, took up the challenge. In 1757, he found that properly ground
combinations of flint and crown glass lenses virtually eliminated chromatic aberration. Dollond died in 1761, but his son, Peter Dollond,
continued to improve the newly named achromatic refracting telescopes, adding a third lens for finer corrections, producing his first
"triplet" in 1763. This revolutionary telescope, with an aperture just
under 4 inches, was 42 inches long. No longer would refractors have to
be built to absurd lengths to avoid chromatic aberration, though the
longer refractors remained the norm for some time, as Dollond held a
patent on the achromatic lenses until 1772.

It was nearly impossible to find enough decent glass to make a lens
larger than a few inches in diameter, so the new achromatic refractors
remained rather small. Still, that wasn't really a drawback, since most

astronomers were primarily interested in finding the precise position of the sun, moon, and planets of the solar system in relation to the stars in order to solve the problem of determining longitude at sea.

The Best Telescopes Ever Made

From the outset, William Herschel wasn't satisfied with the plodding work of the positional astronomers. The solar system was just a tiny, whirring piece of the awesome clockwork of the universe. Where were we located in the midst of this mind-boggling swirl of stars? What was the Milky Way, that mysterious bright swath across the heavens? How had the universe evolved, and what would be its fate?

Until Herschel began to grind ever-larger mirrors in order to peer farther into space, the "fixed stars" outside the solar system were seen, as in Ptolemy's time, as mere pinpricks on the dark fabric of the night sky. They were inconceivably far away, but they seemed equidistant. Astronomers wanted little more than to map their position on the inverted bowl of the sky.

One of Herschel's earliest papers was devoted to the problem of finding the "parallax" (distance) of a star, a quest that became a lifelong obsession. *Parallax* refers to the changing position of a nearer object when viewed from different perspectives against more distant objects. Hold a finger up at arm's length. While looking at it, close one eye, then the other, and you will see your finger move sideways in relation to its background. If you measure the distance between your two eyes and the vertex angle of the long isosceles triangle, simple geometry yields the distance to your finger. That is parallax, which is easier to distinguish the farther apart the "eyes" are set and the bigger and more acute they are.

Stars are so far away that they require a huge distance between observations to discern their parallax. Views from either side of the earth (i.e., from the same spot twelve hours apart on a long winter night) didn't provide sufficient distance between the two telescopic "eyes," but Herschel figured that he might be able to spot a change in position by taking observations at six-month intervals, from opposite ends of the earth's orbit around the sun (some 186 million miles in diameter). To measure a star's parallax, he needed to measure its position against the background of stars so far away that they appear unmoving. To do that, he had to see farther away, so he needed bigger mirrors. Herschel coined the phrase "light grasp" to describe the ability of ever-larger mirrors to gather more light in order to see ever-dimmer stars.

In 1778, after testing different speculum metals and grinding endless mirrors of various sizes (he was to make more than 400 by 1782), Herschel completed a "most capital" 6.2-inch mirror for a 7-foot-long telescope, his finest effort yet.* At first he tried the Gregorian system espoused by James Short, but the necessity for a hole in the primary and a matching concave secondary drove him to adopt the simpler Newtonian design in which a flat secondary mirror reflected the light to a focus at an eyepiece on the telescope's side. With the 7-foot telescope, in August 1779 he began his second comprehensive review of the heavens visible from Bath.

He was able to see stars down to the eleventh magnitude and beyond.** Herschel also found numerous double stars that appeared single to the naked eye, with one member of the pair often fainter than the other. Herschel reasoned that in many cases the brighter star was closer and just happened to appear near the other. In that case, he hoped that he could determine the parallax of the nearer star by seeing it move in relation to the other a half-year hence.

On nights when a full moon glowed in the night sky and ruined stargazing, Herschel turned his telescope on the moon itself and tried to measure the height of its mountains. One night near Christmas in 1779, Herschel was observing the moon on the street in front of his house when a curious stranger asked if he might have a look and expressed "great satisfaction" at the view. He turned out to be Dr. William Watson, a fellow of the Royal Society, who became Herschel's great friend and supporter. Watson encouraged him to join the Bath Literary and Philosophical Society, where Herschel began to submit papers on lunar

---

*Modern telescopes are usually identified by the size of their mirrors, but Herschel usually referred to his by the length of the tube.

**The bigger a star's magnitude, the dimmer the star. Thus, a star of the first magnitude is very bright; a sixth-magnitude star is the faintest that can be seen with the naked eye.

mountains, a variable star (one whose brightness varies with time), and philosophical matters.

In April 1781, Herschel wrote a paper for the Bath society called "An Account of a Comet" in which he described an unusual object he had observed on March 13 as he was conducting his routine sky sweeps. While examining small stars, "I perceived one that appeared visibly larger than the rest." Four nights later, he found that its position relative to the small star near it had changed, and he concluded that it must be a comet because it was moving so near the earth. Herschel was wrong. Within a year, better mathematicians than Herschel had worked out its orbit and declared it to be a seventh planet.

The news galvanized not only the astronomical community but also the general public, since it suddenly expanded the solar system known since antiquity. What other planets might there be? Herschel was catapulted to international fame. He was given the prestigious Copley Medal and made a fellow of the Royal Society in December 1781, a few months after his disastrous effort to cast the mirror for a 30-foot telescope. Better yet, he attracted the attention of King George III, an astronomy enthusiast with his own private observatory.

On May 20, 1782, Herschel packed up his 7-foot telescope and went to London for an audience with the king, who asked him to set up the instrument for three weeks at the Royal Observatory at Greenwich, so that professional astronomers could judge it, and then to bring it to the royal court at Richmond. His telescope performed superbly at Greenwich. "We have compared our telescopes together," Herschel wrote to his sister, "and mine was found very superior to any of the Royal Observatory. Double stars which they could not see with their instruments I had the pleasure to show them very plainly."

Although Herschel enjoyed showing off his telescope, he was itching to get back to his mirror workshop. "I pass my time between Greenwich and London agreeably enough," he wrote to Caroline, "but . . . I would much rather be polishing a speculum." He was pleased but bemused by all the attention he was getting. "Among opticians and astronomers nothing now is talked of but *what they call* my Great discoveries. Alas! this shows how far they are behind, when such trifles as I have seen and done are called *great*. Let me but get at it again! I will make such telescopes, and see such things. . . ."

A Kingly Grant to Penetrate Space

On July 3, Herschel set up his telescope at Richmond for the king, queen, and other royalty. "My Instrument gave a general satisfaction," Herschel wrote. "The King has very good eyes and enjoys Observations with the Telescopes exceedingly." Nothing gave Herschel greater pleasure, he wrote, than to show "those beautiful objects with which the Heavens are so gloriously ornamented."

Soon after their meeting, the king offered Herschel a 200-pound annual stipend, which allowed him to abandon his musical career and devote himself to astronomy full-time. The king also ordered five 10-foot telescopes. The monarch had only six years to enjoy Herschel's telescopes before he suffered his first attack of what was probably a rare medical condition called porphyria. The king with "very good eyes" for astronomy died blind, lonely, and mentally deranged. Although he is remembered as the "mad king" who lost the American colonies, George was an intelligent, well-read monarch fascinated by science. "Perhaps the biggest thing King George ever did," wrote his biographer, "was to patronize Herschel." In gratitude, Herschel named his newly discovered planet Georgium Sidus, although it was eventually called Uranus to match other planet names derived from Roman mythology.

By the end of July 1782, William Herschel, his sister Caroline, and his brother Alexander had moved to Datchet, near Windsor Castle. Caroline was dismayed to find that the house he had rented was falling apart and the grounds swampy. But William was delighted, converting the stables into a mirror-grinding facility and clearing weeds to make room for a 20-foot telescope featuring a 12-inch mirror and an elaborate mounting that allowed Herschel to peer through the eyepiece on the side, even as it swerved up toward the zenith.

By the time he took up astronomy full-time, Herschel was forty-three years old and worried that he would not have time to delve far enough into space to unravel the mysteries of the universe. He threw himself into his full-time occupation, observing every clear night and calling out his observations to Caroline below, who diligently recorded them. "If it had not been sometimes for the intervention of a cloudy or moon-light night," Caroline recalled, "I do know not when my Brother (or I either) should have got any sleep." Herschel himself wrote: "I have many a night, in the course of eleven or twelve hours of observation, carefully and singly examined not less than 400 celestial objects."

The German astronomer J. H. von Magellan visited Herschel at Datchet. "I went to bed about one o'clock," he wrote, "and up to that time he had found that night four or five new nebulae. The thermometer in the garden stood at 13 degrees Fahrenheit; but in spite of this, Herschel observes the whole night through." Sometimes Herschel's feet froze to the ground while he observed. To prevent catching the ague, he rubbed himself all over with a raw onion.

During the day, Herschel ground and polished mirrors and wrote papers. On the first day of 1783, while he was observing on a frigid night, the 12-inch mirror cracked with the sound of a gunshot. Herschel replaced it, but he was already planning to make an 18.75-inch mirror for another telescope of the same length that came to be called the "large 20-foot." Eager to observe with this improved model, Herschel began sweeping the sky before it was finished in the fall of 1783. "Every moment I was alarmed by a crash or fall," Caroline wrote, "knowing him to be elevated 15 or 16 feet on a temporary cross-beam." One windy night, just after Herschel reached the ground, the whole structure collapsed. "Some neighboring men were called up to help extricating the mirror which was fortunately uninjured," Caroline concluded.

Caroline herself—a diminutive woman who barely reached 5 feet—was not so fortunate. On December 31, 1783, while running to turn the big telescope, she fell in slushy snow, snagging her leg on an iron hook and losing a chunk of flesh when she was pulled off. "I had however the comfort to know that my Brother was no loser through this accident," the devoted sister noted, "for the remainder of the night was cloudy."

Although she had not been thrilled by her brother's obsession with astronomy, Caroline quickly caught the bug. In 1783, William gave her a small Newtonian telescope, a bit over 2 feet long. Later her brother completed a much more powerful "sweeper" for her, over 5 feet long, with a 9.2-inch mirror. Caroline was to discover eight comets and become a famous astronomer in her own right. During the daylight hours, she helped grind smaller mirrors for telescopes that William sold to supplement the inadequate stipend from the king.

By 1785, Herschel, who was supporting a large work crew as well as his mother back in Germany, had exhausted his savings. Despite the disastrous attempt to make a mirror for a 30-foot telescope, he still yearned to make a gigantic instrument to pierce the night sky. "The great end in view," he wrote to Sir Joseph Banks, head of the Royal Society, "is to increase what I have called 'THE POWER OF EXTENDING INTO SPACE. . . .' Telescopes have this power of collecting light in proportion to their apertures, so that one with a double aperture will penetrate into space to double the distance of the other." With Banks's support, the king granted Herschel an additional 2,000 pounds to build a 40-foot telescope with a mirror 4 feet in diameter.

By this time, Herschel had a new reason to yearn for a bigger mirror. He wanted to examine nebulae, aptly named fuzzy objects in the night sky. In 1783, the French comet-hunter Charles Messier had published a list of just more than 100 of these mysterious objects, which he considered pests. Like most contemporary astronomers, he had no aspirations beyond the solar system, and these far-off objects kept distracting him. They resembled comets, but they never moved, so he decided to catalog them in order to avoid wasting time on them. Herschel eagerly examined each Messier nebula and quickly discovered many others. With his large 20-foot telescope, Herschel could resolve many of them into individual stars, but others remained cloudy.

In Herschel's telescope, the Milky Way itself turned from a whitish swath into millions of individual stars, clustered in various arrangements. "It is very probable," he concluded in 1784, "that the great stratum, called the Milky Way, is that in which the sun is placed, though perhaps not in the very centre of its thickness." He described our galaxy as a "very extensive, branching, compound Congeries of many millions of stars" and speculated that "it is highly probable that every star is more or less in motion," including the sun along with its dependent planets.

Herschel thought that many of the nebulae were distant galaxies, "island universes," some of which "might well outvie our Milky Way in grandeur." He had the remarkable ability to imagine the view from many light-years away: "To the inhabitants of [other] nebulae . . . our sidereal system must appear . . . as a small nebulous patch." The varied shapes he viewed through his telescope, he thought, might be stars and galaxies in different stages of evolution, comparable to a garden in which some plants were just sprouting while others bloomed or decayed. And just as dying plants provided nutrients for the new, "we ought perhaps to look upon such clusters, and the destruction of now and then a star, in some thousands of ages, as perhaps the very means by which the whole is preserved and renewed. These clusters may be the laboratories of the universe."

By 1786, when the Swiss professor Marc Auguste Pictet visited, Herschel had moved to Slough to avoid the damp climate at Datchet. The immense scaffolding of the new 40-foot telescope was already in place. "In the middle of the workshop there rises a sort of altar," Pictet observed, on which rested the 48-inch metal mirror. Twelve men dressed in numbered overalls polished it according to Herschel's shouted directions. "At one time no less than 24 men (12 and 12 relieving one another) kept polishing, day and night," Caroline wrote, "and my Brother of course never leaving them all the while, taking his food without allowing himself time to sit down to table."

The mirror, too thin to hold its shape, never satisfied Herschel. A second casting cracked, so Herschel reduced the amount of tin to make it less brittle, and this third effort, weighing more than 2,000 pounds, finally produced a usable mirror. Frustrated by the enormous manpower required to polish such a mammoth mirror, Herschel invented an effective polishing machine. Meanwhile, the gigantic telescope tube lay on the ground awaiting its Cyclops eye. One August day in 1787, King George III led the Archbishop of Canterbury through the tube, quipping, "Come, my Lord Bishop, I will show you the way to Heaven."

The great telescope was finally ready to view the heavens in August 1789, and when Herschel turned it on Saturn, he immediately discovered a sixth moon. When he looked at the bright star Sirius, it shone "with all the splendour of the rising sun, and forced me to take my eye from the beautiful sight." Yet the telescope never fulfilled expectations, and Herschel used it only rarely in the ensuing years. Part of the problem lay in the composition of the mirror itself. At their best, speculum mirrors reflected a bit more than 60 percent of the light they received. This giant mirror, cast with too much copper and too little tin, reflected even more poorly. In order to avoid losing any more light from a second reflection, Herschel eliminated the secondary flat mirror and placed his eyepiece at the top of the massive tube—an arrangement now called Herschelian. He had to tilt the primary mirror slightly to the side in order to focus it on the eyepiece, thereby causing some astigmatism.

The mirror in the 40-foot telescope constantly suffered from tarnish despite an annual repolishing. "My brother had reason for choosing the cold season for this laborious work," Caroline recalled, "the exertion of which alone must put any man into a fever." It was also dangerous. "In taking the forty-foot mirror out of the tube," Caroline recorded in her diary on September 22, 1806, "the beam to which the tackle is fixed broke in the middle. . . . Both my brothers had a narrow escape of being crushed to death."

The huge telescope was unwieldy and difficult to use. For the rest of his life, Herschel continued to rely on his trusty 20-footer, while the 40-foot monstrosity served primarily as a tourist attraction.

A Supernatural (But Sometimes Mistaken) Intelligence

As he was preparing the great mirror for insertion into that mighty muzzle in 1788, William Herschel, the workaholic forty-nine-year-old bachelor, surprised everyone by marrying Mary Pitt, a wealthy widow. No one was more shocked than sister Caroline, who felt disoriented and threatened. She continued to assist her brother with his telescopes, but it took years for the new Mrs. Herschel to win her over. Caroline subsequently destroyed her diary entries for the first nine years of their marriage.

In March 1792, Mary Herschel gave birth to John, the couple's only son. When he was five, a visitor found him "entertaining, comical, and promising." The young Herschel grew up peering through his father's telescopes, lending a hand at mirror-grinding, experimenting with chemistry, and wielding a small hammer to help the carpenters.

Meanwhile, William continued to sweep the night skies. He was particularly intrigued with some nebulous objects that refused to resolve themselves into stars and that he named (confusingly) "planetary nebulae" because they presented a round profile like a planet. On November

13, 1790, he observed one that shook his confidence, recording in his journal: "A most singular Phaenomenon! A star ... with a faint luminous atmosphere, of a circular form. ... The star is perfectly in the center and the atmosphere is so diluted, faint and equal throughout, that there can be no surmise of its consisting of stars." The cloudy atmosphere was clearly related in some way to the star. "Perhaps it has been too hastily surmised that all milky nebulosity ... is owing to starlight only," he noted. As a consequence, Herschel pulled back from his assertions about thousands of "island universes" lying far outside the Milky Way. Though he continued to believe that such galaxies existed, he concluded that many of them might be "true nebulosities" much closer to earth.

During the day, Herschel turned smaller telescopes to the sun. Like Newton, he nearly blinded himself, temporarily losing the sight in one eye. He tried grinding a glass mirror and backing it with black velvet, but it didn't prevent the light from being too harsh; neither did colored-glass filters. Finally, he found that a diluted ink solution reduced the heat and light enough for him to make decent observations.

Herschel's experience with glass filters led him to an important discovery. "I used various combinations of differently coloured darkening glasses," he wrote in 1800. "What appeared remarkable was that, when I used some of them, I felt a sensation of heat, though I had but little light." Intrigued, he set up a Newtonian experiment, splitting light with a prism and allowing one color at a time to heat a thermometer. He discovered that red light produced the greatest heat; violet rays, at the other end of the spectrum, produced the least. Then Herschel placed a thermometer slightly *beyond* the red end of the spectrum—and the thermometer spiked even higher!

Herschel had discovered that light stretched beyond the visible spectrum into the infrared. "We cannot too minutely enter into an analysis of light," Herschel wrote, "which is the most subtle of all active principles that are concerned in the mechanism of the operation of nature." Now he concluded that "radiant heat [might] consist, if I may be permitted the expression, of invisible light." Sir Joseph Banks wrote presciently to Herschel: "Highly as I prize the discovery of a new Planet, I consider [your infrared studies] as a discovery pregnant with more important additions to science." He was right.

The following year, German chemist Johann Wilhelm Ritter noticed that silver chloride darkened when exposed to light, particularly toward the violet end of the spectrum—and beyond. He had discovered invisible ultraviolet light. Over the course of the nineteenth century, scientists would discover an ever-broader spectrum.

Herschel was wrong about many matters. He thought the sun was inhabited. He believed that the gassy rings around the planetary nebula were collapsing in to form a new star, but today we know that they represent the remnants of a red giant's death—the twilight rather than the beginning of a star's life span. Herschel ended his own time on earth thinking erroneously that the Milky Way would gradually break up into disparate clusters. He never did measure the parallax of a single star and was forced to conclude that double stars really were connected, since he found that they slowly circled one another.

Even when he was wrong, however, Herschel often pointed the way. By asking the right questions, and by placing us in a vast, evolving universe, inconceivably large and old, he founded modern astronomy and gave mankind a new perspective. His broader conclusions were astonishingly accurate. Galaxies are indeed the "laboratories of the universe," and Herschel was correct that many nebulae are indeed "island universes" similar to the Milky Way, though he would not be proven correct for another two centuries.

In 1813, when William Herschel was seventy-five years old, he spent a Sunday with poet Thomas Campbell. By that time, Herschel's iron constitution had finally broken under the strain of sleepless nights and endless mirror-grinding. Campbell was impressed with the elderly astronomer's "simplicity, his kindness, [and] his readiness to explain." With a "modesty of manner" that conveyed plain facts rather than vanity, Herschel told Campbell, "I have looked further into space than ever human being did before me. I have observed stars of which the light, it can be proved, must take two million years to reach the earth. Nay more, if those distant bodies had ceased to exist millions of years ago, we should still see them, as the light did travel after the body was gone." The overwhelmed Campbell felt "as if I had been conversing with a supernatural intelligence," but Herschel was simply telling him what he had learned by making large paraboloid mirrors and pointing them in the right direction.

John Herschel Sweeps the Southern Skies

Campbell also met John Herschel, whom he described as "a prodigy in science and fond of poetry, but very unassuming." The son had just graduated from Cambridge University, published a paper on mathematical theory, and been made a member of the Royal Society at twenty-one. But he did not want to follow in his father's footsteps. Instead, he entered law school.

John Herschel stuck with his legal studies for a year and a half, but they bored him, so he quit, returning to his old Cambridge haunts as a tutor. Meanwhile, his elderly father struggled to repolish his giant mirror. "His strength is now, and has for the last two or three years not been equal to the labour required for polishing forty-foot mirrors," Caroline wrote in her diary on September 30, 1815.

A year later, with extreme reluctance, John Herschel decided to move to Slough as his father's assistant and heir. William, now nearly seventy-nine, refused to give up mirror-making or observing. The last evidence we have of the elderly astronomer's attempted observations shows him calling once more on his beloved sister: "Lina," he wrote in a shaky hand on July 4, 1819, "there is a great comet. I want you to assist me."

The following year, he made his final mirror by proxy, guiding his son through grinding, figuring, and polishing an 18-inch mirror for John's own 20-foot telescope. Two years later, at eighty-three, William Herschel died. In poor health and not expecting to live long herself, Caroline, seventy-two, went back to Germany after a fifty-year absence to live with her younger brother, Dietrich, with whom she disagreed about everything. She lived an active, cantankerous, unhappy life in Germany—"I find myself, unfortunately, among beings who like nothing but smoking, big talk on politics, wars, and such like things"—until she died, aged ninety-seven, in 1848.

One of the first female astronomers, Caroline Herschel was elected a member of the Royal Astronomical Society, though she demurred: "I did nothing for my brother but what a well-trained puppy-dog would have done. . . . I was a mere tool which *he* had the trouble of sharpening." Even though she clearly prided herself on her comet-hunting, she dismissed it as a "children's game." Up to her death, Caroline kept abreast of new developments in astronomy. At ninety-six, she told a friend that while lying on her couch she had "with her mind's eye, set up a whole solar system in one corner of her room, and given to each newly discovered star its proper place."

Having abandoned the study of law, John Herschel proved as devoted an astronomer as his father. "A week ago I had the twenty-foot directed on the nebulae in Virgo," he wrote to his aunt in 1825. "These curious objects . . . I shall now take into my especial charge—nobody else can see them." Remarkably, Herschel's telescope, designed by his father more than forty years previously, remained the best in the world. It took John only eight years, until 1833, to complete the review of all 2,500 nebulae his father had identified. Although he was assiduous in his task, he was not always as patient as his father. "Two stars last night," he wrote in his journal on a July night in 1830, "and sat up till two waiting for them. Ditto the night before. Sick of star-gazing—mean to break the telescopes and melt the mirrors."

In November 1833, his review of the heavens finished and his mother deceased, Herschel packed up his telescopes, wife, and three children— in 1829 he had married eighteen-year-old Margaret Brodie, exactly half his age at the time—and sailed for the Cape of Good Hope, where he intended to explore the virgin skies of the Southern Hemisphere. When Caroline heard of his trip, she wrote in a passionate mixture of German and English: "Ja! If I was thirty or forty years younger, and could go too? In Gottes nahmen!"

Upon his arrival in South Africa in January 1834, Herschel found it "a perfect paradise," with gorgeous flowers, scenic mountains, and calm, clear nights lit by "the astonishing brilliancy of the constellations." In June he wrote to his aunt: "The twenty-foot has been in activity ever since the end of February, and, as I have now got the polishing apparatus erected and three mirrors (one of which I mean to keep constantly polishing) the sweeping goes rapidly." His father had made one of the mirrors, another he made under his father's direction, and the third he had produced on his own.

Herschel worked with extraordinary speed. Within a year, he wrote to his aunt: "I have already collected a pretty large catalogue of southern nebulae, for the most part hitherto unobserved. . . . I have not had the least difficulty in my polishing work, and my mirrors are now more perfect than at any former time since I have used them." His sweeps of the southern skies revealed 1,202 double stars and 1,708 nebulae, including the two Magellanic Clouds, which he found to be "of astonishing complexity."* A fine artist, he drew sketches of his observations, though he sometimes despaired of recording the "endless details." Declaring that "Science is Poetry," Herschel described one constellation as "a gorgeous piece of fancy jewelry," and he compared the Milky Way to "sand, not strewn evenly as with a sieve, but as if flung down by handfuls."

Herschel's exploits captured the public imagination and inspired a hoax. In August 1835, the *New York Sun* published "Remarkable Discoveries Made at the Cape of Good Hope," claiming that Herschel's telescope had found batmen living on the moon. Lavishly illustrated, the article was translated into several languages.

In 1838, after four years of intense work, John Herschel packed up his telescopes and family—there were now six children—and sailed back to England, where he took up residence at Slough. The colossal 40-foot telescope had become a dangerous dinosaur, so Herschel had the creaky scaffolding dismantled. The next year, on New Year's Eve, the family celebrated the arrival of 1840 by standing one more time inside the giant tube and singing a requiem composed by John Herschel. The chorus tried to be cheerful ("Merrily, merrily, let us all sing / And make the old Telescope rattle and ring"), but it was an elegiac occasion. They sang of the tube, and perhaps of William as well: "Now prone he lies, where he once stood high, / And searched the deep heaven with his broad bright eye."

In a moving verse, John Herschel wistfully recalled his father's cold, lonely, exalted nights, gazing at light emitted long before the first humans existed:

Here watched our Father the wintry night
And his gaze hath been fed with pre-Adamite light,
While planets above him in mystic dance
Sent down on his toils a propitious glance.

---

*Although Herschel did not know it, these are our two nearest galactic neighbors that will someday probably be swallowed by the Milky Way.

Shortly afterward, the Herschels moved from Slough to a large house called Collingwood. John Herschel published a catalog of all known clusters and nebulae, then another of double stars. He wrote a popular science book called *Outlines of Astronomy*, translated Homer's *Iliad*, contributed to the discovery of photography, and followed Newton in serving as Master of the Mint—but he gave up observational astronomy, perhaps to spend more time with his twelve children. He loved gardening and music. He never spent another sleepless night at the telescope or polished another mirror.

Two years before his death at the age of seventy-nine, a humble John Herschel advised a younger astronomer: "In the midst of so much darkness, we ought to open our eyes as wide as possible to any glimpse of light, and utilize whatever twilight may be accorded us, to make out, though but indistinctly, the forms that surround us."

Fraunhofer's Magical Glass

Despite the astonishing success the Herschels had with their large reflecting telescopes, most other astronomers preferred refractors, in large measure because every time a metallic mirror needed to be cleaned it also needed to be repolished and often refigured. In general, that meant that the master mirror-maker who created the speculum in the first place had to maintain it. This explains why few of the telescopes William Herschel sold were effective in other hands.

In the meantime, refractors had improved dramatically, thanks to a self-educated German genius named Joseph Fraunhofer. Born in 1787, the youngest of eleven children, Fraunhofer learned to cut and fit glass from his father, a master glazier, before both of his parents died. Orphaned at twelve, Fraunhofer apprenticed with Philipp Anton Weichselberger, a mean-spirited mirror-maker and ornamental glass–grinder. Two years later, Weichselberger's shop literally collapsed, burying Fraunhofer under the debris. After a dramatic four-hour rescue operation that attracted Elector Maximilian IV Joseph, the boy was pulled out alive. The elector gave him a gift that enabled Fraunhofer to study optics and buy a small grinding machine. By 1807, he had written his first scientific paper on paraboloid mirrors and reflecting telescopes—an ironic subject for the man who was to promote refractors.

That same year, Fraunhofer joined a glassworks in the former monastery at Benediktbeuern, nestled at the base of the Alps, with endless forests to fuel the furnace. The Swiss glass master Pierre Louis Guinand was producing high-quality flint and crown glass. Wood fires did not permit high enough temperatures to produce homogenous melted glass, so Guinand reheated and stirred it with a giant fireclay paddle. Fraunhofer, still dissatisfied with the result, insisted on absolutely pure ingredients, keeping his formula secret. His constant experimenting and note-taking irritated Guinand, who returned to Switzerland in 1814 after the younger man was made his boss. There, Guinand manufactured his own lenses and telescopes.

By that time, Fraunhofer had been made a partner in the renamed Benediktbeuern Optics Institute, and the firm was revolutionizing achromatic refracting telescopes, making high-quality lenses of previously unthinkable sizes, culminating with the 9.5-inch objective for the Dorpat Observatory refractor in Russia, which saw first light in 1824. Not only were the optics superb; Fraunhofer had also designed the first major equatorial mount for a large telescope. To follow a star, William Herschel's telescopes, mounted in the standard alt-azimuth, had to make two adjustments. First, the tube had to move up or down in altitude, and second, the entire telescope had to be wheeled around horizontally on a round track in azimuth. The Dorpat equatorial was mounted on a steel bar set parallel to the earth's polar axis. When fitted with a clock drive, the telescope could automatically follow a star across the heavens merely by turning on one axis.

Refractors made at Benediktbeuern sold quickly to major observatories throughout Europe and were eventually copied by others. Among other luminaries, John Herschel made a pilgrimage to meet Fraunhofer shortly before his death of tuberculosis in 1826 at the age of thirty-nine, but refractors would dominate the field for the remainder of the century. Using a refractor that Fraunhofer was working on when he died, Friedrich Wilhelm Bessel finally measured a star's parallax in 1838, calculating the distance of 61 Cygni to within 10 percent of the modern figure of 10.9 light-years. George Merz, who had worked under Fraunhofer, took over the firm and, in 1838, made a 15-inch refractor for the

Imperial Russian Observatory at Pulkowa, followed by a nearly identical telescope for Harvard College Observatory in 1846.*

The Leviathan of Parsonstown

Given the ascendancy of refractors and the failure of Herschel's 40-foot reflector, it took a mad Irish lord to aim an even bigger mirror skyward. William Parsons, who became the third Earl of Rosse upon his father's death in 1841, lived in Birr Castle in the heart of southern Ireland, a drizzly locale that could scarcely be equaled as an unsuitable site for a telescope. The Parsons nobility had lived there, off and on, since 1620, so it was known as Parsonstown. Born in 1800, William Parsons studied mathematics and engineering at Trinity and Oxford. Soon afterward, he was dividing his time between Parliament and mirror-making at Birr, and in 1828 he published his first scientific paper, "Account of a New Reflecting Telescope." He took issue with those who believed that "since Fraunhofer's discoveries, the refractor has entirely superseded the reflector, and that all attempts to improve the latter instruments are useless."

At first, Parsons (known as Lord Oxmantown until he became the Earl of Rosse) tried to make specula from segments, since he doubted his ability to cast a suitably large disk in one piece. He also experimented with reducing the weight by using ribs on the back, as Herschel had once suggested. By 1840, Parsons had made two 36-inch mirrors, one segmented, with a tinned surface, and one solid. To shape them, he designed and built an ingenious mechanical grinder-polisher, powered by a small steam engine. He kept the speculum metal at a constant temperature by keeping it immersed in a water bath, and he tested it by viewing a watch-dial hung 50 feet over the mirror, blocking different parts of the mirror to see whether the image remained clear.

---

*The improved glass quality also made lenses for compound achromatic microscopes far better. Microscopes had always used lenses, although in 1725 Edmund Culpeper added a concave mirror at the bottom of his microscopes to provide better light for transparent specimens, and such mirrors became standard features. In 1813, Giovan Battista Amici produced the first reflecting microscope, soon followed by others, but mirrored microscopes were difficult to fabricate and use, and they lost ground around 1840.

To prevent the mirror from flexing unevenly in the tube as it swung up and down, Parsons adopted a clever system of weighted support levers invented by Thomas Grubb of Dublin. He built a wooden alt-azimuth modeled after William Herschel's telescopes, installing a Newtonian secondary flat mirror, so that the observer viewed the sky through an eyepiece near the top of the tube.

Reverend Thomas Romney Robinson, the director of the Armagh Observatory farther north in Ireland, came to Birr to help test the two mirrors, but night after night, the weather was uncooperative, the air unsteady. Using eyepieces with high magnification on bright stars, however, Robinson and Parsons concluded that the segmented mirror caused problems. From then on, Parsons made only traditional, solid specula. Robinson was enormously impressed, proclaiming the Birr 36-inch to be "the most powerful telescope that has ever been constructed." But for Parsons it was just a preliminary effort. He immediately began to work on a mirror twice that size, a mammoth reflector 6 feet wide.

On April 12, 1842, Parsons, now the third Earl of Rosse, ordered the crucibles—each 24 feet across—to be preheated for two hours in three separate furnaces. It then took another ten hours for the metal ingots to melt. At 1 A.M., the dramatic casting took place, as the bright metal poured simultaneously from the three crucibles into the metal mold. Robinson, who had come down from Armagh, was exalted by the eerie scene. "Above, the sky, crowded with stars and illuminated by a most brilliant moon, seemed to look down auspiciously on their work. Below, the furnaces poured out huge columns of nearly monochromatic yellow flame, and the ignited crucibles during their passage through the air were fountains of red light, producing on the towers of the castle and the foliage of the trees, such accidents of colour and shade as might almost transport fancy to the planets of a contrasted double star."

The new mirror, which weighed four tons, was immediately dragged on rails into an annealing oven, where it was allowed to cool gradually for sixteen weeks, then ground. After all that effort, it cracked, and the whole process was repeated, using higher copper content to make it less brittle. This one took a good figure and polish, but Parsons knew that he needed two good mirrors to avoid the fate of Herschel's ever-tarnished

40-foot dinosaur. The third and fourth castings failed for various reasons, but the fifth finally produced a replacement mirror.

By the time the first mirror was usable, in 1845, Parsons's workmen had built two masonry walls 24 feet apart, each 56 feet high, between which he mounted the 56-foot wooden tube, held together with iron rings. The box containing the mirror at the bottom was bolted to a massive iron universal joint. Two workmen could raise and lower the tube along the meridian—the north-south line—with windlasses, pulleys, and chains. Another assistant moved it from side to side in "right ascension" between the pillars. The observer could make finer adjustments from the platform. Although the tube could move about 15 degrees between the stone walls, it could not follow stars for longer than an hour. Instead, observers had to wait for the heavens to pass over the huge mirror.

They also had to wait for clear skies. For the first two weeks of February 1845, Robinson and Parsons waited for a break in the clouds. Finally, on February 15, they saw Castor in Gemini, a glorious double star, before losing the sky again. Even on a night without clouds, however, atmospheric turbulence usually ruined the "seeing," as astronomers referred to visual conditions. Bright stars, one observer noted, were seen in the massive telescope as "balls of light, like small peas, violently boiling." Only occasionally would a magical calm allow good seeing, and then the stars were well-defined pinpricks in the black velvet of the night sky.

On such a rare night in April 1845, William Parsons viewed M51, the fifty-first object in Messier's catalog, which the French comet-hunter had called "a very faint nebula without stars." In the reflected light of his huge mirror, M51 resolved into a majestic spiral "pretty well studded with stars," Parsons recorded. It was subsequently called the Whirlpool nebula. Parsons was astonished at this discovery—a giant nebulous pinwheel whose light took millennia to arrive in his telescope mirror, probably composed of millions of stars. "That such a system should exist, without internal movement, seems to be in the highest degree improbable," he wrote. He thought it must be slowly turning, though he could never prove it.

Just as Parsons's appetite to look for more spirals was whetted, the Irish potato famine of 1845 forcibly brought his attention back to earth. For the next three years, the earl threw himself into relief work. Nonetheless, by 1850 he had identified fourteen spiral nebulae, some spread nicely like the Whirlpool, but most at an angle like the Andromeda.

Romney Robinson jumped to the conclusion that "no *real* nebula seemed to exist. . . . All appeared to be clusters of stars." Oddly, Robinson and most other astronomers concluded that these were not external galaxies but were all part of the Milky Way. Lord Rosse felt that the spirals might indeed be island universes, and although he incorrectly felt that all nebulae, including the Orion Nebula and Ring Nebula in Lyra, might be resolvable into stars with a bigger telescope, he withheld a final judgment.

The Leviathan's revelations generated enormous excitement. In his 1851 book *Architecture of the Heavens*, the Scottish astronomer John Pringle Nichol wrote, "The magnificent and most exact instruments of Parsonstown have converted what was twilight into daylight, and penetrated into regions of space formerly enveloped in utter darkness." Still, Nichol recognized that the Leviathan could not resolve most of the nebulae. He admitted, "Even the six-feet mirror, after its powers of distinct vision are exhausted, becomes in its turn simply as the child, gazing on those mysterious lights with awful and solemn wonder."

At first, it seemed that the Leviathan would revolutionize astronomy. In the ensuing years, however, it failed to perform further miracles, largely because of the perennial tarnishing problem. Even with two mirrors, keeping a bright, well-figured mirror pointing heavenward was an enormous challenge in the damp climate of southern Ireland. "I am sorry to say [the specula] very often [are] not as bright as they should have been," Lord Rosse admitted. The 3-foot mirror usually worked better than the 6-foot.*

---

*Two other wealthy British amateurs also made substantial reflectors, with Lord Rosse's encouragement. In 1849, James Nasmyth, a Scottish ironmaster and inventor of the steam hammer, made a "comfortable telescope" with three mirrors, allowing the observer to sit without moving, but he received only a quarter of the available light because of the triple reflection. William Lassell, a Liverpool brewer, made excellent metal mirrors and was among the first astronomers to seek better seeing conditions, mounting a 48-inch equatorial in Malta in 1861.

Silver-on-Glass and the Great Melbourne Failure

By the time Lord Rosse made this admission in 1861, a new technology should have made solid metal mirrors obsolete. The advance was motivated by health issues facing looking-glass manufacturers, with their workers suffering from weakness, irritability, tremors, and delirium associated with mercury erethism. In 1835, the German chemist Justus von Liebig first discovered that silver was deposited by a chemical reduction of a silver nitrate solution, but his method involved boiling the solution. In 1843, Thomas Drayton, a British gentleman tinkerer, patented a silver deposit method that didn't require heat.*

In 1856, von Liebig developed a more sophisticated method for slowly depositing an even silver film on glass. Although silver nitrate could explode, it did not poison workers as mercury did. Moreover, silver reflected light much more efficiently and tarnished more slowly. Commercial mirror-makers quickly switched to the new method, protecting the rear-surfaced mirrors with paint and varnish. Household mirrors became even better and cheaper.

Later that year, Carl August von Steinheil, who worked in Munich near von Liebig, used the process for telescope mirrors, applying the silver to the front surface, since putting it behind the glass would have produced no image due to double refraction through the concave glass surface.

The following year, the French physicist Léon Foucault independently silvered a 4-inch paraboloid telescope mirror, which he mounted on an equatorial stand. By that time, Foucault was already famous, having demonstrated the earth's rotation with his huge pendulum. He had already used a metallic concave mirror in a projecting microscope, and he had been forced to use the new silver-deposit

---

*London's 1851 Crystal Palace, which attracted some 6 million visitors, featured more than 1 million square feet of glass. Among the 13,000 exhibits were glass globes and vases made reflective with a silver nitrate and grape sugar process, patented by Messrs. Varnish and Mellish, but it was treated only as a novelty. "We cannot think that the pure white glass," wrote a contemporary journalist, "is in any respect improved by silvering."

technique on a rapidly spinning flat glass mirror (used the determine the speed of light) because centrifugal force flung off the traditional tin amalgam.

In 1857, Foucault traveled to Dublin to deliver a paper ("A Telescope Speculum of Silvered Glass") to the British Association for the Advancement of Science. He got the distinct impression that no one was listening. At the same meeting, a Dublin mirror-maker named Thomas Grubb described his plans for a proposed "Great Southern Telescope," the brainchild of Romney Robinson. It was to be a mammoth equatorial that would be a Leviathan for the Southern Hemisphere, where it would reexamine the nebulae that John Herschel had seen at the Cape. To avoid the awkward, dangerous observing platform that a large Newtonian necessitated, Grubb would make it a Cassegrain, so that observers could remain closer to the ground.*

But the Great Melbourne Telescope, named after the Australian site chosen for it, would have a solid metal mirror. "It seemed imprudent to risk the success of the undertaking by venturing on an experiment [silver-on-glass] whose success was not assured," Romney Robinson wrote later. "It was not known whether the silver could be uniformly deposited on so large a scale; some facts appear to show that glass is more liable to irregular action than speculum metal."

The advantages of silver-on-glass mirrors were immediately obvious to John Herschel. "Glass," wrote Herschel, "is incomparably stiffer than metal; so that a glass speculum, to be equally strong . . . need weigh only one-fourth of a metallic one." In addition, glass could be cast, annealed, and ground "with infinitely less labour, hazard, and cost." He pointed out that although silver, like speculum metal, tarnished over time (though more slowly), "the reproduction of the polish is the work of a few minutes, and is performed without any chance of injuring the figure." Finally, Herschel noted that silver reflected 91 percent of available light, as opposed to 67 percent for the very best speculum.

---

*Thomas Grubb's engineering firm made banknotes for the Bank of Ireland. Robinson had commissioned him to make his first telescope—an equatorial reflector—in 1835, and he soon established a reputation that he passed on to his son, Howard Grubb.

Robinson would have none of it. He, Lord Rosse, and Thomas Grubb formed a kind of Irish astronomical Mafia, and they were hostile to meddling by British and French proponents of glass mirrors.

Foucault sensed this hostility. "For the English [and Irish]," he wrote, "[my telescope] does not exist. It has been, it is, and for some time yet will be as if it had never happened."* Neither was the French physicist impressed with the Leviathan, which he visited in Birr. "Lord Rosse's telescope is a monstrosity," he asserted.

Foucault earned the right to his opinion. The following year, he invented what is known as the Foucault knife-edge test, allowing much better astronomical mirror-testing than had ever been done. It is easiest to picture how it works for a perfectly spherical mirror. A pinhole light set in the central focus will reflect directly back to itself. While peering at the mirror from near the light source, cut the light beam slowly with a knife. If the mirror is perfectly spherical, the light will shut off uniformly over the entire mirror. Under similar circumstances, a paraboloid mirror presents distinctive patterns, so that problem areas can be identified and refigured.

In 1862, Foucault figured a 31-inch glass blank from Saint-Gobain using his new test. He also abandoned the traditional copper mirror-grinding tools, instead working glass against glass to achieve an initial spherical shape, then transforming it to a parabola by applying local corrections with rouge. Mounted on a forked equatorial, the silver-on-glass mirror gave superb results. Glass is much easier to grind than metal, so with a deeper concavity, Foucault made a mirror with a relatively short focal length of 14 feet. The telescope continued to do useful astronomy for more than a century.

That same year, Thomas Grubb was making a 48-inch metal mirror weighing 2,200 pounds for the Great Melbourne Telescope. He ground it shallowly, with a high focal ratio that demanded a 30-foot tube. The

---

*Foucault was only partially correct. Amateur telescope-makers throughout Europe and North America quickly realized the advantages of silver-on-glass mirrors. Glass was much cheaper, more widely available, and easier to work. Cobblers, railway porters, blacksmiths, and others in the working class now ground their own telescopes and formed astronomy clubs.

heavy mirror skewed the center of gravity so that the tube had to be supported near the bottom, leaving the long skeletal tube to be buffeted by the wind. Whereas the moving parts for Foucault's telescope weighed 1.5 tons, those for the Melbourne monstrosity weighed 8.3 tons when it was finally erected and ready for observing in 1869.

The telescope came with two mirrors and an assistant trained by Grubb in how to repolish them. Unfortunately, after one repolishing in 1870, he quit, and the mirror was not polished again for another seventeen years, when the sixty-two-year-old director tried to teach himself through long-distance correspondence with Howard Grubb, who had taken over the family firm when his father died. The Melbourne telescope, a failure from the outset, never revealed any new cosmic laws.

Nonetheless, Howard Grubb remained an advocate of large reflectors composed of speculum metal. He recommended that "a person should be sent out with the telescope duly instructed in the art of figuring the [metal] mirror." Either that, or three or four speculum mirrors should be provided, "and each mirror as it becomes tarnished, sent back to the maker to repolish."

Grubb gave these opinions to Richard S. Floyd when he met him in 1876. Floyd, the president of the Lick Trust, was touring European and U.S. astronomical facilities, trying to decide whether to commission a refractor or a reflector for the new Lick Observatory, to be built atop Mount Hamilton in California—the first observatory to seek a mountaintop for better seeing conditions. The prospect of dragging a two-ton metal mirror off the mountain and shipping it back to Ireland for repolishing could not have thrilled him. After protracted debate, the Lick board opted for a 36-inch refractor from Alvan Clark and Sons, the premier U.S. telescope-maker.

Grubb was bitterly disappointed, the Clarks triumphant. For $50,000, the Clarks ground the lenses for the world's biggest refractor. In 1893, Alvan G. Clark gave a speech. "It is my idea that the great telescopes of the future will be refractors, not reflectors," he told his Chicago audience, bragging that his firm was even then building a 40-inch refractor for the University of Chicago. "*Large* reflecting telescopes have never accomplished much except in the hands of the opticians who made them."

Light Through the Slits

Clark was wrong. Ever-larger reflecting telescopes—not refractors—would allow astronomers of the twentieth century to explore the secrets of the universe, though the way they examined the light gathered in the massive mirrors—and their understanding of the nature of that light—changed dramatically. To understand why, we need to go back to the beginning of the nineteenth century, when Thomas Young looked at light streaming through two adjacent slits.

Chapter 8

KALEIDOSCOPIC LIGHT WAVES

Although I had thus combined two plain mirrors, so as to pro-
duce highly pleasing effects, from the multiplication and circu-
lar arrangement of the images of objects . . . yet I had scarcely
made a step towards the invention of the Kaleidoscope.

DAVID BREWSTER, 1819

T HE 1800S WERE THE century of light. Scientists of the
nineteenth century not only made ever-larger telescope mirrors
but also studied, parsed, divided, interfered, redefined, reflected, re-
reflected, and expanded light, leading up to Einstein's magical act of
turning it into the glue (and speed limit) of the relativistic universe.

Interfering with Light

In January 1800, when the London physician Thomas Young declared
his faith in the wave theory of light in a Royal Society paper, no one
took him seriously. Isaac Newton had ruled that light was composed of
infinitesimal particles, and that was that.*

In 1807, Young published his classic experiment that demonstrated
"the interference of light." He cut two narrow slits a fraction of a millimeter apart in a board, then put a candle in front of it. On the screen
behind, he saw a pattern of alternating light and dark bands that he
called interference fringes. He explained that light, like sound, was

---

*Actually, Newton was not nearly so dogmatic as his followers. Although he be-
lieved in particles, he acknowledged evidence for wave behavior as well.

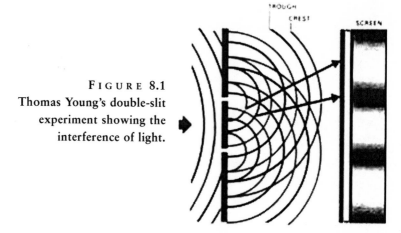

FIGURE 8.1
Thomas Young's double-slit
experiment showing the
interference of light.

composed of waves. Where two beams' peaks and valleys coincided, the light was reinforced, but where the crests of one beam hit the valleys of another, the interference produced a dark line (see Figure 8.1).

Young realized that the colorful Newton's Rings were interference fringes created by light partially reflecting back on itself between thin layers of glass. Colors were due not to different types of light particles but to different wavelengths of light, which consequently refracted differently. Using Newton's meticulous measurements, Young deduced the wavelengths of different colors of light. People saw red when light had 39,180 waves per inch, and they saw violet with 59,750 waves per inch. Young explained the greater refractive index of violet by its shorter wavelength and higher frequency. In glass or water, violet light slowed down more than red, he hypothesized, so it bent farther.

Having failed to convince his fellow scientists that light arrived in waves of *any* length, Young went on to decipher Egyptian hieroglyphics on the Rosetta Stone and to write on astronomy, magnetism, electricity, pumps, steam engines, firearms, hydraulics, engraving, and drawing before his death at fifty-five in 1829.

The French engineer Augustin Jean Fresnel pursued the wave theory and light interference in a much more convincing manner, creating mathematical equations that accurately predicted diffraction fringes down to one-hundredth of a millimeter. Fresnel thought that light, heat, and electricity might be aspects of a single universal fluid of some sort.

To disprove Newton's theory that light particles were somehow pulled or pushed around corners of a slit, Fresnel *bounced* light from mirrors in order to produce the same interference fringes. He set up a tiny light source directly in front of two flat mirrors placed together at an angle of just less than 180 degrees, and sure enough, the reflection produced alternating light and dark bands.

Fresnel also pondered the mystery of polarization, in which light is given a directional quality by passing through a type of crystal or by being reflected repeatedly. Fresnel concluded that light must travel in transverse waves, like a stretched rope being shaken, rather than longitudinal waves, such as the way sound moves by air compression. It seemed indisputable that light did travel in waves, but if so, what was waving? Through what medium did it travel so quickly? There must be an "ether" such as Newton and Huygens had proposed, but no one could find it.*

Dark Lines, Lucid Cameras

William Hyde Wollaston, a British doctor, quit seeing patients in 1802 in order to devote all his time to scientific pursuits. Wollaston let sunlight through a slit 1/20th of an inch wide, then stood 10 feet behind a flint-glass prism to observe it. Because the beam was narrow, the colors in the resultant spectrum did not overlap as Newton's had, and Wollaston saw several dark bands cutting between the different colors, with the two darkest lines in green and blue.

Twelve years later, Joseph Fraunhofer rediscovered Wollaston's dark lines while studying the refractive indices of different wavelengths of

---

*Before his death of tuberculosis at the age of thirty-nine in 1827, Fresnel also designed a lighthouse lens—a bull's-eye convex lens surrounded by concentric circles of prisms—to beam more light than the traditional concave metal mirrors. The outer prisms acted as mirrors, bouncing the light in total internal reflection. The enormous Fresnel lenses, up to 12 feet high and containing more than a thousand prisms, sent powerful light beams 18 miles to sea. In other nineteenth-century applications, parabolic mirrors were used in streetlights and searchlights. Yankee soldiers used the first military searchlight in an 1863 night attack. Dentists used small mirrors to peer into oral recesses, and doctors used the ophthalmoscope's partially reflecting mirror to direct light into the eye.

light. By examining sunlight through a slit, prism, and telescope, Fraunhofer saw what he called *"an infinite number of vertical lines of different thicknesses,"* mapping them carefully and giving letter labels to the most prominent ones, beginning with "A" at the red end of the spectrum and ending with "I" in the violet. These letters are still applied to what are now called the Fraunhofer lines.

Because shorter wavelengths such as violet are refracted more than the red end, prisms produce elongated spectra in which the red lines are crammed together. Fraunhofer discovered that he could produce more regular and detailed spectra using equally spaced thin wires instead of a prism to intercept the light. Through constructive interference, these diffraction gratings produced spectra as well, and the closer together the wires, the more detailed the results. Fraunhofer applied gold film to a glass plate, then ruled fine parallel lines on it with a diamond stylus. From the superbly detailed spectrum this produced, he determined precise wavelengths mathematically.

Still, no one knew what Fraunhofer's mysterious lines meant. They appeared in the same place when he observed Venus, because he was just observing reflected sunlight, but bright stars such as Sirius produced a variety of different spectra. Fraunhofer also noticed that his dark "D" lines in the solar spectrum coincided with two bright yellow lines he found in the spectrum of his sodium lamp. Unfortunately, his early death in 1826 cut short his experiments. The mystery was not unraveled for more than three decades.

In the meantime, a William Wollaston invention provided a new optical aid for artists. In 1807, Wollaston described how he had "amused myself with attempts to sketch various interesting views, without an adequate knowledge of the art of drawing." Inspired one morning by a crack in his glass shaving mirror, in which he saw an oddly reflected image, Wollaston invented the camera lucida, a small, four-sided prism held on a vertical stick. The tiny prism acts like two mirrors, internally reflecting light twice, so that the scene is not viewed in reverse. With it, an artist can look down at his piece of paper and, by carefully placing half of his pupil over the prism, see his subject projected on the paper.

The camera lucida, less cumbersome than the camera obscura, was quickly adopted by artists. Jean-Auguste-Dominique Ingres probably used a camera lucida to draw many of his pencil portraits, for instance.

The artist who produced more camera lucida drawings than any other was not a professional, however, but John Herschel, for whom William Wollaston was a mentor.

Following Fraunhofer, Herschel studied both dark and bright spectral lines. Impressed by Young and Fresnel, he espoused the wave theory of light, explaining the shifting opalescence of mother-of-pearl by the theory of light-wave interference. And as he tramped around Europe in 1830 with his new bride, scaling volcanoes and descending "at full gallop," he used his camera lucida to make exquisite drawings. "We go about sketching, reading, & hammering the hills," he wrote to a friend.

It was not such an easy instrument for others, such as Herschel's friend William Henry Fox Talbot, another scientific polymath. In October 1833, "I was amusing myself on the lovely shores of the Lake of Como in Italy, taking sketches with Wollaston's Camera Lucida, or rather I should say, attempting to take them," Talbot later recalled. "For when the eye was removed from the prism—in which all looked beautiful—I found that the faithless pencil had only left traces on the paper melancholy to behold."

Giving up on the camera lucida, Talbot reconsidered the camera obscura, in which he saw "fairy pictures, creations of a moment, and destined as rapidly to fade away." But what if he could find a way to "cause these natural images to imprint themselves durably, and remain fixed upon the paper"? He thought of silver nitrate, well known for its sensitivity to light.

Although Talbot figured out a way to make pale permanent silhouette photos within the next two years, he didn't publicize the process until 1839, when Louis Daguerre revealed his extraordinarily detailed pictures on metal plates. By that time, John Herschel had returned from South Africa, and he and Talbot began a frenzied collaboration to improve the paper process. Herschel recalled his experiments years before on hyposulfites of soda, realizing that they would be the perfect "fixer." Herschel even coined the term *photography* in naming the new field, along with the terms *negative* and *positive* for the initial impression, which reversed black and white, and the subsequent re-reversal. It would take many more years, however, for photography to mature.

The Kaleidoscope Craze

Before he invented photography, Henry Talbot dabbled in spectral analysis, writing a paper called "Experiments on Colored Flames" in 1826 and asking his friend John Herschel to send it to the famous Sir David Brewster for his comments. Talbot and Brewster subsequently became friends, with Brewster cheering Talbot's invention of photography. By that time, Brewster had achieved fame for his studies of polarized light and a curious optical toy involving mirrors.

As a child, David Brewster hung around the shop of an ingenious young blacksmith in the small Scottish town of Inchbonny. In addition to hammering out ploughs, James Veitch cast, figured, and polished his own telescope mirrors, and in 1791 he helped the ten-year-old boy make his own telescope.* Brewster went on to study at the University of Edinburgh. In 1806, he plunged into optics, studying the reflection and refraction of light from and through every imaginable substance.

In 1815, while reflecting polarized light between two plates of gold and silver, Brewster remarked on the "succession of splendid colours." The next year, while studying the behavior of light in fluids, he made a trough from two glass plates, blocking the ends with glass and using cement to hold the whole thing together. When he looked down the trough, he noticed that the cement lines were reflected in a regular circular pattern. Within a few months, he had invented a new "philosophical toy," naming it the *kaleidoscope*, meaning "to see beautiful forms."

Brewster discovered that if he rotated a cell with colored bits of glass at the end, the resulting view gave a spectacular rose-window effect as the pie-shaped colors were reflected and re-reflected until they formed a circle. By opening and closing hinged mirrors, Brewster realized that any angle that divided evenly into 360 degrees would produce a regular figure. A 20-degree angle, for instance, yields eighteen pie slices, arranged in alternate reflections in a circle that creates a nine-pointed star. The kaleidoscopic views—unique, symmetrical, and lovely as snowflakes— enchanted Brewster. It was almost inconceivable that the twisted wire,

---

*James Veitch tested the figure of his telescope mirrors by their ability to resolve the sparkle in the eye of a bird perched in an oak tree a half-mile away. Despite his increasing fame, Veitch chose to remain a humble blacksmith.

colored glass, lace, and beads clunking around in the object-box could be converted to the gorgeous patterns seen between the two mirrors.

Brewster applied for a patent in 1817, but then "the gentleman who was employed to manufacture them under the patent, carried a kaleidoscope to show the principal London opticians, for the purpose of taking orders from them." Almost overnight, a kaleidoscope craze ensued, quickly jumping the English Channel to France. Within three months, 200,000 kaleidoscopes were sold in London and Paris. In May 1818, Brewster wrote to his wife: "You can form no conception of the effect which the instrument excited in London. . . . No book and no instrument in the memory of man ever produced such a singular effect. . . . Thousands of poor people make their bread by making and selling them."

The kaleidoscope may have provided much-needed income for these needy entrepreneurs, but Brewster fumed that he himself made only modest profits, because most of the kaleidoscopes were pirated copies. "The mortification is very great," he complained to his wife.

Hoping to generate more sales, Brewster commissioned a "polyangular" model in which viewers could adjust the angle between the two mirrors. In others, he added a third mirror to complete the triangle, producing patterns reflected not just in a circle, but filling the view. Finally, he introduced the telescopic kaleidoscope, or teleidoscope, with a convex lens at the end of the tube instead of an object-box. With it, viewers could turn the world they viewed—houses, people, trees, dogs—into symmetric kaleidoscopic wonders.

Brewster had grandiose plans for the kaleidoscope. "It will create, in a single hour, what a thousand artists could not invent in the course of a year." Because humans naturally love symmetry in "forms of animal, vegetable, and mineral bodies," the kaleidoscope designs (copied with a camera lucida) could be incorporated in architecture, sculpture, paintings, stained glass, carpets, books, and jewelry.

Stereoscopic Snits

Brewster's kaleidoscope indirectly inspired Charles Wheatstone, an acoustics expert, to make a connection between sound and light, music

and mirrors. Wheatstone worked in the family business of music pub-
lishing and instrument manufacturing, inventing the concertina, among
other instruments.

In 1826, Wheatstone created the kaleidophone, naming it after
Brewster's device. His direct inspiration came from a Thomas Young
experiment in which Young wrapped a piano string in silver wire, then
shone a light on it as it was struck, so that the wire acted like a vibrating mirror. "The luminous point will delineate its path like a burning
coal whirled round," Young had written in 1800.

Taking advantage of this persistence of visual memory, which makes
the eyes see a line of light where a single spark moves rapidly, Wheatstone's kaleidophone created much bigger and more varied lines of reflected light with a vibrating metal rod, held vertically in a wooden
base. With a leather-covered hammer, Wheatstone hit the rod, tipped
with a silvered glass bead, either in bright sunlight or indoors near a
lamp or candle, yielding a low tuning-fork sound and a bright fast-
moving mirrored light. "By striking the rod in different parts and with
different forces, very complicated and beautiful curvilinear forms may
be obtained," Wheatstone wrote.*

A few years later, Wheatstone's fascination with light and mirrors
produced another invention, the stereoscope, although he didn't publicize it until 1838. This toy's serious purpose was to reveal "some remarkable, and hitherto unobserved, phenomena of binocular vision."
Although Euclid had noted that the left and right eye see slightly different parts of an object, and the well-known notion of parallax relied on
the differing perspective of two views, no one before Wheatstone had
realized that depth perception relied on the brain's combining and interpreting the two disparate visions.

---

*The nineteenth century was rife with mirrored toys with long Greek names. In
1833, Belgian physicist Joseph Plateau, 32, invented the Phenakistoscope (called
a Phantascope in England). The viewer looked into a mirror through regularly
spaced slots on the edge of a spinning wheel. On the other side of the wheel,
there were pictures of a dancer in slightly varying positions, so that in the mirror
the viewer saw her performing a pirouette through persistence of vision, as in
modern movies. At age forty-one, Plateau blinded himself by an experiment in
which he stared for 25 seconds at the sun. Thus, he could not see his dancing illusion for the last 40 years of his life.

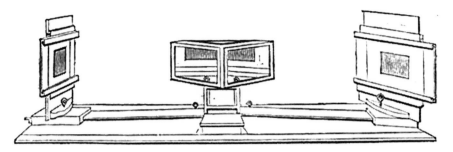

F I G U R E 8.2   Charles Wheatstone's reflecting stereoscope.

Wheatstone asked himself: "What would be the visual effect of simultaneously presenting to each eye, instead of the object itself, its projection on a plane surface as it appears to that eye?" He made wire outlines of cubes, pyramids, and other objects, set them nearby, and, closing one eye, drew what he saw. Then, closing the other eye, he drew the same objects again. He rigged up a device to force the left eye to look at one drawing while the right viewed the other. He did this by positioning two mirrors facing out at right angles, with the corner placed between his eyes. With the pictures pinned to viewing boards on either side, Wheatstone's eyes united the two views into the illusion of a three-dimensional object (see Figure 8.2).

Six months after Wheatstone published his stereoscope paper, Daguerre revealed his photographic process, quickly followed by Talbot. Wheatstone immediately realized that a camera could take pairs of stereoscopic photographs that would work in his device, and he commissioned several from Talbot and other photographers. But nothing commercial came of it until 1849, when David Brewster created a smaller, cheaper version. Instead of mirrors that reflected separate pictures held on either side, Brewster's lenticular stereoscope used a lens-prism so that the two photographs could be mounted on a single card and slipped into a slot at the front. This time, Brewster made a tidy profit from the ensuing craze.

In 1852, Wheatstone introduced his pseudoscope, which reflected light so that the right eye viewed what the left eye would normally see, and vice versa. The pseudoscope creates optical illusions in which depth perception is reversed. The inside of a teacup appears to be a convex bulge. Objects in front of a wall seem to be imbedded behind it. "With the pseudoscope," Wheatstone wrote, "we have a glance, as it were, into another visible world." The strange device, along with its alternative reality, did not appeal to the general public, however.

In some ways, Wheatstone and Brewster were mirror images of one another. Both men were phobic about public speaking, suffered from migraines, were superb historians of science, and studied a wide range of subjects. They were knighted, lived to a ripe age, married late in life, and were fascinated by mirrors, light, and vision.

Yet they were bitter foes, with Brewster disputing Wheatstone's stereoscope priority. Their main differences were intellectual and temperamental. Brewster, who clung to the particle theory of light, resisted theories that he regarded as too speculative, whereas Wheatstone loved mind-play. For Brewster, vision stopped at the retina, whereas for Wheatstone perception ultimately resided somewhere in the brain. The prolific Wheatstone also invented the gyroscope, the rheostat, an early typewriter, a crude "talking" machine, and the first working electric telegraph, which made him a wealthy man.

Marrying Electricity, Magnetism, and Light

Michael Faraday, a blacksmith's son who rose from the London slums to become the chief experimenter at the Royal Institution, became one of Charles Wheatstone's best friends. Faraday frequently presented the shy Wheatstone's experiments to the public, as he did one April evening in 1846. As an afterthought, he tacked on his own speculations about "ray vibrations."

Faraday suggested that light might consist of transverse waves in an electromagnetic field. In 1820, the Danish scientist Hans Christian Oersted—another Wheatstone friend—first noticed that an electric current moving along a wire produces a magnetic field. Faraday found that the converse was also true: A moving magnet could induce electricity in a loop of wire. This discovery led not only to the electric motor but also to a revolution in the conception of light. Perhaps light, magnetism, and electricity were somehow connected.

Wheatstone, who experimented with all three phenomena, tried to measure the speed of electricity in 1834 by looking at the images of sparks produced by an electric discharge sent through quarter-mile lengths of thin copper wire. By viewing the separate sparks in a rapidly rotating mirror, he tried to figure the speed of electricity, deducing it from the mirror's rate of rotation and the measured displacement of the spark's reflection. He came up with a speed of 250,000 miles per second, which is much too fast.* But he also suggested using a similar method to measure the speed of light more accurately than the astronomical methods then employed.

Léon Foucault and his friend Armand-Hippolyte-Louis Fizeau collaborated on experiments with daguerreotypes, taking the first photographs of the sun in 1845. Soon afterward, they tried to modify Wheatstone's rotating mirror to measure the speed of light, partly in order to test the wave theory of light, which predicted that it would move more slowly in water than air. Foucault and Fizeau fell out, so they pursued parallel experiments.

In 1850, Foucault designed a method that verified the wave theory. He sent a beam of light through a "beam-splitter," a flat piece of glass held at a 45-degree angle so that part of the light reflected to the observer and part of it went straight on to the spinning mirror, where it was reflected to a stationary flat mirror, then back to the spinning mirror and thence to the beam-splitter. In a second run, the light was bounced through a water-filled tube before hitting the flat mirror (see Figure 8.3).

Foucault proved that water did indeed slow the light down. Twelve years later, using a similar but more accurate setup, he came very close to the modern figure of 186,000 miles per second in air.

Light moved unbelievably quickly, but what *was* it, if it didn't consist of tiny particles? In 1862, James Clerk Maxwell, the brilliant Scottish physicist, mathematically derived the hypothetical speed of Faraday's "ray vibrations" in an electromagnetic medium. To his astonishment, the figure was very close to Foucault's speed of light. When he published his findings, he used italics to drive home his point: "We can scarcely avoid the inference that *light consists in the transverse undulations of the same medium which is the cause of electric and magnetic*

---

*Ideally, electricity travels at the speed of light, but that is assuming no resistance in the wire. In practice, the rate varies depending on the conductor and other factors.

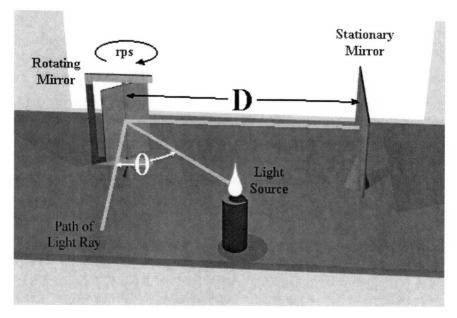

FIGURE 8.3 Using a spinning mirror, Foucault devised an ingenious method of measuring the speed of light.

phenomena." He subsequently defined light as "an electromagnetic disturbance propagated through the field."

Maxwell couldn't account for *how* light moved or precisely what it was, but it clearly had something to do with electricity and magnetism, and it obeyed his equations. There must be an invisible ether to carry light, he thought, even though his math didn't demand it. In a private letter, however, he called ether a "most conjectural scientific hypothesis." In an 1875 *Encyclopaedia Britannica* entry, Maxwell envisioned a way to test for ether by measuring the variations in the velocity of light on a double journey between two mirrors. Because the earth is traveling through this hypothetical substance, light should travel slightly slower against the flow of the "ether wind." Maxwell died of abdominal cancer in 1879 at the age of forty-eight.

The following year, Alexander Graham Bell invented the photophone, which appeared to give credence to the ether by transporting the human voice on light beams. Bell bounced sunlight off a mirror, through a lens, and onto a thin, dime-sized mirror attached to a speaking tube. As he spoke, the mirror vibrated, causing minute fluctuations in the light reflected from it. Across the room, a paraboloid dish-mirror concentrated the light waves onto a selenium cell and a telephone receiver. It worked.*

Eventually, Bell was able to hear light-bearing messages up to 700 feet away. "I have heard articulate speech produced by sunlight!" the thirty-two-year-old wrote to his father. "I have heard a ray of the sun laugh and cough and sing!" Because the device only worked in the line of sight for short distances—and then only on sunny days—it never replaced the telephone, which Bell had already invented.

The year after Bell's photophone invention, twenty-eight-year-old Albert Michelson took up the challenge of trying to prove the existence of a light-carrying ether, shining a beam of light through a 45-degree-angle beam-splitter, a partially silvered mirror that reflected half the light while letting the other half go through. The two beams were then reflected from flat mirrors back through the beam-splitter and to a screen where the light created interference fringes. Then he moved the entire apparatus around 90 degrees and repeated the experiment. If there were an ether wind, the interference fringe pattern should have changed. This mechanism, the interferometer, allowed elegant measurements to within a fraction of a wavelength of light.

The experiment failed because nearby movement—even a passing pedestrian on the street outside—vibrated the instrument. Six years later, Michelson repeated the experiment in collaboration with Edward Morley. This time, to insulate it, they mounted the entire apparatus on a two-ton sandstone slab atop a wooden support floating in mercury. In order to lengthen the light path, they used multiple mirrors to bounce the beams back and forth four times. But no matter how they turned the device, the interference fringes looked exactly the same. There was no ether, though Michelson continued to half-believe in it until he died four decades later.

Invisible Light Every Way You Look

In 1887, while Michelson and Morley were failing to find ether, Heinrich Hertz, a physicist in Karlsruhe, Germany, was also extending

---

*When Bell's wife, Mabel, gave birth to a baby girl in 1880, he wanted to name her Photophone, but Mabel objected, so he settled for Marian.

Maxwell's legacy by exploring how electromagnetic waves were propagated. He produced electric waves with an induction coil connected to two rods between which he produced a spark, detecting them by looking at fellow sparks induced at a distance in a simple unclosed loop of wire. He could ascertain the wavelengths by moving the detector around a darkened lecture hall. Then, knowing the frequency of the oscillation, he could deduce the speed of the waves. Hertz was thrilled when this came out to be the speed of light.

The wavelengths Hertz studied were much longer than infrared. Some could be measured in meters. He had discovered radio waves, and he proceeded to prove that they behaved like light, refracting them through huge prisms of hard pitch and reflecting them from the walls of the room to obtain interference fringes. Because of their long wavelengths, they did not need highly polished surfaces to reflect them—any smooth surface would work. Hertz focused the radio waves with huge concave mirrors and cast "shadows" with conductive obstacles. Until then, most European theorists believed that electromagnetism acted instantaneously, through "action at a distance." The experiments with mirrors clearly disproved that hypothesis.

It wasn't the radio waves that were setting off the secondary spark in the detector, however. After a great deal of experimentation, Hertz determined that it was ultraviolet light that somehow caused the flash of light. He called this the photoelectric effect but couldn't explain it.

Hertz conducted his experiments in constant pain from a bone condition that his doctors could not understand. In 1889, he had all his teeth pulled out to stop his horrible toothaches and underwent several head operations that offered only temporary relief. He died of blood poisoning in 1894 at the age of thirty-six, leaving his name to posterity: radio frequency is measured by the number of hertz (cycles per second). Two years later, a twenty-two-year-old Italian physicist, Guglielmo Marconi, succeeded in sending a message with radio waves to someone a few miles away.

The pace of discovery as the world approached the twentieth century was extraordinary. In 1895, Wilhelm Conrad Röntgen, a Würzburg physics professor, suspected that high-voltage electrical discharges in a glass vacuum tube might send waves of some sort outside the tube, so he tested it in a darkened room, where a sheet of paper glowed after being treated with the salt barium platinum-cyanide. When Röntgen held something in front of the paper, he was amazed that he could see the bones in his fingers in the shadow it cast. He had discovered X-rays, with wavelengths even shorter than ultraviolet, so powerful that they plowed right into regular mirrors.

In 1897, J. J. Thomson, who headed the Cavendish Laboratory (founded at Cambridge University by James Clerk Maxwell), performed a series of cathode ray–tube experiments proving the existence of the electron, a tiny negatively charged constituent of all atoms. Thomson incorrectly believed that this was the *only* fundamental particle, but eight years later Ernest Rutherford, a former student of Thomson's, discovered a relatively heavy nucleus, which was later found to consist of protons and neutrons. Atoms, it appeared, were something like tiny solar systems, with electrons swirling around the nucleus. Rutherford also discovered gamma rays, with even shorter wavelengths than X-rays, emitted by radioactive substances like uranium.

Although a true definition of light remained elusive—after all, what *is* electromagnetic radiation, and why do most wavelengths reflect from mirrors?—clearly that definition had to be much broader than anyone had previously thought. The electromagnetic spectrum ranges from gamma rays, with extremely short waves and high frequencies, down to long radio waves with low frequencies (see Figure 8.4).

If you consider the entire range as a piano keyboard, visible light occupies only an octave, from about 400 to 700 nanometers (a nanometer equals one-billionth of a meter). Why are our eyes and those of other species sensitive to this particular range? Because earth's atmosphere blocks most of the adjacent wavelengths, life-forms evolved to take advantage of the light coming through this narrow window. The only alternative—assuming eyesight developed at all—would be for us to have eyes the size of satellite dishes to focus the long wavelengths of radio waves, which also get through the atmosphere.

Light, the Universe, and Everything

William Herschel had assumed that heat and light were related but separate. Now scientists realized that heat—infrared radiation—was

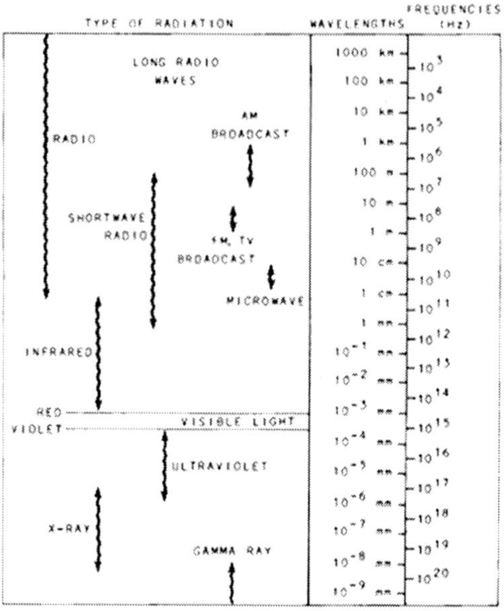

FIGURE 8.4
Visible light is only a small part of the electromagnetic spectrum.

simply another form of light at a longer wavelength. All light seemed to be a form of energy of one sort or another, most of which could be refracted and reflected. But what accounted for the way different kinds of energy were absorbed or emitted from different kinds of matter? In 1900, the German physicist Max Planck came up with a formula that described the intensity of radiation emitted at any wavelength at various temperatures. Planck's formula worked beautifully, but he didn't know why.

Five years later, a twenty-six-year-old examiner in the Swiss Federal Patent Office offered an explanation in his paper, "Concerning a Heuristic Point of View of the Creation and Transformation of Light." The examiner, Albert Einstein, also published his Special Theory of Relativity and a seminal paper on Brownian motion that year. In his "heuristic" article (for which he won the Nobel Prize), he took Planck's figures and applied them to an atom, assuming that the total energy radiation would match the formula or a multiple of it—but never a fractional amount. Einstein called these little parcels of energy quanta, though they were later renamed photons.

Einstein reasoned that light (electromagnetic radiation) is a mechanism for transferring energy. The shorter the wavelength of light, the higher the energy, which explains why X-rays can penetrate materials that visible light does not. In Einstein's theory, light is emitted as a photon, a discrete bundle of energy, a particle. Newton was right after all! Then, as it traverses space, light acts like a wave. Huygens and Young were right! When light hits something and is absorbed, it turns back into a photon, a particle. Somehow, light is both wave and particle.

In his Special Theory of Relativity, Einstein produced his most famous equation, $e = mc^2$, which simply means that energy *is* mass, multiplied by a huge number, the speed of light squared. In other words, light (energy) can become matter, and matter can be transformed into energy—an incredible amount of energy, as humans were to discover four decades later with the detonation of the first atomic bomb. Robert Grosseteste was right! Light is the glue that binds the universe, its energy delivery system. Everything is made out of it. We are all forms of light, at least potentially.

It took a few more years before Neils Bohr, a Danish physicist working with Ernest Rutherford in England, developed more of the theory of quantum mechanics, as it eventually came to be known. Electrons spin in variously layered orbits in different atoms. In hydrogen, the simplest atom, one electron spins around the nucleus.* When a photon with exactly the right amount of energy hits it, the atom absorbs it, and the electron jumps to a higher, more energetic orbit. Conversely, when it falls back to a lower orbit, it emits a photon of the same energy/wavelength. Bohr traced the hydrogen emission lines, and they conformed to his theory.

Now Hertz's photoelectric effect made sense. The ultraviolet rays knock electrons off conductive metal, whose electrons are held more loosely than other forms of matter, giving one metal surface a small positive charge, leading to a spark across the gap. Quantum theory also explains vision. Light of a particular wavelength/energy hits a particular receptor in the retina and is absorbed, setting off electrons, sparking neuronal firing and chemical reactions between neurons, and eventually registering in the brain as a specific color in one spot. Wheatstone was right!

Finally, the theory explains reflection. It is convenient to speak of light "bouncing" from a smooth metallic surface, but in fact the pho-

---

*It is actually inaccurate to speak of an electron as if it were a simple particle. Like light, electrons should be thought of as both particle and wave.

tons are interacting with loose electrons in the metal, which absorb and then reemit photons. In his fascinating book *QED: The Strange Theory of Light and Matter*, the physicist Richard Feynman explained that we can speak only of the *probability* of light's interactions at a mirror surface. Yes, the angle of incidence equals the angle of reflection, but in fact some photons are hitting all over the mirror, then are re-emitted to our eyes at unequal angles. Some are simply absorbed—about 9 percent in the case of silver—while a few others take longer paths. The overall effect yields the familiar central reflection, unless we have turned the mirror into a diffraction grating by scratching tiny parallel grooves, and then, from the proper angle, we see a particular color reflected from all over the mirror, not just the central portion. "Isn't it wonderful?" Feynman asks. "You can take a piece of mirror where you didn't expect any reflection, scrape away part of it, and it reflects!"

Partial reflection from glass or water is even more confusing, as explained by quantum mechanics and wave interference. From a solid chunk of glass, about 4 percent of the light is reflected. But from a thin piece of glass, with two parallel surfaces, the reflection can increase up to 16 percent—or it can be reduced to zero. It all depends on slight variations in thickness that produce constructive or destructive interference as the light ricochets through the glass.

The New Astronomy

Quantum mechanics explains *why* particular atoms absorb and emit light at certain wavelengths, but in 1859 two German scientists broke the code of the spectrum itself, revolutionizing the study of the stars. Finally, Fraunhofer's dark solar lines made sense. Robert Bunsen, a forty-eight-year-old chemist, was using his newly invented burner to analyze salts by their distinctively colored flames. He first tried viewing the flames through colored glass or liquids. Then his friend Gustav Kirchhoff, a physicist thirteen years his junior, suggested examining their spectra through a slit and prism.

Kirchhoff, a cheerful, energetic investigator despite an accident that had left him unable to walk, studied the dark solar "D" lines and discovered that when he interposed a sodium flame the lines got much darker. By itself, the sodium flame produced *bright* lines at the same points in the spectrum. Kirchhoff concluded that a substance capable of emitting light at a certain wavelength and temperature could also absorb it and that the sun's atmosphere must therefore contain sodium. Bunsen immediately recognized that spectroscopy, as the new science was named, offered a new way to identify chemical elements, and within a year he had discovered cesium and rubidium, named respectively after the blue and red spectral lines that revealed their presence.

While Bunsen toiled on earthly matters, Kirchhoff mapped the solar spectrum, identifying elements by the lines they produced. Besides sodium, he proved by 1861 that iron, magnesium, calcium, chromium, copper, zinc, barium, and nickel were present in the sun's atmosphere.

Word of the Bunsen-Kirchhoff discovery spread quickly. In England, the thirty-five-year-old William Huggins made a prismatic spectroscope and attached it to his 8-inch refracting telescope, set up at Tulse Hill, then a rural suburb of London. The excited Huggins soon found evidence for iron, sodium, calcium, magnesium, hydrogen, and other elements in stellar spectra. On the evening of August 29, 1864, he turned his telescope to the Cat's Eye Nebula in Draco. "I looked into the spectroscope. No spectrum such as I expected! A single bright line only!" Although he soon found a few other bright lines, it was clear to Huggins that he was looking at a luminous gas with only a few emission lines, rather than stellar matter, with its numerous dark absorption lines.

Those who felt that all the nebulae could ultimately be resolved into stars were wrong. These "planetary" nebulae were, as William Herschel had suspected, composed of "shining fluid" of some sort. Huggins identified hydrogen but couldn't figure out the other lines, which he considered a new element, nebulium. Years later, these lines were identified as ionized oxygen and nitrogen. Thanks to Huggins, the universe shrank a bit. These cloudy nebulae were not composed of stars at such immense distances that they could not be seen but were much closer cosmic clouds of gas. In the next four years, Huggins examined the spectra of sixty nebulae, one-third of which were gaseous.

Meanwhile, he struggled to determine stellar velocity through spectroscopy. Fizeau and Austrian physicist Christian Doppler had both theorized that wavelengths would be slightly shifted for objects speeding toward or away from us. If they were approaching, the wavelengths would be compressed and, if receding, would be stretched toward the red end of the spectrum. In 1868, Huggins finally succeeded in measuring this Doppler effect for Sirius, concluding that its radial velocity (speed along the line of sight) as it rushed away from the earth was 29.4 miles per second, larger than the correct figure. Nonetheless, he had pioneered the redshift method that would later prove pivotal in exploring the structure and evolution of the universe.

Thus far, Huggins had used a refractor for his spectroscopic work, but he realized that mirrors could theoretically yield superior results, since even achromatic refractors produced some spectral aberrations, thereby blurring spectroscopic lines. Glass also absorbs ultraviolet and infrared rays, so only reflecting telescopes could allow ultraviolet and infrared exploration. In 1870, the Royal Society paid for a new tandem reflector-refractor fixed together on an equatorial mount, made by Howard Grubb and on permanent loan to Huggins. The 18-inch Cassegrain reflector featured a metal mirror. With it, he began to explore the ultraviolet range. He also tried to photograph spectral lines, but his initial efforts were too blurry.

A U.S. surgeon, Henry Draper, produced the first successful photographs of spectral lines. While the inspired William Huggins was beginning to explore spectroscopy, Draper was amputating limbs in the Civil War, before his own poor health forced his discharge at age twenty-five late in 1862. He immediately returned to his observatory on his father's estate at Hastings-on-Hudson, New York, where he resumed mirror-making.

Draper's father, John William Draper, was a physician-scientist who had taken the first astronomical photograph, a daguerreotype of the moon in 1840, and passed on a passion for both astronomy and photography to his son. Something of a child prodigy, Henry Draper graduated from medical school in 1857 at the age of twenty, then went abroad for a year, where he was inspired by Lord Rosse's Leviathan. Back in the United States, he attempted to grind a metal mirror, but it cracked in frigid temperatures. In an 1860 meeting with John Herschel, John William Draper mentioned his son's frustration, and Herschel strongly suggested trying a silver-on-glass mirror.

Henry took the advice and made his first successful mirror just before leaving for battle in 1861. After he returned the following year, he made 100 more mirrors, perfecting a silvering technique that produced "bright, hard and in every respect perfect [silver] films." In 1867, Draper married Anna Mary Palmer, a wealthy heiress who was quickly swept up in her husband's intellectual pursuits. They spent their honeymoon in New York City selecting a glass blank for a 28-inch mirror, and Anna helped grind and polish it over the next five years. She also helped with the coating and developing of wet-plate photographs. In August 1872, with a camera, spectroscope, and telescope holding the newly installed 28-inch mirror, the Drapers successfully photographed the stellar spectrum for the first time.

In 1879, Draper visited England, where William Huggins showed him that dry photographic plates had become more sensitive—and therefore required less exposure time—than the messy, awkward, wet collodion process. With the new photographic process, Draper took photos of spectra of bright stars, the moon, Mars, and Jupiter. With an excellent clock drive installed on his telescope, Draper could use long exposures to photograph the Orion Nebula and the moon. "We are on the point of photographing stars fainter than we can see with the same telescope," he wrote, but in 1882 he died of pneumonia at forty-five. His widow funded the Henry Draper Memorial research project at Harvard to carry on his spectrographic work.

That same year, Henry Rowland was perfecting a machine to make extraordinarily fine lines on a mirror in his basement laboratory at Johns Hopkins University. His fellow American, Lewis Rutherfurd, had pioneered the use of reflective diffraction gratings, proving that they were superior to prisms in producing detailed spectra. Not only were the lines more detailed; by reflecting the light rather than passing it through glass, they made it possible to study infrared and ultraviolet wavelengths. Rutherfurd had managed to produce 17,000 lines to the inch, but his rulings were often irregular, as were the resulting spectra.

Rowland, a brilliant physicist who had studied magnetism and electricity with James Clark Maxwell and Hermann von Helmholtz, now applied himself to creating better diffraction gratings. He realized that in order to produce tiny, regular grooves—the key to greater spectral dispersion—he had to make a perfect, regular screw to drive a diamond stylus over the surface of a metallic mirror. He succeeded in 1882, and in his basement laboratory, isolated from all vibrations, he ruled up to 43,314 lines to the inch. To make the reflected spectrum self-focusing, he then made his gratings from spherical concave mirrors more than 5 inches wide. Rowland sold his jewel-like precision gratings at cost to astronomers worldwide. He married at forty-two but died of diabetes ten years later. As he wished, his ashes were interred in the wall of his basement laboratory, near his machine with its perfect screw and finely lined mirrors.

William Huggins also married later in life. In 1875, the fifty-one-year-old Huggins married Margaret Lindsay Murray, half his age. Like Anna Draper, Margaret Huggins quickly became a collaborator and colleague. The following year, they first photographed ultraviolet spectral lines, using the 18-inch reflector, and the couple worked together until Huggins's death in 1910.

Draper and Huggins—along with other early photographers and spectroscopists—were pioneers in what Huggins named the New Astronomy. "This really involves a whole world of change," wrote Agnes Clerke in the fourth edition (1902) of her *Popular History of Astronomy During the Nineteenth Century*. As a result of photography, for instance, "in a small part of one night stars can now be got to register themselves more numerously and more accurately than by the eye and hand of the most skilled observer in the course of a year."

Clerke stressed that spectroscopes and cameras relied on telescopes to provide them with light and that reflectors were "specially adapted" for that purpose. She singled out one reflector in particular. "A new era in its employment west of the Atlantic opened with the transfer . . . to Mount Hamilton of the Crossley reflector." She referred to a 36-inch mirror completed in 1877 by George Calver for A.A. Common, a British sewage engineer who preferred looking to the heavens. With it, Common took a superb photograph of the Orion Nebula in 1883, then went on to grind and polish his own large mirrors. Common sold his original 36-inch telescope to Edward Crossley, a wealthy textile manufacturer.

After Howard Grubb refigured the mirror, in 1895 Crossley gave the telescope to the Lick Observatory, which thus received a free 36-inch reflector to match its 36-inch refractor, on which so much money and agonizing decisionmaking had been lavished. In the hands of James Keeler at the Lick Observatory, the Crossley reflector would outperform the expensive Lick refractor.

Magic Mirrors Shine Again

Edward Crossley gave away his telescope because he had developed strong religious beliefs that soured him on trying to delve into the mysteries of the universe.* Once more, as in John Dee's time, mirrors reflected conflicts between science and religion. In 1802, when William Herschel discussed the wonders of the heavens with Napoleon, the French leader interrupted to ask, "And who is the author of all this?" When Pierre-Simon Laplace tried to explain it all as a "chain of natural causes," Napoleon objected. God must be given the credit, not evolution.

Following Charles Darwin's 1859 publication of *The Origin of Species*, that argument became more heated, as science appeared to threaten traditional religion and values. In 1874, J. W. Draper, Henry's father, published *History of the Conflict Between Religion and Science*, defending scientists against attacks by theologians. In the late Victorian era, as technological and scientific innovations produced unprecedented changes at a dizzying rate—steam engines, electricity, telegraphs, railroads—many people sought alternate forms of religious solace. Spiritualism, with its mediums, séances, and crystal balls, became popular. Scrying and magic mirrors enjoyed an extraordinary revival, even as rational scientific progress appeared triumphant.

John Melville's *Crystal Gazing and Clairvoyance*, published in 1896, gave a pseudoscientific patina to scrying. "The crystal or mirror should frequently be magnetised by passes made with the *right hand*, for about five minutes at a time," Melville explained, though he also thought that intense staring helped. "The *Magnetism* with which the surface of the

---

*Not all believers spurned astronomy. Many of the endorsement letters George Calver received were from ministers. In 1880, for instance, the Reverend Conybeare W. Bruce of Cardiff wrote to Calver: "The mirror is a *beauty*. Canon Beechey and I got Jupiter and Saturn at 2:30 A.M. yesterday morning—and they were *superb*."

mirror or crystal becomes charged, *collects there from the eyes of the gazer*, and from the universal ether, the Brain being as it were switched on to the Universe."

David Brewster was fascinated by magic mirrors (including the Chinese variety), though he explained them scientifically. "If a fine transparent cloud of blue smoke is raised, by means of a chafing-dish, around the focus of a large concave mirror," Brewster observed, "the image of any highly illuminated object will be depicted, in the middle of it, with great beauty. A skull concealed from the observer is sometimes used, to surprise the ignorant."

Thus originated the phrase *smoke and mirrors*, as magic shows and theatrical effects thrilled Victorian audiences. Equipped with concave mirrors, lenses, and bright lime- or magnesium lights, magic lanterns projected "terrific heads [with] awful eyes and tremendous jaws" onto an invisible gauze screen; the images "then suddenly vanished," as Eusebe Salverte wrote in his 1847 book, *Philosophy of Magic, Prodigies, and Apparent Miracles*.

In 1863, John Henry Pepper, the director of the Royal Polytechnic Institute in London, demonstrated an ingenious theatrical device in which live actors apparently interacted with ghosts on stage. The illusion, which relied on the beam-splitter qualities of glass, could be achieved through advances in plate-glass technology. A large sheet of glass was hung at a 45-degree angle with its foot near the front of the stage and its top extended toward the audience. With the stage lit and the audience in darkness, the glass was invisible. In a pit below the glass the "ghost" was hidden on an inclined platform. When a bright limelight in the pit illuminated the ghost, its reflection magically appeared to the audience, for whom the apparition appeared upright, translucent, and the same distance away as the onstage actors, who couldn't see the reflection at all but had to act as if they did.

The Pepper's Ghost illusion, as it came to be known, appeared in five London stage productions within a year. It wasn't an easy life being a ghost. The pit earned the nickname "the oven" because of the hot lights and black drapes required for areas invisible to the audience. One boy performing in *The Death of Little Jim* was supposed to ascend to heaven at the appropriate moment, but he fell asleep in the pit and appeared to be diving in the other direction when the lights came on. In another humorous production, an audience member was called onto stage, where (invisible to him) a sexy female Pepper's ghost made alluring gestures toward him.

Mirrors also appeared onstage with Victorian magicians, although viewers generally weren't aware of them. In 1865, Joseph Inglis, appearing as Colonel Stodare, world traveler and explorer, introduced *The Sphinx*. He carried a small box onstage, set it on an undraped, three-legged table in a curtained recess about 10 feet wide, then opened the front of the box to reveal a disembodied, dark-skinned head wearing an Egyptian turban, its eyes closed. It looked remarkably lifelike. Stodare went out into the audience and said, "Sphinx awake!" The head slowly opened its eyes, staring straight ahead without expression. The audience gasped. Then, as if only gradually coming to life, it looked to one side, then the other, its head turning slightly. Later in the act, the head recited a lengthy oracle in verse, before Stodare closed the box again.

How did he do it? The three table legs concealed the edges of two flat mirrors joined at right angles, so that the audience saw reflections of the curtains to either side of the stage, identical with the curtains at the back. The empty space under the table was illusory. A turbaned accomplice squatted behind the mirrors and put his head up through a trap door in the tabletop and into the box. In the following years, variations on the same theme enabled magicians to place a live woman's head on a giant spider's body and to make people, pigeons, and donkeys appear or disappear. The legendary Harry Houdini once dematerialized an elephant onstage.

Mirror mazes were popular Victorian entertainments, disorienting visitors in an early incarnation of the funhouse. In Paris, the Musée Grevin built a huge hexagonal mirror room in 1882, later adding swiveling cornices that magically changed the decor from Turkish to tropical in a few seconds in the dark. Packed into the room, viewers were treated to an eerie light show of multiple reflections. In 1896, at the Swiss National Exhibition in Geneva, architect Heinrich Ernst created an extraordinary mirror maze based on the Spanish Alhambra. Whichever way maze-walkers looked, it appeared they could walk down a long corridor, but when they tried, they bumped their noses on a mirror. In a relatively small space, they became utterly disoriented. After the exhibition, the popular maze was moved to Lucerne.

The theme of illusory or magical reality also struck a chord with writers of the era, beginning with Alfred Tennyson's *Lady of Shallott*, cursed never to view the world directly but only reflected in a mirror. "I am half sick of shadows," she complains in the poem, eventually looking out the window to see Sir Lancelot directly, whereupon the mirror cracks and the lady dies.

Many writers portrayed "doubles," as Robert Louis Stevenson did in *The Strange Case of Dr. Jekyll and Mr. Hyde* (1886), where mirror images of good and evil reflected popular fears of mad scientists who were discovering too much too fast. In Oscar Wilde's *The Picture of Dorian Gray* (1890), the evil protagonist sees himself eternally youthful and unchanged in his mirror while his hidden, hideously aging portrait is the *real* mirror of his soul.*

Bram Stoker's *Dracula* (1897) established the quintessential mirror-horror motif, hearkening back to the ancient belief in the mirror of the soul. Visitor Jonathan Harker is surprised to find that there are no mirrors anywhere in Count Dracula's castle. One morning as he is peering into his portable shaving mirror, Dracula enters unannounced. "I started, for it amazed me that I had not seen him," Harker says, "since the reflection of the glass covered the whole room behind me." The vampire's image does not appear in a mirror, presumably because he has no soul.

In *Also Sprach Zarathustra* (1891), Friedrich Nietzsche had Zarathustra dream of a child holding a mirror up to him, in which the hero sees a grimacing devil. For Russian symbolists like Andrey Bely, mirrors provided an ambiguous intersection between two worlds of reality and illusion, sanity and madness. "We may turn out to be not people, but only their reflections," Bely's 1904 character Evgeny Handrikov ponders. "And it is not we who approach a mirror, but it is the reflection of someone unknown who approaches me from the other side." Handrikov drowns in a mirrorlike lake, as he tries to join his double.

Not all literary mirrors were so frightening. For the shy British mathematics don Charles Dodgson, mirrors provided the entry into a world

---

*In Oscar Wilde's "The Fisherman and His Soul," the Mirror of Wisdom in an Eastern temple reflects "all things that are in heaven and on earth" except the observer's face.

of childish wonder and topsy-turvy logic. Using the pen name Lewis Carroll, he had written *Alice in Wonderland* for Alice Liddell in 1865. When he met another Alice, Alice Raikes, the playful logician placed an orange in her right hand, then put her in front of a mirror and asked her in which hand the child in the mirror held the orange. It was the left. How did she explain that? "Supposing I was on the other side of the glass, wouldn't the orange still be in my right hand?" she asked. Delighted with her answer, Carroll wrote *Through the Looking-Glass* in 1872.

In the book, Alice tells her cat about "Looking-glass House," reflected in a large overmantel mirror. "First, there's the room you can see through the glass—that's just the same as our drawing-room, only the things go the other way." The books look the same, but the writing is reversed. "I wonder if they'd give you milk in there?" she asks Kitty. "Perhaps Looking-glass milk isn't good to drink." Yearning to explore this alternative world, Alice pretends that "the glass has got all soft like gauze, so that we can get through. Why, it's turning into a sort of mist now, I declare!" And with that, she climbs onto the mantel and steps easily through the mirror, where her curious adventures commence.

George Stratton's Mirror Worlds

George Malcolm Stratton, a California experimental psychologist, took the Victorian fascination with mirror worlds, new possibilities, and altered perceptions to extremes. While studying visual perception for a Ph.D. in Leipzig in 1896, the thirty-one-year-old Stratton devised a pair of glasses (he used lenses, although he could have achieved the same effect with mirrors) that made him see the world upside down through his right eye, with his left eye masked. For three days, he wore his inverting spectacles, blindfolding himself at night. "All movements of the body at this time were awkward," he wrote. "Knocks against things in plain sight were more or less of a surprise."

The next year, back in California, Stratton once more strapped on his spectacles, this time for eight days. Again, "the wrong hand was constantly used to seize anything that lay to one side." The first day, he experienced "signs of nervous disturbance" and "mild nausea." One

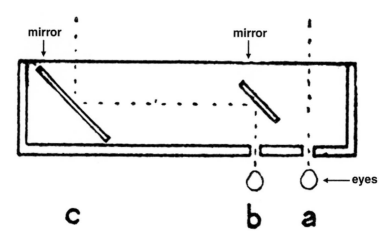

F IGURE 8.5   George Stratton's drawing of his telestereoscope.

evening he sat mesmerized, watching flames lick downward from his fireplace, "without seeing that one of the logs had rolled far out on the hearth and was filling the room with smoke." Gradually, Stratton became more accustomed to living in this upside-down mirror-world. By the seventh day, "there was perfect reality in my visual surroundings, and I gave myself up to them without reserve."

In 1898, Stratton invented what he called a telestereoscope, using two mirrors that, in effect, moved the eyes farther apart (see Figure 8.5). The resulting exaggerated stereoscopic vision gave "abnormal relief to objects in the foreground," Stratton noted.

Intrigued with the distortion of depth perception and distance, Stratton strapped on another device for three days in 1899. A shoulder harness held a facedown 24-by-20-inch mirror 10 inches above his head. A smaller mirror, 4 inches square, was held aslant before his eyes so that it reflected his vision to the upper mirror, and a dark cloth blocked any other visual input. In effect, Stratton saw himself from the vantage point of a bird hovering just overhead (see Figure 8.6).

"There was slight dizziness at first," he found, "and it was difficult to direct visually the movements of my feet and hands." By the third day, however, he moved with "comparative freedom and precision," though he needed someone to guide him so that he didn't run into things. "I sometimes felt myself strangely tall, as if my body had been elongated." At times, "I had the feeling that I was mentally outside my own body."

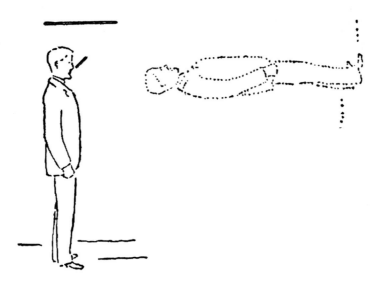

FIGURE 8.6　George Stratton's illustration of his shoulder harness device.

Stratton concluded from his experiments that "a correspondence, point by point, between touch and sight, is built up associationally." People could adapt to just about any situation. "Upright vision, in the final analysis, is vision in harmony with touch and motor experience." There was "nothing absolute."

In the ensuing years, Stratton applied this philosophy to culture and world affairs, trying to broaden the vision of the general public. In 1908, he wrote *Experimental Psychology and Its Bearing Upon Culture*, explaining that "out of the depths of the mind, new powers are ... always emerging," though the mind is "universally subject to illusions." For the rest of his long life, Stratton strove valiantly to remain optimistic, to hold the mirror of his humanistic, relativistic vision up to the light in a world that seemed to be going mad. As war loomed, he wrote *International Delusions* (1936), noting that nations "act as though they were insane." Every nation feels self-righteous, superior, and peace-loving, whereas foreigners alone are violent and to blame for everything.

In 1952, at the age of eighty-seven, Stratton wrote his last book, *Man: Creator or Destroyer.* "One morning soon after the explosion of atomic energy over Japan had stricken the world," he observed, "I met a friend, an astronomer, on a path leading through pine and redwood from his observatory. Witlessly I asked him what the stars were thinking of us. 'We and our earth,' he answered, 'are nothing but a mote in their immensity. They have no interest in us.'" But Stratton refused to give in to nihilism or despair. He thought that "the actual transformation of man's mind" was imminent and that creative energy would win out over destructive tendencies. Just as he had once forced himself to adjust to a new worldview with his odd mirror contraptions, Stratton strove to broaden others' vision. He died at ninety-two in 1957, at the height of the Cold War and the arms race.

Chapter 9

BIG EYE

The figure of a mirror is rather like the figure of a woman—a curve that is not only beautiful but exactly adapted to its purpose. No sculptor ever smoothed the contours of a statue with more loving care than the optician who works a lustrous surface upon glass down to the perfect parabolic curve which will capture not only the admiration of his fellows but the secrets of the stars.

DAVID O. WOODBURY,

THE GLASS GIANT OF PALOMAR, 1939

ALTHOUGH REFRACTING telescopes dominated most of the nineteenth century, reflectors revolutionized the first half of the twentieth, thanks largely to George Ellery Hale, a manic-depressive solar astronomer with a genius for extracting money from millionaires, and George Willis Ritchey, a brilliant, egotistical optician who made ever-larger mirrors. Their collaboration resulted in extraordinary discoveries about the universe.

Hale was the son of a wealthy Chicago businessman whose elevator company made the city's skyscrapers possible. As a child, Hale read Jules Verne's 1863 classic, *From the Earth to the Moon*, with its fictional 280-foot-long telescope set atop the Rocky Mountains. Verne's telescope featured an inconceivably large mirror, 192 inches in diameter, nearly three times the width of Lord Rosse's Leviathan. The book made a deep impression on young George Hale.

Hale became obsessed with solar research and, by the time he ran into Ritchey at an 1890 astronomical gathering, Hale, then twenty-two and a recent MIT graduate, owned the private Kenwood Observatory and had invented the spectroheliograph, a clever device with which he could take photographs of the sun in the light of a specific wavelength. Ritchey, twenty-six, taught woodworking during the day but was grinding telescope mirrors in his home shop. The two quickly became friends and allies.

On his honeymoon that year, Hale dragged his new bride, Evelina, to the top of California's Mount Hamilton, where he abandoned her at night to peer through the new 36-inch Lick refractor. Two years later, Hale learned that the 40-inch glass blanks ordered for a larger refractor were languishing in Alvan G. Clark's workshop, since the funding had failed. With cooperation from the new University of Chicago, Hale swiftly moved to raise money from Charles Yerkes, a streetcar tycoon, to purchase the glass and make a telescope. In August 1893, Hale and Ritchey listened to Alvan G. Clark deliver his speech, "Great Telescopes of the Future," in front of the gigantic mounting and tube for the Yerkes telescope displayed at the Chicago World's Columbian Exhibition. Clark predicted that refractors would dominate, but the telescope whose 40-inch lenses he was then polishing would in fact be the largest refractor ever made.

Even as they listened to Clark's speech, Hale and Ritchey believed that the future belonged to reflectors. Two years later, Ritchey was grinding a 24-inch Saint-Gobain glass blank to make his own reflecting telescope. Hale convinced his wealthy father to buy a much larger glass disk that Ritchey could make into a mirror. In 1896, with specifications provided by Ritchey, Hale ordered a 60-inch blank from Saint-Gobain and hired the ecstatic Ritchey to work for him full-time.

That fall, as the Yerkes Observatory neared completion on the shores of Lake Geneva near Williams Bay, Wisconsin, 80 miles north of Chicago, Hale moved there. A few months later, Ritchey joined him at Yerkes, taking charge of the lengthy basement optical laboratory, big enough for Foucault tests on the 60-inch mirror he was to make.

In 1897, George Hale published a seminal paper that mapped out the future of twentieth-century astronomy, which he had christened astrophysics. He wrote in the *Astrophysical Journal,* which he had founded in 1895, of "the more important advantages for many classes of astrophysical work which the reflecting telescope seems to possess." Mirrors eliminated chromatic aberration, he explained. In addition, "no combination of lenses yet devised can compare with a paraboloidal mirror in the capacity to unite in a single focal plane all wave-lengths from the extreme infra-red to the ultimate limit of the ultra-violet." Not only that, reflectors were cheaper, more compact, and no one knew how large a mirror might be made. Thus far, no one could use one of Henry Rowland's precision gratings on stars, because no existing telescope collected enough light to make the widely dispersed spectral lines distinguishable. Hale hoped that this would soon change.

"As regards the future development of telescopes in the direction of increased light-grasping power," he concluded, "the reflector promises far greater gains than the refractor, especially for spectroscopic work in the so-called photographic region." For the rest of his life, Hale would pursue ever-larger mirrors to grasp ever-more light.

Keeler Revives the Crossley

The year after Hale published this article, James Keeler, a spectroscopist who had studied under Henry Rowland, took over as director of Lick Observatory, where he found the rickety old 36-inch Crossley reflector, which had hardly been used. Keeler immediately resilvered its mirror and remounted the reflector so that he could follow stars and nebulae for photographic exposures up to four hours long. Using the refurbished Crossley, he began to make incredible photographs of the mysterious spiral nebulae. Near the bigger spirals, little ones appeared, so that Keeler estimated that there must be 120,000 of them within range of his reflector.

In 1899, Keeler sent some of his photos for display at Yerkes, where the Astronomical and Astrophysical Society was meeting. According to Hale, the photographs "created a genuine sensation and showed to many who had been skeptical regarding the advantages of reflectors what the instrument is capable of doing in the right hands."

Then, as the dynamic Lick director prepared to push for further results, he died of a stroke on August 12, 1900, just short of his forty-third birthday. "My principle is, do the best you can, and then laugh if you fail," Keeler once said. He had not failed. From then on, all major research telescopes would have mirrors.

The Master Mirror-Maker

Soon after Keeler's death, George Ritchey climbed with a friend to the top of the Yerkes dome and stared at the midnight sky. "Even the largest mirrors," Ritchey said, "will still be puny, pygmy instruments" for plumbing the depths of space, but he planned to make the biggest, best reflectors possible. Back on the ground, Ritchey began to take photographs with his newly completed 24-inch telescope, producing even better pictures than Keeler. Ritchey's mirror, though smaller, was perfectly shaped and solidly mounted, allowing for longer exposures. It was also ground to a deeper parabola, with a short focal ratio of f/4. Such a "fast" mirror concentrated light from a narrower field, focusing on a smaller piece of sky.

Nova Persei was blazing a new pinpoint in the sky, and Ritchey rushed to document it. His first plate, taken on September 20, 1901, showed a few cloudlike wisps near the new star. Two months later, another photo displayed a much larger wispy area. At first, Ritchey and others thought that they were observing the birth of a nebula, but they then realized that the nebulous wisps were already there—they were simply being lit up by the star. Because of the huge distances involved, it took the light several months to reach out that far.

Like a proud father, Hale distributed Ritchey's photos of the nova, spiral nebulae, and gaseous nebulae to astronomers around the world, and he arranged for the photos to be shown at the American Astronomical Society reception at the end of 1901.

Ritchey, the descendent of Scotch-Irish craftsmen, was never regarded as an equal by astronomers, who considered him a working-class technician. The increasingly self-confident Ritchey nonetheless began to give popular slide lectures throughout the Midwest. The Smithsonian commissioned him to write *On the Modern Reflecting Telescope and the Making and Testing of Optical Mirrors*, published in 1904. "Without egotism I can say that the article will be an important one in the scarce literature of the subject," Ritchey wrote, "and will be regarded and used as authoritative for years to come."

He was right about everything except the lack of egotism. Emphasizing the need for cleanliness and perfection, Ritchey described how to grind and polish glass to a spherical curve—the natural result of systematic, random grinding between two surfaces—and test it with the knife-edge Foucault test. Once that was accomplished, he explained how to make a perfectly flat mirror, which could be tested with reflected light from the spherical mirror. Only then could the master optician parabolize the sphere by removing a little more glass in just the right places, testing it periodically with the flat mirror.*

Ritchey had begun grinding the 60-inch mirror blank as soon as it arrived at Yerkes in 1897, with Hale assuring him that he would somehow find the money to mount it in a telescope. Meanwhile, Ritchey ground a very shallow 24-inch mirror with a long f/82 focal ratio to help Hale pursue his solar research. If a regular reflecting telescope were pointed at the sun, the concentrated light would melt the secondary mirror, and the image would be too bright and too small to study easily. A long focal length provided a much bigger view of the sun without too much heat. It was impractical to try to follow the sun with a long tube, so Ritchey made a horizontal telescope inside a 170-foot-long wooden house on stilts to minimize heating and cooling effects of the earth, which would distort the air and result in poor seeing. To bring the sunlight into the tube, he made a flat revolving coelostat mirror to track the sun, reflecting the light to another fixed flat mirror that sent it down the tube, where Hale's mirror grating dispersed it into a wide spectrum.

Ritchey finished the solar telescope in the fall of 1902, only to see a fire destroy it that December. Hale immediately commissioned him to replace it with funds donated by Helen Snow, a wealthy Chicagoan who wanted a memorial to her father. The same year, the steel tycoon Andrew Carnegie set up the Carnegie Institute in New York, and Hale wasted no time requesting funds for the 60-inch telescope, sending along Ritchey's most spectacular pictures of Andromeda, the Orion Nebula, and the Pleiades. Hale suggested building the telescope "at a high elevation in southern California or Arizona."

---

*Making a perfectly flat mirror is extremely difficult, because two surfaces rubbed together tend to produce spherical curves. It requires three pieces of glass, constantly interchanged. Ritchey used the flat mirror to test the parabola, shining a light from the concave mirror's focal plane. The parabola reflected the light out in parallel lines to the flat mirror, which sent it back to be perfectly refocused by the parabola. By looking at the imperfections in the returning light, the optician could determine where the mirror still needed work.

In 1903, the Carnegie Institute provided seed money for astronomer W. J. Hussey of Lick Observatory to climb mountains in the United States, Australia, and New Zealand in search of good seeing. In California, Hussey was much impressed with Palomar Mountain, "a hanging garden above the arid lands," but since it was too inaccessible, he settled on Mount Wilson, west of Pasadena, as the ideal site. Hale climbed the mountain on a glorious June day in 1903 and was smitten. Within a few months, he decided to move to Pasadena and set up an optical shop with his own money, hoping that Carnegie would fund a telescope.

George Ritchey was equally enthralled with the "superb climate" and "wonderfully clear and transparent . . . sky" when he arrived the next year to set up an instrument shop in Pasadena with five men working under him. Ritchey had just returned from Lynn, Massachusetts, where he met with Elihu Thomson, an imaginative General Electric inventor-scientist who thought he could make fused quartz mirrors. If he could succeed, Ritchey realized, it would revolutionize telescopic mirrors. Fused quartz, almost unaffected by heat or cold, would maintain its shape and optical qualities. The trouble was that it required enormous temperatures to melt and fuse quartz.

At the end of 1904, the Carnegie Institute came through with $30,000, and Ritchey escorted his precious 60-inch mirror blank from Yerkes to Pasadena. He recommended work on it in August 1905. For the next two years, Ritchey and his assistants ground and polished the mirror in a temperature-controlled room with double-sealed windows, all air filtered to keep out dust particles that might scratch the glass. After half a year, they had achieved a spherical curve. At that point, the opticians removed the mirror and grinding machine, cleaned them thoroughly, scrubbed every inch of the shop, and varnished the walls.

Then Ritchey began the final parabolizing. Wearing a surgical cap and gown, he began to remove the last few millionths of an inch of glass, using jeweler's rouge. The closer he got to the perfect figure, the more time it took to work on the mirror, because the rubbing heated and distorted the glass. He would have to wait for hours, set the mirror on edge to test it again using a flat mirror and the Foucault knife-edge, and then begin again, knowing that if he overcorrected he could ruin the work of years. In August 1907, he finally finished polishing the mirror. George Hale recognized Ritchey for having produced a shining surface "with no error in form exceeding two millionths of an inch."

Ritchey also designed the entire telescope, bringing his seventy-year-old father to Pasadena as his draftsman. To facilitate smooth motion, the entire telescope floated in a pool of mercury. The paraboloid f/5 mirror required a 25-foot tube. With a flat diagonal secondary mirror, it could be used as a Newtonian telescope. A hyperboloid secondary mirror converted it to a Cassegrain, reflecting the light back through a hole in the primary mirror. A third flat mirror diverted the light down the polar axis of the telescope in a *coudé* arrangement (from the French word for "elbow"). The coudé, with its 150-foot focal length, led the light down to heavy, immobile spectroscopes where Rowland mirror gratings could be used on starlight.

The indefatigable Ritchey even supervised the widening of the Mount Wilson road and planned the transport of the telescope and mirror up the mountain, where the telescope saw first light in December 1908. Turning the telescope on the moon, Ritchey wrote, "I was amazed to see the astounding, indescribable wealth of details shown." In the next few years, using the 60-inch mirror, Ritchey would take extraordinary long-exposure photographs (tracking the same piece of sky for several nights in a row) that revealed large, well-defined spiral nebulae surrounded by many smaller nebular objects invisible in previous telescopes.

Even as he polished the 60-inch disk, however, Ritchey was planning bigger mirrors. Percival Lowell, a wealthy Bostonian obsessed with Mars—he thought he saw irrigation canals there and believed that Martians were highly intelligent beings struggling to survive on an arid planet—had established the Lowell Observatory in Flagstaff, Arizona, with a large refractor. In 1906, he summoned Ritchey to discuss making a 7-foot (84-inch) reflector for him. When those plans fell through, John D. Hooker, a wealthy Los Angeles businessman, stepped into the breach.

Both Hale and Ritchey had befriended Hooker, an amateur astronomer who kept a telescope in his elaborate rose garden. A collector of all fine things—books, paintings, musical instruments, blossoms—Hooker fell in love with Ritchey's photographs, telling him, "Neither literature, art, music, nor even flowers can sound the greatest depths of me, as these marvelous pictures of the heavens do." Ritchey made a special illuminated cabinet for Hooker to enjoy glass transparencies of his astronomical photographs. Learning that plans for the 7-foot Lowell mirror had fallen through, Hooker grandly said that he would pay for an 8-foot (96-inch) mirror, then upped it to 100 inches. In September 1906, Ritchey cabled the order to Saint-Gobain in France.

Two years later, after months of annealing in a gigantic manure pile, the disk arrived in Pasadena in December 1908 just as the 60-inch telescope saw first light. But Ritchey and Hale were dismayed when they pried open the box and saw hundreds of air bubbles trapped in the glass, the result of three separate pots poured into the mold one after the other. They rejected the disk, asking Saint-Gobain to try again. Ritchey ventured to Paris for a few months as an adviser, but all subsequent efforts to cast the 100-inch blank were failures, and Hale eventually ordered a reluctant Ritchey to begin grinding the original flawed disk in October 1910.

Hale's Demons, Ritchey's Frustration

By then, both Ritchey and Hale were suffering mental meltdowns, and the former allies had become bitter enemies. Though he could practically run up Mount Wilson in his exuberant manic modes, the high-strung Hale sank into periodic depressions that became more frequent as he aged. Hale complained about his "head" as though it were a dangerous foreign object; he feared falling into a "neurasthenic quagmire."

Despite his mental problems, Hale continued his research on the sun, designing two solar telescope towers of 60 and 150 feet, erected on Mount Wilson in 1907 and 1912, respectively. By capturing the sun in thick flat mirrors high above the ground and sending the light straight down to the spectroheliograph, Hale was able to obtain sharper images than the horizontal Snow telescope, which suffered from heat waves and mirror distortion. He spent much of June 1908 scrambling up and down the ladder of the new tower telescope. Photographing the alpha-line of hydrogen, he saw eddies that resembled tornadoes swirling around sunspots, which he had already proved to be somewhat cooler than the rest of the solar surface.

"I have been carried away by the solar whirlwinds," Hale wrote to a friend. He suspected that solar magnetic fields were affecting the light. In his lab, he showed that spectroscopic lines split under the influence of magnets, then found the same effects surrounding sunspots. This was the greatest scientific discovery of Hale's life, even though he pursued solar research sporadically for another three decades.

By the fall of 1910, worried over the failed 100-inch disks and Hooker's vacillation over additional payments, Hale fell apart completely, departing for an extended overseas trip with his wife. The couple arrived in Egypt in January 1911. Even there, however, he was plagued by ringing ears, tingling feet, and an inability to concentrate. In addition, he couldn't sleep and sometimes tried to climb the picture frames on the wall in his torment.

Meanwhile, back in Pasadena, Ritchey, too, suffered from insomnia, though his troubles were more understandable, caused by his archenemy, Walter S. Adams, the research astronomer Hale left in charge. Adams regarded Ritchey as an upstart technician who needed to be put in his place. In early 1910, without consulting Hale, Ritchey asked Hooker to fund a special astronomical photographic laboratory to produce ever-more sensitive emulsions. When Hale heard about it, he exploded, because he didn't want to divert Hooker's money to ancillary causes. From then on, he viewed Ritchey as a turncoat, and Adams took every opportunity to reinforce that attitude. "I really think," Adams wrote to Hale in 1910, "we shall have to consider the question of Ritchey's connection with the observatory in the near future."

Frustrated and ambitious, Ritchey began his own mirror-making business for amateur astronomers, cutting back to two-thirds time at the observatory. Then Henri Chrétien, a brilliant young French mathematician and astrophysicist, arrived at Mount Wilson in 1910 to conduct research, and Ritchey taught him how to take long-exposure photographs. They worked together to take spectacular pictures of Halley's Comet that year, but they also invented a revolutionary new telescope concept that they dubbed the "new-curves" design, since the mirror surfaces of both primary and secondary were unique.

Now called the Ritchey-Chrétien telescope, it was designed to cut down on coma, an aberration inherent to paraboloid primary mirrors.

Parallel, perpendicular light waves coming from the object in the center of the field of view are focused perfectly by a paraboloid mirror, but light from stars at the edge of the field is distorted so that they resemble small comets rather than points. The bigger and deeper the primary mirror, the worse this coma problem becomes. In 1910, with Ritchey providing the ideas and Chrétien working out the math, the two men designed a new type of Cassegrain telescope in which the primary and secondary were specially designed hyperboloids—concave and convex, respectively. As an added advantage, such mirrors featured fast focal ratios, which meant that they could be relatively compact.

Naturally, Ritchey wanted to make the 100-inch a Ritchey-Chrétien. He also wanted to manufacture it as a sandwich of three thin disks separated by cemented spacers, rendering it lightweight, inexpensive, and easily ventilated for temperature control. In his home workshop, he produced a sample 20-inch cellular disk. Ritchey wanted to mount the big new telescope high above the ground in open air, with only a windscreen and rollover shelter for foul weather.

Hale didn't want to risk failure with the 100-inch, and he refused to make it a prototype for an untested design such as the Ritchey-Chrétien, which existed only on paper. Neither did he want a cellular mirror or a tube open to the air. Because Hale was not interested, Ritchey described his plans to Hooker, who was intrigued. Then Hooker died in May 1911, leaving Ritchey friendless and forlorn. The forty-six-year-old Ritchey feared he would be fired and wrote pathetically to the head of the Carnegie Institute, describing his "frequent fainting spells, continued insomnia, and extreme mental depression."

Ritchey wasn't fired, but he was demoted from a regular member of the observatory staff to a mere optician. He was to make the 100-inch mirror but would have no part in the telescope's design, which was given to Francis Pease, Ritchey's former assistant.

Ritchey recovered from his depression. He turned the 100-inch mirror up on its side for a test every morning, then lost himself in work, along with two assistants, even though he privately doubted that the bubble-infested glass would work well. "It required the greater part of my time and intense care for five years," he recalled later. In the end, he concluded that "the front, concave surface of the glass is almost perfectly free from even small bubbles," and his tests revealed a perfect paraboloid curve to within .000003 of an inch. To comprehend the minuscule size of that variation, Ritchey envisioned the mirror enlarged 250,000 times, so that it would have a 400-mile diameter. On that scale, the largest error would amount to three-fourths of an inch.

Ritchey and his assistants silvered the mirror in July 1916, then began work on the hyperboloid secondary mirrors for the Cassegrain and coudé arrangements. By November 1917, the mirrors were in the telescope—which, like the 60-inch, floated on a pool of mercury—ready for their first test. Ritchey was not invited. Early in the evening, Hale and Adams looked through the eyepiece at Jupiter and were appalled to see blurred, overlapping images. After a sleepless night, they tried again at 2:30 A.M. Jupiter had departed, so they swung the telescope toward the star Vega, which, to their relief, blazed as a bright pinpoint. It turned out that workmen had left the dome open during the day, and the hot mirror was severely distorted in the cooling night air until it reached equilibrium in the early morning.

War Optics

By this time, the United States had entered World War I. The Great War, with its machine guns, tanks, submarines, poison gas, and airplanes, produced unprecedented carnage. Mirror-makers found themselves contributing with range finders, periscopes, and searchlights.

Hale, whose depression lifted whenever he was busy organizing and planning, thrived in Washington, D.C., as a founder of the National Research Council, a group of engineers and scientists enlisted to aid the war effort. He secured contracts for Ritchey, who mobilized his optical shop, hiring and training new assistants to mass-produce small mirrors, prisms, and lenses for optical range finders designed by Albert Michelson. In Germany, opticians for the Zeiss and Schott optical and glass firms turned out similar items for the Kaiser.

In Dublin, the aging Howard Grubb had been making prisms and mirrors for submarine periscopes as early as 1902, and his firm produced 95 percent of British surveillance instruments. During the war, as Irish political unrest seethed, the British insisted that Grubb move his shop to England.

Grubb yearned to get back to making telescopes. He had begun to make a reflector for Russia in 1912, but that was put on hold, and the Russian Revolution of 1919 didn't help matters. Finally, in a last gasp, the economically strapped firm finished the telescope, but it was its final job before bankruptcy. Charles Parsons—the inventor of the steam turbine and the youngest son of the Third Earl of Rosse (who made the Leviathan telescope)—purchased the remains of the firm in 1925, renaming it Grubb Parsons and carrying on the optical tradition.

Searchlights, with mirrors made from glass, aluminum, nickel, silver, gold plate, porcelain enamel, or even flat white paint, came into their own during World War I. French, German, British, and U.S. engineers scrambled to provide searchlights that would pierce the night sky to reveal enemy airplanes. Huge mobile searchlights—first 36 inches, then 60 inches wide—were mounted on trucks or wagons, along with their own electric generators. "This not only renders him [the enemy pilot] open to attack," the American engineer W. F. Tompkins observed, "but the searchlight beams dazzle and confuse the aviators, and, by contrast, hide objectives."

The Expanding Universe

Despite George Ritchey's heroic efforts in the optical shop, where he was supervising sixty assistants by November 1918, the end of the war signaled the end of his mirror-making for the Mount Wilson Observatory. Hale heard that Ritchey had predicted that the 100-inch telescope would be a "total failure" because of a poorly designed mount, and that did it. In his speech dedicating the telescope in June 1919, Hale even neglected to mention Ritchey, who had spent five years making the all-important mirror.

Hale fired him in October 1919, and for the next five years Ritchey tended his citrus trees. But the fifty-five-year-old mirror-maker was not defeated. He continued to dream of huge telescopes, and his ego remained intact. "Some astronomers say, so easily, that the making of very large and very accurate optical surfaces is 'merely a mechanical operation,' whatever that may mean," Ritchey wrote later. "We note that these astronomers have never made any."

Ironically, Ritchey was happier than the troubled Hale, who soon retired as an "honorary director" to his private Pasadena solar observatory, where he became a virtual hermit. Hale entombed himself in a basement laboratory, where he studied the sunbeams reflected down to him by a series of mirrors. Headaches and insomnia—he named the maelstrom in his head "the whirligus"—continued to plague him.

Meanwhile, the 100-inch telescope, which had finally wrested the title of the world's largest from the derelict old 72-inch Leviathan at Birr, was revolutionizing the view of the universe. Ever since William Herschel, astronomers had pondered the true nature of the spiral nebulae. Were they relatively nearby cloudy nebulae, part of the Milky Way? Or were they "island universes," each composed of billions of stars so distant that they could not be seen distinctly? No one knew, and in April 1920 two astronomers, Harlow Shapley and Heber Curtis, squared off for what came to be called the Great Debate on this issue.

Like most astronomers at the time, Shapley thought that the spirals were gaseous nebulae, not distant galaxies, even though they did appear to have some stars in them. In 1902, Agnes Clerke had relegated the notion of island universes to "the region of discarded and half-forgotten speculations," largely because two bright novae had appeared in spirals in the late nineteenth century. No one could conceive of such extraordinarily bright stars appearing at the immense distances required by the island-universe theory. "Imagination recoils from contemplating" the gigantic suns that would account for such novae, she wrote.

In addition, in comparing photographs of the same spiral nebulae taken over a number of years, the Dutch astronomer Adriaan van Maanen thought he saw evidence that they had rotated slightly, at the rate of 1/100,000 of a revolution per year. If that was so, they could not be millions of light years away, or the observed rotation would have been faster than the speed of light.

Finally, Shapley estimated that the Milky Way was 300,000 light years across, much larger than anyone had thought, and that it contained the entire universe. He reached this conclusion by studying special variable stars—those that grow brighter and dimmer—called Cepheids. At Harvard, Henrietta Leavitt, an underpaid "computer" who studied photographs of variable stars, noticed in 1908 that the longer it took for Cepheids to dim and brighten again, the brighter they were. Because she was studying stars that were all part of the Small Magellanic Cloud, she knew that they were roughly the same distance away. Here, then, was a "standard candle" that could serve as a yardstick for the universe. Once you found a Cepheid and tracked its period of variability, you could then determine its relative distance by how bright it was.

Using the 60-inch mirror at Mount Wilson, Shapley found Cepheids in globular clusters—circular swarms of thousands of stars—in the Milky Way. He determined that most of them lay toward the center of the galaxy and that our solar system lay somewhere far out on the rim. He overestimated the galaxy's size (it is really about 100,000 light years across), primarily because he didn't take account of interstellar gas and dust that dimmed the Cepheids.

Shapley presented his arguments at the debate. Heber Curtis countered with photographs George Ritchey had taken in 1917 with the same 60-inch telescope. On a long-exposure photo of a spiral nebula, Ritchey had noticed a dim 14th-magnitude star that hadn't been there on previous photos. It was a nova, and when its light was dispersed by a spectroscope, it showed a continuous spectrum with a few bright emission bands, typical of such new stars. Soon, Ritchey found several other novae in spirals. Curtis, who for years had been taking similar pictures with the Crossley 36-inch mirror at Lick, quickly reexamined his own plates and found more new stars in spirals. He concluded that spirals were indeed separate and very distant galaxies.

The Great Debate did not settle the issue, but Edwin Hubble soon did, using the new 100-inch mirror on Mount Wilson. The son of a Missouri insurance agent, the imposing, athletic Hubble had attended Oxford University as a Rhodes Scholar, then served briefly in World War I. An unabashed self-promoter and snob, he annoyed his fellow astronomers by affecting a British accent and telling tales of heroic exploits that stretched the truth to the breaking point. Hubble arrived on Mount Wilson in October 1919, three weeks after the 100-inch telescope was put into service, and he used it to peer deeper into space, probing the mystery of the nebulae. On October 5, he hit pay dirt with a forty-five-minute exposure of M31, the Andromeda Nebula, where he noticed three new stars. But when he compared the plates with Andromeda photos taken by Ritchey and others, he found that one of the stars wasn't new after all—it just grew dimmer and brighter. It was a Cepheid variable, the first to be identified in a spiral. Hubble estimated its distance as 1 million light years, putting it definitively outside the Milky Way.

Hubble triumphantly notified Shapley, who was now director of the Harvard Observatory. Shapley, also a Missouri native, loathed the pretentious Hubble, but he could not fight facts. "Here is the letter that has destroyed my universe," he told a grad student. Over the next year, Hubble continued his assault on the heavens, finding twelve more variables in Andromeda as well as Cepheids in other spirals. The Great Debate was over. Spiral galaxies were indeed island universes, each composed of billions of stars. The remarkably bright new suns from which Agnes Clerke's imagination had recoiled were supernovae, and they really were millions of light-years away. Van Maanen was wrong—the spirals' rotations were a product of his expectation and imagination, just like Lowell's illusory Martian canals.

Now Hubble turned his attention to stellar velocities as documented by their redshifts. As William Huggins first demonstrated, the distinctive spectroscopic lines for different elements in stars were often shifted toward the red end of the spectrum, and the faster the stars were fleeing in relation to the earth, the bigger the redshift. Using a 24-inch refractor at the Lowell Observatory, Vesto Slipher had studied spectroscopic shifts of spirals over a fifteen-year period. He found that our neighboring Andromeda spiral is blueshifted, moving *toward* the earth, but that all other spirals he could measure were redshifted, rushing away at speeds as high as 1,200 kilometers per second. (*Note:* Throughout this book, measurements are given as cited in the literature.)

But Slipher was limited by his refractor. The 100-inch mirror allowed astronomers to see dimmer objects, but even with its better light grasp, it required nightlong exposures to get their absorption lines to appear on spectroscopic photographs. For this demanding work, Hubble turned to Milton Humason, who began his career atop Mount Wilson as a bellboy at an old resort hotel, then graduated to mule driver, observatory janitor, and finally astronomer. The night assistant with the eighth-grade education turned out to be a meticulous, skillful observer and photographer, despite occasional snorts from his bottle of "panther juice."

On two successive frigid nights in 1927, Humason kept the big mirror trained on a distant nebula out of Slipher's range, then developed the plate, where he soon found the H and K Fraunhofer lines produced by calcium. They were indeed redshifted, and Hubble calculated the spiral's speed as 3,000 kilometers per second. Over the next year and a half, Humason and Hubble worked in tandem. Humason tracked speed through redshifts while Hubble sought Cepheids in the same galaxies to determine distances. In 1929, Hubble published a revolutionary six-page paper, "A Relation Between Distance and Radial Velocity Among Extra-Galactic Nebulae," in which he documented that the farther away a galaxy was, the faster it fled.

The conservative Hubble resisted interpreting his data. He called the redshifts *apparent* velocities, believing that there might be some other explanation. But for nearly everyone else, Hubble and Humason had proved that the universe was expanding. For Albert Einstein, the expanding universe produced both relief and humility. His theory of general relativity had actually predicted that it must be either expanding or contracting, but because most astronomers assured Einstein that the universe was static, he had added a fudge factor to his equation called the cosmological constant. Now he called this his "biggest blunder."

Ritchey's Last Hurrahs

By this time, George Ritchey was once more a force to be reckoned with. In 1924, he had left his California lemon farm, sailing for France. A wealthy couple—Assan Farid Dina, a French-Indian engineer, and his wife, Mary Shilito Dina, a Cincinnati department-store heiress—were going to pay for a 104-inch mirror to be mounted in a telescope in the Alps. It would be a Ritchey-Chrétien design. Ritchey took over a lab at the Paris Observatory as well as an office at the Institute of Optics, where Henri Chrétien was now a professor.

The rejuvenated fifty-nine-year-old threw himself into designs and experiments, soon announcing that he had no intention of producing a mere 104-inch mirror; instead, Ritchey would make a 200-inch or 240-inch reflector with a built-up, cellular disk. Charles Fabry, the director of the Institute of Optics, thought Ritchey was *maboul* (cuckoo), but the self-confident American persisted, making progressively larger cellular prototype mirrors, although the cement holding the dividers caused problems. Writing to a friend in February 1925, Ritchey expressed grandiose, near-messianic views. His project was "surely providentially managed—*divinely* planned." Ritchey said that he expected to build a 40-foot (480-inch) telescope mirror before he died.

The grand project sputtered, however, when the 60-inch cellular mirror cracked in April 1926. Dina fired Ritchey, and the entire project fell apart after Dina's death in 1928. Ritchey hung on at the Paris Observatory, thanks to other French supporters, finally completing the first working Ritchey-Chrétien telescope with a solid 20-inch mirror. It mostly eliminated coma, but because of the poor mounting and the necessity to grind curved glass photographic plates and coat each one individually with emulsion, his pictures were unimpressive. Desperate to return to the United States, he tried to secure commissions there but was repeatedly blocked by Hale and Adams behind the scenes. Only French and Canadian journals would publish his articles.

The more frustrated Ritchey became, the more grandiose his dreams, which he detailed in a six-part series. In the first article, published in the May 1928 issue of the *Journal of the Astronomical Society of Canada,* Ritchey described his plans for a "fixed universal telescope." At the top of the fixed tower, a huge, flat mirror, 5 meters (197 inches) across, would act as a coelostat, tracking the stars and reflecting them to another large flat mirror, which would direct the light down the tower to a 4-meter (158-inch) Ritchey-Chrétien hyperboloid, then up to a secondary, and down through a hole in the primary or off to the side in a coudé arrangement. The telescope would be "universal" because a variety of primary mirrors could be wheeled quickly into position at the bottom of the tower to create different focal lengths. Three would be Ritchey-Chrétiens, and two would be Schwarzschilds—invented independently by Karl Schwarzschild to lessen coma with similar "new curves," but with a concave rather than convex secondary mirror.

All of the mirrors would be cellular. "It seems so ridiculous," Ritchey stated, "to insist upon using for a great optical mirror a heavy, solid disk, expensive, almost prohibitively difficult to cast and to anneal, and weighing several tons, in order to support in perfect optical form a silver film weighing only a very small fraction of an ounce!"

Ritchey acknowledged that his coelostat design would limit the amount of sky an observer could view, but with three such telescopes— one at the equator and the other two at 33 degrees north and south latitude—astronomers could cover all of the heavens. He wanted to build the first one on the edge of the Grand Canyon. It was a fantastic plan, and it kept getting bigger. By the time Ritchey finally left France at the end of 1930, he was talking about a fixed supertelescope as high as the Eiffel Tower with a mirror 80 feet (980 inches) across.

Some day, Ritchey predicted, "we shall look back and see how inefficient, how *primitive* it was to work with thick, solid mirrors, obsolete mirror-curves, [and] equatorial telescope-mountings of antiquated types requiring enormous domes and buildings." He urged people not to allow their imaginations "to be limited to one small world—to one microscopic corner of the *universe* of worlds." Instead, they would look out with his telescopes to see the Milky Way, with its "*tens of billions of suns*," and beyond that to millions of galaxies just as large.

Alas, although Ritchey's soaring vision was essentially sound, his prophetic powers were not. No one funded any of his enormous supertelescope fixed towers. His first potential American patron died soon after he returned to the United States in 1930. Then the U.S. Naval Observatory funded a 40-inch Ritchey-Chrétien, completed in 1935 when Ritchey was seventy years old. Because of the poor seeing conditions in Washington, D.C., it performed poorly. Ritchey retired to his California lemon farm, where he died in 1945.

Yet Ritchey has had the last laugh. Ten years after his death, the 40-inch Ritchey-Chrétien telescope was moved from Washington to a site near Flagstaff, Arizona, where it produced spectacular galactic photographs. It inspired an entire generation of major telescopes with Ritchey-Chrétien mirrors. Other Ritchey concepts have also been vindicated, including cellular mirrors, though the fixed vertical telescope has yet to be built.

Toward Palomar

Meanwhile, despite his whirligus, George Hale remained an engaged if mostly invisible presence. From his solar laboratory he churned out a series of articles and books of popular science. In April 1928, just as Ritchey was preparing his visionary series of articles, Hale published his most important, influential article, "The Possibilities of Large Telescopes," in *Harper's Magazine*. "Starlight is falling on every square mile of the earth's surface," he wrote, "and the best we can do is to gather up and concentrate the rays that strike an area 100 inches in diameter."

Hale compared astronomers to cosmic explorers, explaining that they had now "worked beyond the boundaries of the Milky Way," peering at distant galaxies. "As yet we can barely discern a few of the countless suns in the nearest of these spiral systems. . . . While much progress has been made, the greatest possibilities still lie in the future."

Hale sent a galley proof of the article to the head of the Rockefeller Foundation, and within a few months the foundation announced a gift of $6 million to build a 200-inch telescope. It would be placed on Mount Palomar, whose isolation now made it much preferable to Mount Wilson, where light pollution from greater Los Angeles had become a problem.

Septuagenarian Elihu Thomson, the head of General Electric's research laboratory, convinced the Mount Wilson astronomers and the Rockefeller Foundation that he could make the big mirror of fused quartz—the optimal material. The trouble was that fused quartz required incredibly high casting temperatures. Even though General Electric offered to make the mirror at cost, taking no profit, two years of experiments costing $600,000 failed to produce a successful 60-inch test blank.

In October 1931, Hale called a halt, switching to his second choice— a Pyrex disk made by Corning Glass in upstate New York. Pyrex, a borosillicate glass used in cookware, was affected by temperature change less than regular glass, although it expanded and contracted much more than fused quartz. Arthur Day, a former Corning executive, suggested casting the disk with a ribbed waffle-like back in order to save weight, and Corning special projects manager George McCauley took charge of production.

McCauley encountered various problems as he produced progressively larger test blanks. The viscous Pyrex congealed in the mold before it flowed into all corners, meaning that the mold itself had to rest inside a furnace. Corning built what they called an "igloo" over it. Overcoming other problems, in October 1933 the team cast a successful 120-inch blank, and on Sunday, March 25, 1934, in front of a crowd of dignitaries and journalists, McCauley supervised the casting of the 200-inch. Five crews of men would simultaneously maneuver giant ladles to pour sixty-five tons of molten Pyrex.

The pour began at 8 A.M., with ladle after ladle—each holding 750 pounds of glass—stuck through the door of the white-hot igloo. Just before noon, one of the workers passed out from the heat. In the afternoon, one of the silica brick cores that produced the hollows between the ribs broke loose and bobbed to the surface, then several more. The disk was ruined.

By the end of the year, with a special cooling system installed for each core, McCauley was ready for another attempt, and on December 2, 1934, with few spectators, the second casting succeeded. At a temperature of 1,525 degrees Celsius, the great disk—the largest piece of glass ever made—was moved into an annealing oven, where it would be gradually cooled for ten months. The huge piece of Pyrex survived a near-disastrous flood, emerging with three gashes in the surface where the overheated oven top had bent into it. Fortunately, they were shallow enough to be ground out.

Encased in a steel box with "PYREX 200" TELESCOPE DISC MADE BY CORNING GLASS WORKS" in bold letters on the side, the ribbed glass was ready for its meticulously planned cross-country trip in March 1936. Insured for $100,000, it was then the most valuable single object ever carried by rail. For its 3,000-mile, two-week journey, the future mirror was a celebrity. Schools let out as the train rumbled past at twenty-five miles per hour. Crowds lined the track to cheer Corning's rolling 17-foot billboard.

In Pasadena, the optician Marcus Brown and his assistants took possession of the disk at the huge windowless optics lab. The two-story doors swung open for the mirror blank, which would not reemerge for eleven years.

Russell Porter: A New Leonardo

The art deco optics lab was designed by a multifaceted Vermonter named Russell Porter, who was also the father of the U.S. amateur telescope-making movement. In 1928, George Hale summoned the fifty-seven-year-old Porter to California to help with the 200-inch project. The two had met at a lunch in New York City, where Porter covered napkins and the menu with quick sketches of large telescope designs. Porter's ideas and beautiful drawings were to have a fundamental impact on the 200-inch Mount Palomar observatory.

In the first half of his life, Porter's only contacts with astronomy were the star transits he timed with frozen fingers during his many arctic expeditions, on which he was charged with determining longitude and latitude. Porter studied architecture at MIT, but the arctic bug caught him in 1894. He participated in several assaults on the North Pole, and though he never got there, he recorded the desolate beauty in fine landscape pencil drawings and watercolor portraits of Eskimos, as well as an expressive journal in which he described the scenery as "brutal, unreal, mystic."

Though Porter never completed a college degree, he ended up teaching at MIT just before World War I. By that time, with failing hearing and poor eyesight, he had settled down, married, and had a daughter, but his adventurous spirit was never stilled.

Porter became fascinated with mirror-grinding through his friendship with James Hartness, who owned a machine-tool company in Porter's home town of Springfield, Vermont. Hartness had his own home observatory, and in 1913 he sent Porter two 16-inch glass blanks, with which the artist made a clever stationary telescope in which a rotating flat mirror captured starlight and brought it down to the paraboloid primary mirror.

Among Porter's many innovative telescope designs, in 1918 he proposed (in articles in *Popular Astronomy* and *Scientific American Supplement*) a yoke to hold the tube for a large telescope so that the center of mass could fall within its three supports. Such a design would solve a problem that encumbered the 100-inch telescope at Mount Wilson, where the support system blocked access to part of the sky. Eventually, Porter helped design the 200-inch Palomar mount according to this plan, with a gigantic "split-ring" horseshoe riding on a thin film of pressurized oil.

In 1919, he moved to Vermont to work for James Hartness, who called him his Leonardo da Vinci because of his artistic and mechanical skills. Porter inspired a group of Springfield residents—mostly blue-collar male mechanics, but including one woman—to form a telescope-making club, and he taught them all to grind and test their own mirrors. The Springfield Telescope Makers met atop nearby Breezy Hill on land Porter had inherited. There they built a clubhouse on whose steep gable were inscribed words from the Psalms: "The Heavens Declare the Glory of God." Porter named it Stellar Fane (Shrine to the Stars), soon truncated to Stellafane.

In November 1925, *Scientific American* editor Albert Ingalls published an account of his visit to Stellafane—stressing the camaraderie, all-night observing, and strawberry shortcake for breakfast—and concluded that almost anyone could make a telescope. "You must be handy, of course, but you do not have to be a genius."

The article launched a movement. Within months, Porter got sixty-one letters from thirty states, the West Indies, and Hawaii. Ingalls got 368 letters requesting specific instructions, so he asked Porter to write two articles explaining how to grind a mirror and make a telescope. "As time went on," Ingalls recalled, "the telescope-making hobby enlisted the interest and keen enthusiasm, sometimes almost fanatical, of more and more of the readers of the *Scientific American*," which published an expanded 500-page book, *Amateur Telescope Making* (the first of three volumes), in 1935.

By that time, the annual summer Stellafane convention of amateurs, who camped out and stayed up all night pointing their mirrors skyward, attracted stargazers from New England and beyond, and other clubs popped up in the United States and elsewhere. Porter himself attended infrequently, since he lived in Pasadena, making models and drawings for the 200-inch telescope. His evocative cutaway drawings of the future telescope are still better than photographs of the real thing.

Men in White

Marcus Brown, the man in charge of grinding the Pasadena 200-inch mirror, grew up on a California chicken farm and took a job driving a delivery truck up Mount Wilson. "Brownie," as he was called, admired the fastidious men grinding glass at the optical shop, so he took a pay cut to work there. Many workers couldn't take the gut-wrenching routine, boredom, and confinement, not to mention the stress of knowing that months of work could be ruined by one small mistake. Brown loved it, and by the time the mirror blank arrived in March 1936, he was a master mirror-grinder.

"Glass won't ever do what you expect," he warned his new assistants. "It has as many moods as a movie star." He reminded them, "If you don't know what you're doing, don't do anything until you find out." Every day his team donned all-white uniforms, including white sneakers, before grinding glass. Floors and walls were swept and washed every day, a magnet rolled across the floor to pick up stray metal specks that might scratch the disk.

Using a slurry of carborundum, they took more than a year to grind the front and back surfaces flat and install the mirror supports. Then, in the summer of 1937, they began to cut a spherical curve into the huge hunk of Pyrex, with the automated grinding tool sweeping in the prescribed lissajous patterns.* After the rough carborundum took it near the proper shape, the workers cleaned for three and a half months, threw away their old clothing and sneakers in favor of uncontaminated new outfits, and switched to fine polishing rouge.

In September 1938, on a Saturday when they could turn off the ventilators and air-conditioning to still the air, Marcus Brown and his physicist boss, John Anderson, first tested the mirror, swinging it to a vertical position and using a Foucault knife-edge test, reflecting light from a pinhole 120 feet away. The tests revealed areas that needed more work, but they also showed an inexplicable astigmatism, as if giant fingers were squeezing the glass gently from the sides, no matter how they turned it. Anderson eventually corrected the problem with a slight adjustment of the mirror supports. Still, it took another three years until they achieved a nearly flawless polished spherical curve and were ready to begin the final push toward a parabola.

George Ellery Hale didn't live to see himself reflected in the giant mirror. A few days before he died at sixty-nine in February 1938, he

---

*The overlapping, sweeping patterns were named for Jules Antoine Lissajous (1822–1880), who found the patterns in reflected light from a mirror atop a struck tuning fork, a version of Wheatstone's kaleidophone.

looked wistfully out the window and said, "It is a beautiful day. The sun is shining and they are working on Palomar." In an editorial, the *New York Times* suggested that the huge mirror and observatory be named for Hale.

The One-Armed Maniac

When Bernhard Schmidt died of pneumonia in an insane asylum in Hamburg, Germany, three years before Hale's passing, nobody paid much attention.* The eccentric, misanthropic optician had, however, invented a remarkable telescope that would soon revolutionize astronomical photography, beginning on Mount Palomar.

Born on the small Estonian island of Nargen in 1879, young Bernhard combed the beaches for items that he could use for inventions, and at night he memorized the constellations. He built himself a fiddle, wove a net from spider webs, and made a pipe bomb with homemade gunpowder, accidentally blowing off two fingers of his right hand. The doctors amputated his arm above the elbow, leaving only a stump.

One winter day, Schmidt noticed that a transparent piece of ice magnified dried leaves and straw in exquisite detail, and he soon focused on optics, making his own camera and (according to legend) using sand to grind the bottom of a broken bottle into a lens. In 1895, the sixteen-year-old Schmidt took a job as a night telegraph operator, where he found time to observe the sky with homemade telescopes. In 1901, he was one of the first to see the Nova Persei that George Ritchey photographed. Later that year, Schmidt moved to Mittweida, a small German town near Jena, an optical center. There he set up a mirror-grinding shop in an abandoned bowling alley, where for a quarter-century he proved that a one-armed optician could turn out astonishingly accurate telescope mirrors.

---

*In his final days, Schmidt was delirious, suffering from horrible headaches, and crying out as he was restrained in a straitjacket and forced into cold baths. Until recently, it was assumed that he died an alcoholic, but his nephew and biographer, Erik Schmidt, believes that his uncle's symptoms stemmed from undiagnosed meningitis.

Dressed formally in striped trousers and a cutaway coat, Schmidt disappeared into his shop for days and nights, neglecting meals but fortified by cigars and corn liquor. He scorned grinding machines, preferring to work by hand or stump. "My hand is more sensitive than the finest gauge. . . . If the hand encounters friction, you have to stop work at once until the temperature evens out." To test his mirrors, Schmidt hung silvered glass balls in trees at a nearby park, then shined a searchlight (made with one of his mirrors) on them at night to create artificial stars. Another of his testing methods utilized interference fringes.

Around 1909, Schmidt made a horizontal fixed telescope with a 16-inch primary mirror and 36-foot tube. Light was reflected down the tube from a flat mirror turned to follow a star with an ingenious water clock. With it, Schmidt took impressive photographs of the moon and planets. During World War I, local police suspected that the secretive optician was using his telescope to flash light signals to Russian aircraft, and he was briefly interned as an enemy alien.

Schmidt, who took a dim view of humankind in general, wanted no part of the war. "Only one man alone is worth anything," he once said. "Put two men together and they quarrel. A hundred of them make a rabble, and if there are a thousand or more, they'll start a war." At least the carnage helped Schmidt in one way—with so many veterans missing limbs, he became less self-conscious about his own disfigurement.

Schmidt's business dried up in the inflationary postwar years. In 1927, he accepted an offer from the Hamburg Observatory in Bergedorf to serve as an unpaid optician with free room and board, setting up a basement workshop. Two years later, he accompanied astronomer Walter Baade to the Philippines to photograph a solar eclipse with his horizontal telescope. On the long voyage home, Schmidt pondered a problem that had been occupying him for several years. His astronomical photographs were sharp only at the center of the field. Toward the margins, coma smeared his stars into teardrops. How could he achieve photographs in focus over the entire field of view?

As he stared out at the ocean swells, it suddenly came to him. Schmidt confided in Baade, one of the few colleagues he trusted, that he might be able to use a spherical mirror—the simplest type to make. As every astronomer knew, however, such a mirror produced spherical aberration, yielding blurry images. But what if Schmidt could somehow

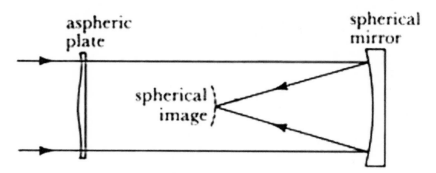

FIGURE 9.1   Schmidt telescope.

grind a corrector plate that fooled the light, refracting it so as to correct the aberration? The excited Baade saw that it might actually work.

Back in his basement lab, Schmidt came up with a brilliant solution—part science, part inspired intuition. He sealed a thin, 14-inch disk of glass to the top of a pot, then drew down a vacuum, sucking the disk down somewhat. With the glass held in this distorted position, he ground and polished it flat. When he released the vacuum, the glass popped into the correcting lens he needed, which he placed through the center of curvature of the 17-inch mirror.* Then, using circular negatives clipped to an internal holder shaped to the spherical focal curve (halfway between the corrector plate and mirror), Schmidt and Baade tested the 1930 prototype, snapping a photo of a distant cemetery. Amazingly, the entire picture remained in sharp enough focus that they could read the names on the tombstones (see Figure 9.1).

The following year, when Baade left Hamburg for a job at Mount Wilson Observatory, he took that picture with him, along with plans for the Schmidt telescopic camera. An impressed John Anderson realized that such a camera would provide the perfect complement for the new 200-inch giant. Even with a corrector lens designed by optician Frank Ross, which helped remove a bit of coma, the mammoth eye would be able to peer at only a tiny bit of the sky, albeit highly magnified. A Schmidt could act as a "finder scope" to identify interesting objects to zero in on. Russell Porter designed a sleek 18-inch Schmidt cam-

---

*For a Schmidt telescope, the corrector plate must be somewhat smaller than the mirror, so the smaller aperture is used in speaking of the telescope's size.

era, resembling a howitzer, which went into operation on Palomar in 1936, the year after the Estonian optician died a virtual unknown at age fifty-six.

The camera, which could take exquisite long-exposure photographs of big chunks of the sky, permitted Fritz Zwicky, an abrasive, brilliant German, to find supernovae in distant galaxies. The results were so impressive that the astronomers asked Porter to design a much bigger 48-inch Schmidt. Don Hendrix, a young wizard at the Mount Wilson Observatory optical shop, began work on it. He also made tiny 2-inch Schmidts for spectrographic work.

Finishing the Big Eye

Both major projects—the big Schmidt and the 200-inch—were delayed by World War II. Hendrix and an assistant had completed the 72-inch spherical mirror for the big Schmidt (the lens aperture was 48 inches, the mirror considerably larger) in only five months before the war, but the complicated corrector plate took much longer. Marcus Brown and his team in Pasadena finally began parabolizing the 200-inch mirror in September 1941. Original plans called for an entire year to make a flat test mirror from a 120-inch Pyrex blank, but John Anderson devised an ingenious testing method using a much smaller, partially silvered flat mirror. Brown's team had just reached the crucial last stages of parabolization when the Japanese bombed Pearl Harbor on December 7.

As in the previous war, opticians turned to making mirrors and prisms for periscopes, searchlights, and range finders, to which they now added Schmidt-type reconnaissance cameras for airplanes. When the war ended, the aging Brown uncovered the 200-inch mirror and resumed polishing, diluting the rouge with talcum powder to reduce abrasion even further. By 1947, Anderson's Saturday tests revealed that the mirror surface was within one-millionth of an inch of a true parabola. Brown still didn't want to let go, but in November Anderson finally released the mirror. The optics shop doors swung open. Brown, Anderson, and his men posed in front of the vertical disk, one worker sitting in its hole, where he noticed letters scratched inside the hole on the glass: *Marcus H. Brown 1947 A.D.* Brown was retiring. He had devoted eleven years of his life to this piece of glass, and now he had to turn it over to Don Hendrix, who had finished the big Schmidt corrector plate and now took over as chief optician.

Once the mirror reached Palomar Mountain—under heavy guard against religious cranks, who threatened to shoot the glass to prevent forbidden glimpses into God's heavenly sanctuary—it was put into a custom-made vacuum chamber for its reflective coating, which would be aluminum rather than the traditional silver. In 1932, physicist John Strong had first vaporized aluminum in a vacuum, allowing its molecules to settle evenly on a glass surface. Although aluminum doesn't reflect visible light quite as well as silver, it performs better in the ultraviolet range. Better yet, its oxidized coating provides a transparent protective surface, so that tarnishing is not a problem.

In 1947, Strong arrived on Palomar with a case of Wildroot Cream Oil hair tonic, which he mixed with powdered chalk before coating the mirror. "In order to get glass clean," he told the horrified astronomers, "you first have to get it properly dirty." After wiping the goop off with felt, Strong burned off the residue and aluminized the mirror in the vacuum chamber, depositing a thin coating, one-thousandth of a millimeter thick.

Finally, just before Christmas 1947, with the new mirror in place, the first-light ceremony took place. As with most new telescopes, there were problems. Among other things, the mirror support system needed tweaking. A slight astigmatism was cured by hanging four small spring scales—purchased from a hardware store—from strategic points on the back of the mirror.

Ira Bowen, the new director at Palomar and an optical expert himself, ruled that the mirror needed final figuring, based on Hartmann-screen tests he conducted.* In the spring of 1949, Don Hendrix and his assistant stripped off the aluminum coating and corrected several small zones using tiny pitch-covered tools, or a piece of cork, or simply their thumbs. They kept begging for just one more week to get it perfect, but finally Bowen called a halt. The mirror was realuminized, and in November 1949 the astronomers finally turned the Hale Telescope to the

---

*A Hartmann screen consists of a large metal circle with regular holes in it, so that individual portions of a mirror can be checked. The test was devised by German astronomer Johannes Hartmann in 1900.

heavens. For the next few decades—in tandem with the big 48-inch Schmidt, which surveyed the sky—it would reveal redshifts, hitherto unknown galaxies, and mysterious quasars that changed our ideas of the universe and its evolution.

Like the 100-inch on Mount Wilson, the Hale featured a variety of secondary mirrors so that it could function as a Cassegrain or coudé, but because of the gigantic mirror, astronomers could also work directly at the prime focus without blocking too much light. They rode up the side of the dome in a small elevator, then climbed into the prime focus cage, where their instruments could capture the light after only one reflection. While there, they could also gaze directly down into the mirror where the stars swam—pollen in a celestial fishpond.

George Ellery Hale, a manic-depressive astronomer with a genius for extracting money from millionaires, planned ever-larger telescope mirrors. In his final years he became a virtual hermit in his underground solar laboratory, shown here.

With Hale's sponsorship, George Ritchey became a superb optician, mirror-maker, and astronomical photographer. Hale eventually fired the brilliant, abrasive Ritchey, who is shown here in Paris with one of his lightweight cellular mirrors.

During World War I, mobile searchlights with paraboloid mirrors hunted for enemy airplanes. "This not only renders [the enemy pilot] open to attack," a contemporary strategist wrote, "but the searchlight beams dazzle and confuse the aviators, and, by contrast, hide objectives."

With so many Great War veterans missing limbs, one-armed optician Bernhard Schmidt became less self-conscious. Schmidt invented a telescopic camera with a wide field of view, using a spherical mirror and a corrector plate. The coat sleeve over the stump of his right arm is worn where he used it to polish mirrors.

Arctic explorer, artist, architect, and mirror-maker, Russell Porter launched the amateur telescope movement from his home in Springfield, Vermont, where the annual Stellafane convention takes place.

On Breezy Hill in Vermont, Russell Porter and his group built this Turret Telescope in 1930. A mirror placed in the opening to the right reflects the stars into the telescope. On the eaves of the clubhouse in the background they inscribed, "The Heavens Declare the Glory of God."

The modern Stellafane convention attracts amateur mirror-makers and observers from across the country.

On its 3000-mile, two-week rail journey across America in 1936, the 200-inch Pyrex mirror blank was a celebrity and a rolling billboard for Corning.

Marcus Brown (far right) and his team of opticians pose with the mirror blank before dressing in all-white clothing and getting to work.

In 1949, an unidentified worker (probably Don Hendrix) puts the final touches on the 200-inch mirror.

"Our minds are full of windows," John Wanamaker observed in 1916. "Show windows are eyes to meet eyes." In the shop windows, people could look at their own reflections as well as the merchandise.

At night, millions of electric light bulbs transformed Coney Island into a glittering wonderland. "A fantastic city all of fire suddenly rises from the ocean into the sky [and is] mirrored in the waters," observed Maxim Gorky.

Unlike their fathers, most men of the early 20th century were clean-shaven, often using a mirror and throw-away razor blades. This 1910 Gillette ad featured baseball stars who were "clean men—clean of action and clean of face."

In the 1920s and 1930s, it became acceptable to apply cosmetics in public while looking into a compact mirror — and ignoring pesky boyfriends.

The modern sensibility valued youth and image, and advertisements lie this played on the fear of aging.

Hollywood celebrities helped sell make-up, and vice versa. This still from the 1924 film *Men* was used in a drug store tie-in campaign.

Even during World War II, Rosie the Riveter couldn't do without her lipstick and mirror, as this 1942 *New York Times* drawing indicates.

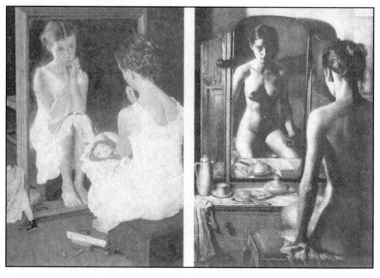

Norman Rockwell's 11-year-old innocent looks wistfully at herself, wondering whether she will ever grow up to be a sophisticated beauty, while Gerald Brockhurst's 16-year-old gazes with solemn wonder, fear, and awe at her newly sensual body.

Searching for radio static with his "merry-go-round" antenna in 1932, Karl Jansky found radio waves from outer space.

In 1957, Bernard Lovell's 250-foot Jodrell Bank radio dish was saved from bankruptcy by Sputnik, when it was the only facility capable of tracking the carrier rocket. The radio dish has now been resurfaced and upgraded.

The world's largest radio mirror, a thousand feet across, lies at Arecibo, Puerto Rico, in a natural bowl. Frank Drake and Carl Sagan used it to listen for messages from extraterrestrial civilizations.

With this strange-looking horn reflector (left), Arno Penzias and Robert Wilson accidentally discovered the background microwave radiation from near the Big Bang.

As the diagram to the right indicates, the bottom of the "scoop" serves as the mirror and is part of a paraboloid curve.

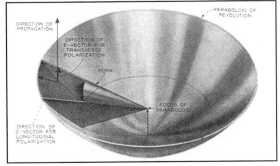

When grad student Jocelyn Bell (now Burnell) heard regular radio pulses in 1967, she and her colleagues thought they might be signals from aliens, so they dubbed them LGM for "little green men." They turned out to be pulsars, fast-spinning neutron stars.

Completed in 1981 on the Plains of San Augustin near Socorro, New Mexico, the Very Large Array is an interferometer of 27 mobile parabolic radio mirrors. They can be spread to give the resolution of a single mirror 22 miles across with the sensitivity of a dish 426 feet wide.

Riccardo Giacconi helped send x-ray mirrors into space before switching to the Hubble Space Telescope and other projects.

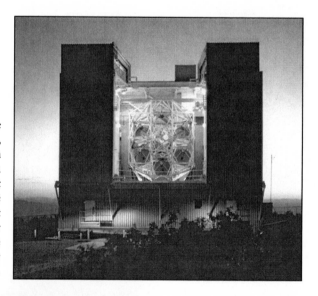

The innovative Multiple Mirror Telescope, which first saw light on Mt. Hopkins in Arizona in 1979, was the first major step beyond the 200-inch telescope. It used lightweight cellular mirrors originally made for secret military spy satellites.

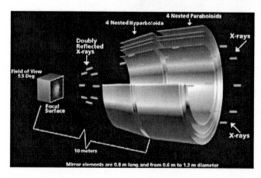

The *Chandra X-ray Observatory*, launched in 1999, has nested mirrors almost four feet wide. Coated with evaporated iridium, they are incredibly precise.

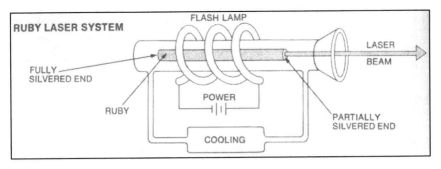

Invented in 1960, the first laser excited molecules in a ruby rod, where the polarized light reflected back and forth between two mirrors before exiting from the partially silvered end.

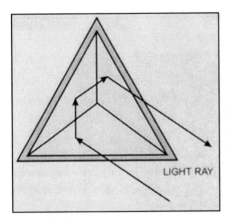

As illustrated here, a corner cube reflector always sends light directly back to where it came from. Astronauts left corner reflectors on the moon, but they are also on the back of bicycles.

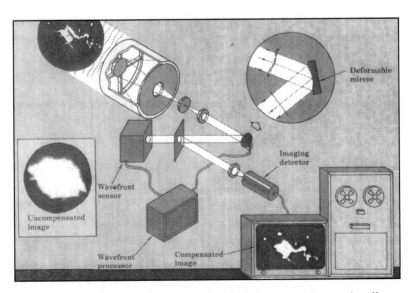

Adaptive optics uses small deformable "rubber" mirrors to correct the effects of atmospheric turbulence on light. The system illustrated here was installed for "dark side" military applications at an Air Force site in Hawaii in 1982.

The Hubble Space Telescope mirror being tested in 1981 at Perkin-Elmer in Danbury, Connecticut. It was touted as "the most precise large mirror ever made," but it was precisely wrong.

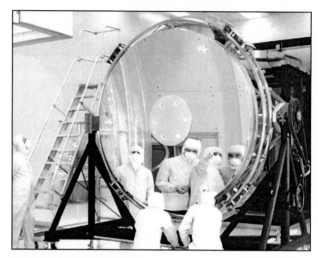

When the Hubble mirror finally flew in 1990, it sent back this fuzzy image of a star, the result of spherical aberration traced back to a missing fleck of paint that allowed a reflection in the wrong place, thus misplacing a testing lens by 1/20th inch.

Cartoonists had a field day blaming the opticians for the Hubble mirror error, but pressure from NASA was largely to blame.

Inspired by the lightweight MMT mirrors, Roger Angel pioneered honeycomb mirrors made by spin-casting a mold inside a giant furnace underneath the University of Arizona football stadium.

The entire Mirror Lab staff stands proudly in front of one of the 8.4 meter honeycomb mirrors that will go into the Large Binocular Telescope. Roger Angel is among them.

Schott, the German glass company, spun-cast this 8.2 meter mirror made of Zerodur. The first three efforts shattered, but this meniscus-style mirror is now in the Very Large Telescope in Chile.

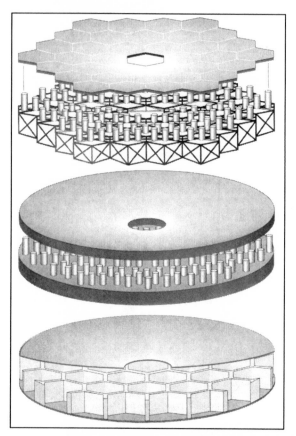

This illustration summarizes the three methods used in the past two decades to surpass the Palomar mirror. Jerry Nelson's segmented mirror is on top, with the meniscus-style VLT mirror in the middle. Both show actuators beneath to control the mirror surface. Roger Angel's stiffer honeycomb mirror is shown at the bottom.

Of the many proposed larger telescope mirrors on earth, the 100-meter Overwhelmingly Large (OWL) telescope planned by Roberto Gilmozzi is the most outrageous and intriguing. It would be made of 2,000 identical spherical mirrors. Here is an artist's conception, peering over the edge of the mirror. Note the size of the man to the right of the ladder.

Two girls watch their fingers elongate into witches' digits in the distorting Prague mirrors.

Pseudoscience thrives in the 21st century, with claims that people with the mythical "multiple personality disorder" see different "alters" in their mirrors.

In the spooky hexagonal mirror room of the Museé Grevin in Paris, a crowd sees lights recede into infinity.

The front side of the eight-story Odeillo building in the French Pyrenees is a giant paraboloid mirror, fed by 63 flat sun-tracking heliostat mirrors on a sloping hillside in front of it. The surrounding mountains are reflected upside down in the building.

Guido Barbini, shown here entering his display room on the island of Murano, is the descendant of Gerolamo Barbini, one of the mirror-makers lured to France in 1665.

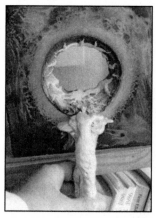

The Odeillo solar furnace melted this hole in solid iron in seconds. Director Gabrielle Olalde holds the solid tears in his hand.

John Dobson was thrown out of the Vedanta monastery for making telescope mirrors out of jug bottoms. Now he tours the world promoting Sidewalk Astronomers and laughing at the Big Bang.

After the death of her son, Cozy Baker found solace in kaleidoscopes. Now kaleidoscopes grace every surface of her sprawling Maryland home.

Don Doak takes a picture of himself in the midst of constructing his dodecahedron kaleidoscope at Catskill Corners. "Picture diving off the edge of a dock and looking up and seeing someone looking down, but it's you, and it repeats itself forever." To avoid feeling ill, he covered the mirrors with blankets as he worked.

Doak's star dodecahedron, produced by three precisely cut tapering mirrors, is a twelve-sided illusion encompassed by swirling yellow lines, a celestial fantasy hanging in space.

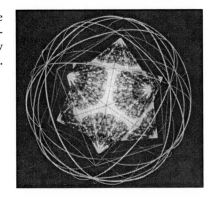

In 1930 William E. Benton patented the "Duality Mirror," which showed how faces would look with two perfectly symmetrical right or left halves. Edgar Allen Poe's real photo is in the middle, but the two symmetrical portraits are amazingly different.

Siblings John and Catherine Walter see themselves unreversed in the True Mirror they market as a way to see yourself as you really are. You may want to change your hair part, too.

The True Mirror features two flat front-surface mirrors at right angles, as shown here. If you turn the mirror sideways, however, it turns your head upside down.

Babies enjoy playing with their reflected companion, but most human infants learn to recognize themselves in mirrors just before their second birthday. This ability is associated with logic, empathy, and introspection.

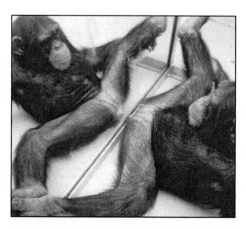

For a long time, researchers thought only humans and higher apes could recognize themselves in mirrors. Curious chimpanzees use mirrors to look at otherwise inaccessible body areas.

The Biami of New Guinea reacted emotionally to their first look into a mirror, with terror, wonder, and comprehension. Within days, however, they were using them to groom themselves. (This is actually a photo of a Tiwi Islander from Australia.)

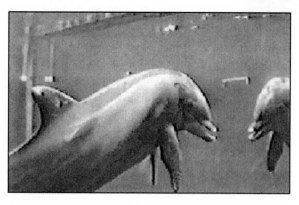

Now it appears that dolphins and elephants may also know they are looking at themselves in the mirror. Here the dolphin Presley gazes into his own eyes.

Chapter 10

THE VANITY BUSINESS

We like to see ourselves. . . . It is seldom enough to place mirrors in dressing rooms, bathrooms, entrance, halls, and bedrooms. There is a place for a mirror in almost any room.

WILLIAM LAWRENCE BOTTOMLEY,
"MIRRORS IN INTERIOR ARCHITECTURE," 1932

Every girl of fifteen has put the same question to a mirror: "Am I beautiful?"

ANAIS NIN DIARY ENTRY, 1937

As THE CELESTIAL variety were reflected by the 200-inch Palomar mirror, Hollywood stars found their images in flat, everyday reflectors, as did just about everyone else in the developed world. Once a luxury of the rich, mirrors were everywhere, thanks to industrial methods of glass production and more efficient silvering processes. "Not until the last few decades," wrote a commentator in 1932, "had there been any appreciable advance in the production of reflecting surfaces. . . . The art of manufacturing mirrors, once as closely guarded as an alchemist's secret, is today a science of quantity as well as quality production."

In the middle of the nineteenth century, mirror quality was still notoriously uneven. A young woman named Maria Daly, after staying overnight in a room with a bad looking glass, complained in her diary, "I looked so old and ugly that I felt distressed." A contemporary photographer observed, "How few of us have a perfect idea how we look." For those with imperfect mirrors, that was true.

In the span from 1850 to 1950, mirrors improved dramatically as the mass production of large sheets of glass advanced. Here is a contemporary description of glassmaking at Chance Brothers for London's Crystal Palace of 1851:

> When the requisite [molten glass] is taken from the furnace by the blower, it is blown into a spherical form. . . . It is then, after having been reheated in the furnace, swung above the head and below the feet of the workman, until it assumes the form of a cylinder. The workman stands upon a stage opposite the mouth of the furnace, with a pit or well beneath his feet, 6 or 7 feet in depth. He swings and balances . . . first above and then beneath him, until it gradually expands.

Once the glass was 4 feet long, it was converted to a tube, then slit down the side and "ironed" smooth with a wooden tool before annealing. To make larger pieces, manufacturers poured molten glass onto a table mold, rolling it out and producing rough glass that then had to be laboriously ground and polished. In other words, very little had changed in glass production since the time of Louis XIV.

As the process was industrialized in the late nineteenth century, compressed air and machinery replaced the sweaty laborer, but the basic process remained the same until 1896, when a Pittsburgh inventor named John Lubbers made a machine that could pull a cylinder of glass as high as 50 feet tall and 3 feet wide. A few years later, the Belgian engineer Emile Fourcault and an American, Irving Colburn, independently developed a process to draw molten glass up in thin sheets, using a metal "lure" and rollers. Such sheet glass was imperfect and relatively thin, but it was useful for windowpanes and small, cheap mirrors, which flooded the market. Larger plate glass still had to be ground on both sides, though this process, too, was industrialized. Pittsburgh Plate Glass and Toledo Glass Company (later Libby Owens-Ford) led the way in the United States, while Pilkington Brothers flourished in England.

Silver was deposited on the glass by mixing silver nitrate and ammonia with caustic soda such as sodium hydroxide and a reducing agent, usually inverted sugar. "Rocking tables" vibrated the solution so that it deposited silver evenly on the glass. Although mercury no longer poisoned mirror workers, explosions were a common hazard, because a silver-ammonia and sodium hydroxide solution can form volatile "fulminating silver." Consequently, all ingredients were usually combined only at the last minute, but mirror-makers still occasionally lost an eye to an unexpected blast. When a spray method was developed in 1940, the chemicals came simultaneously from separate nozzles. Once the silver was deposited on the glass, it was covered with protective copper, then paint or shellac.

Windows and Mirrors for the Modern World

The mass production of glass and mirrors furthered a major cultural shift exemplified by the Roaring Twenties in the United States. This modern sensibility valued consumption, leisure, entertainment, thrills, upward mobility, youth, image, and sex. Although humans had always enjoyed such pleasures, they were usually balanced by a self-restrained work ethic and recognition of the wisdom gained by a long life. Increasingly, in a society glittering with reflective surfaces, superficial values triumphed. Artists of the era frequently portrayed women facing their full-length mirrors.

By the turn of the century, every household in America had at least one and probably several mirrors. The 1897 Sears Roebuck catalog offered 10-by-10-inch mirrors in oak frames for 50 cents apiece, 16-by-16-inch mirrors for $1.35. "No house is complete without a number of small mirrors which are handy in so many rooms," the copy advised. The catalog also offered a heavy 18-by-40-inch French plate-glass mirror in a more ornate frame for $6.50, a toilet table with an oval mirror for $8.75, elegant hall trees (combination hat racks/mirrors/seats) for entranceways (up to $17.50) and sideboards and ice refrigerators, all with large built-in mirrors at eye level, for as much as $50. Sears customers could also buy a "Search Light" kerosene lantern with a concave tin mirror projector for a couple of dollars.

In the early twentieth century, advertisers used mirrors to tout their goods. Oval pocket mirrors, usually with attractive women pictured on the back, promoted shoes, dry goods, gloves, chocolates, books, and (of course) Coca-Cola. Naked women appeared on pocket mirrors advertising products for men, such as cigars and digestive aids. The back of one clever mirror promoting International Life Insurance Company showed a mother comforting her child, with the words: "IF THE MAN ON THE OTHER SIDE SHOULD DIE WOULD HIS FAMILY BE PROVIDED FOR?" Other large mirrors featured advertising painted right on their surfaces.

The availability of bigger, cheaper glass meant a proliferation of shop windows, in which people could look at themselves as well as the merchandise. "Our minds are full of windows," John Wanamaker observed in 1916. "Show windows are eyes to meet eyes." Wanamaker, who made his first fortune supplying Civil War army uniforms, was a department-store pioneer, opening consumer emporiums in Philadelphia and New York, where goods enticed shoppers through "towers of glass," as a distressed Henry James put it in 1904. James objected to "window upon window, at any cost."

Two years earlier, his fellow novelist Theodore Dreiser had described the feverish yearnings shop windows inspired: "What a stinging, quivering zest they display, stirring up in onlookers a desire to secure but a part of what they see, the taste of a vibrating presence, and the picture that it makes." Part of that "vibrating presence" was the self-reflection of the shoppers, who could see their translucent mirror images projected onto the wonderland of saleable goods.

Fifth Avenue in New York became a consumers' paradise. "It is not the things which Fifth Avenue contains that give its greatest interest," a journalist wrote in 1906. "It is the moving, pulsating life which it bears along in its great current. It is like a splendid river filled with all sorts of craft engaged in ministering to the pleasure or the needs of the world." That river of humanity gawked and gazed at its own reflection in the shop windows.

With the addition of real mirrors, the store window became a "splintering maze of glittering crystal," as one retailer described it. In 1897, three years before he published *The Wizard of Oz*, L. Frank Baum put out the first issue of *The Show Window*, a magazine devoted to the art of enticement, full of information on gadgets such as moving electrical displays of revolving stars, incandescent lamp globes, and mechanical butterflies, along with ads for everything from corsets to "Frink's window reflectors." The creator of Emerald City understood the connection between mirrors, illusions, and human desires.

By 1912, lifelike mannequins with real hair and adjustable limbs mirrored the shoppers outside. "Associate the goods with people and events," one retailer advised. They should imply luxury, glamour, and adventure. Some female mannequins, wearing nothing but underwear, attracted crowds and even sparked street riots as people viewed their fantasy-reflections through the glass.

In department stores' interiors, as well as in restaurants and hotels, mirrors multiplied along with innovative electrical lighting. Mirrored elevator doors allowed people to admire themselves while they waited. Imitating Parisian shops, John Wanamaker, and later other retailers, installed glass cases with mirrored backs to display goods. Mirrors sheathed columns and covered walls. Children's departments sported huge green dragons and giant clowns reflected in multiple mirrors and shiny stars suspended from the ceiling. As one Wanamaker decorator noted, "People do not buy the thing, they buy the effect. . . . Make the whole store a brilliant showplace."

Meanwhile, paraboloid searchlight mirrors were turning the outdoors into a brilliant showplace as well. Floodlight impresario W. D'A. Ryan began in 1906 with experiments in Nahant, Massachusetts, where he lit up clouds of steam and a huge U.S. flag. The next year, he played searchlights over Niagara Falls, and in 1908 he lit the Singer Building in New York City, at that time the tallest skyscraper in the United States.

Paris and Berlin, already known as "cities of light," used floodlights on public buildings and monuments. Soon afterward, New York directed searchlights on the Statue of Liberty, and when World War I ended in 1919, colored floodlights played on a triumphal jeweled arch thrown over Fifth Avenue.

Downtown streets became "white ways" in the evening as modified searchlights were installed as streetlights. In combination with regular electric light bulbs illuminating shop windows and theaters, these transformed the urban twilight. "Ah, the promise of the night," Dreiser wrote at the turn of the century. "Says the soul of the toiler to itself, 'I shall soon be free. I shall be in the ways and the hosts of the merry. The streets, the lamp, the lighted chamber . . . are for me.'"

The paraboloid mirror also provided light for an ever-more-mobile society. In 1917, two writers surveying the use of searchlights wrote,

"The automobile headlamp is another phase of light projection upon which much time and energy has been devoted during the last few years." But while headlights used paraboloid mirrors, few cars came equipped with flat rearview mirrors. In her 1909 book, *The Woman and the Car*, Dorothy Levitt advised female drivers to carry a mirror with a handle in the side pocket of the car "to occasionally hold up to see what is behind you"—and just in case you didn't like what you saw, she suggested carrying a small revolver as well.

Rearview mirrors received favorable publicity in 1911 when Ray Harroun employed one on his single-seater Marmon Wasp during the Indianapolis 500 while other drivers relied on their mechanics riding with them to warn of approaching competitors. After Harroun won the race, rearview mirrors became much more popular, and by the 1920s they were standard equipment on most cars. Great Britain made them compulsory in 1932, the United States not until 1966.

New Thought at the Funhouse

As Americans traveled more by rail, streetcar, and automobile, they flocked to a series of world's fairs that dazzled them with more mirrors and light spectacles. The Chicago World's Columbian Exposition of 1893—where the Yerkes telescope tube and mounting were exhibited—marked a new era in American life. Like the Paris Exposition of 1889, the fair, which attracted 14 million people, featured a Hall of Mirrors, as would fairs held in the next twenty-five years. The Chicago fair catered to the American fascination with glittering Oriental themes. "Little Egypt" wore a scanty sequined outfit that threw off sparkles of light as she danced the hootchy-kootchy.

As a character visiting the Chicago fair in a contemporary novel put it, "You're feelin' bewildered with the smells and sounds and sights, always changin' like one o' these kaleidoscopes." For rural folk who couldn't make it to the funhouse, it traveled to them in 1896, when "The Crystal Maze" toured by rail.

The expositions led directly to the full-time public amusement parks on Coney Island, where factory workers and secretaries could enter fantasy mirror-worlds like Steeplechase, Luna Park, and Dreamland, all opened by 1904. At night, millions of electric light bulbs transformed Coney Island into a glittering wonderland, as the Russian writer Maxim Gorky observed: "A fantastic city all of fire suddenly rises from the ocean into the sky . . . Golden gossamer threads tremble in the air. They intertwine in transparent, flaming patterns, which flutter and melt away in love with their own beauty mirrored in the waters."

Coney Island reveled in illusion. In the distorting mirrors of its funhouses, everyday reality was suspended, strict societal norms turned upside down, and pleasure became its own end. Jets of air blew women's skirts over their heads. Violent rides threw men and women together. In the mirror mazes, laughing couples clung to one another as they stumbled through the confusing hallways, startled to confront themselves in a mirror they didn't know was there.

Around the country, other amusement parks—originally quiet getaways sponsored by streetcar companies to encourage weekend travel to the end of the line—turned into mini–Coney Islands. Americans pursued pleasure as never before, and a new brand of religion/psychology called New Thought encouraged them. "To the emancipated soul there is nothing common or unclean," wrote New Thought evangelist Eugene Del Mar in 1903. "There is no necessary postponement of happiness." He railed against the "idea of duty or self-denial."

The Sexy, Made-Up 1920s

Self-denial fell completely out of favor in the 1920s, when casual sex became acceptable and exciting. One motion picture advertised "brilliant men, beautiful jazz babies, champagne baths, midnight revels, petting parties in the purple dawn, all ending in one terrific smashing climax." By this time, most automobiles had rearview mirrors, which proved useful for trysting couples to watch for interlopers. A juvenile-court judge called the auto a "house of prostitution on wheels," and there were 23 million of them by the end of the decade.

New Thought philosophy reached its apotheosis in a self-help craze that swept the country in 1923. The French guru Emil Coué established institutes across the United States, teaching people to repeat the mantra,

"Day by day in every way I am getting better and better," while gazing earnestly at themselves in mirrors.

And they could find mirrors almost everywhere. In clothing stores, men and women could see themselves from all sides in angled three-part floor-length mirrors. By the middle of the decade in the United States, there were more than 20,000 moving-picture theaters, many of them known as "pleasure palaces," with sumptuous, mirrored entrance halls, lobbies, and lounges replete with marble, chandeliers, and gold leaf.

George Rapp, who designed many of these theaters, called them "shrines to democracy," where the lowest ticket-holder could feel like royalty. He gloried in "the cloistered arcades, the depthless mirrors, and the great sweeping staircases," all part of "a celestial city—a cavern of many-colored jewels, where iridescent lights and luxurious fittings heighten the expectation of pleasure."

When they weren't admiring themselves in movie-theater mirrors, flappers could watch themselves do the Charleston, the bunny hug, or the grizzly bear in the wall mirrors of huge dance halls, filled with "bright lights, jazz music, and continuous novelty [to] attract by the thousands young people," as a social worker observed in 1924. Stenographers and mechanics, mill workers, and domestics could forget their daily routines and see themselves reflected as happy dancers in a wild party.

New cultural attitudes had been bubbling since the turn of the century, but the changes of the 1920s were so dramatic that in 1931 Frederick Lewis Allen wrote a book called *Only Yesterday: An Informal History of the Nineteen-Twenties*. In the "Prelude," he painted a portrait of pre-1920s life. "Although the use of cosmetics is no longer, in 1919, considered *prima facie* evidence of a scarlet career, and sophisticated young girls have already begun to apply them with some bravado, most well-brought-up women still frown upon rouge. The beauty-parlor industry is in its infancy."

That changed radically during the 1920s, when the pursuit of beauty became a weekly ritual even in rural America. By the end of the decade, there were nearly 40,000 beauty shops operating across the country. In row upon row, women sat in front of big mirrors where they could watch themselves being remade. The beauty-parlor culture provided a safe place for women to share joys and sorrows without any real face-to-face interaction. Everything took place in the mirror, including conversations and eye contact. The beautician became a counselor as well as hairstylist.

More and more, the ideal of beauty was determined by celebrities, fashion shows, and advertisements. The first Miss America pageant took place in 1921. Allen noted that cosmetics were sold with "glowing testimonials—often bought and paid for—that the advertised product was used by women of fashion, movie stars, and non-stop flyers."

By the 1920s, women could literally "make up" what they wanted their faces to look like. "The vogue of rouge and lipstick," wrote Allen, "which in 1920 had so alarmed the parents of the younger generation, spread swiftly to the remotest village. Women who in 1920 would have thought the use of paint immoral were soon applying it regularly as a matter of course and making no effort to disguise the fact." In addition, "a strange new form of surgery, 'face-lifting,' took its place among the applied sciences of the day." Like many other fashion trends, face-lifts originated in France. By 1930, cosmetics was a $2-billion-a-year industry, and on average every adult American woman applied a pound of face powder annually, along with eight rouge compacts.

The application of makeup became a public ritual. In her pocketbook, every woman carried a compact that snapped open to reveal a small mirror in which she could powder her face, apply lipstick, and adjust her hair. Compacts were made to imitate cigarette cases or golf balls; some were made as detachable belt or shoe buckles. Though some critics objected to such public displays of vanity, Dorothy Cocks defended it in *Etiquette of Beauty* in 1927. "Immorality is doing something one is ashamed to be caught doing. If we rouge our cheeks and powder our noses before every mirror we meet in public, there can be no turpitude in that!"

Women used their small portable mirrors for brief touchup efforts, but they spent hours in front of their boudoir mirrors applying their basic daily makeup. In 1923, Celia Caroline Cole described the elaborate procedure in *Delineator*, one of many women's magazines that offered beauty tips:

> First the cleansing cream, soft, soft as whipped cream, to keep the skin smooth and fine-grained and clean. Next a bottle of skin tonic to tone up the skin and bring out the natural freshness. . . . Then skin-food cream to

pat under the eyes . . . and all over thin faces and necks. Then a small jar of astringent cream for the pores. . . . Then the foundation cream. . . . Next the little box or bottle of rouge. Oh, yes you do! Most of you, anyway. . . . Now an eyebrow-brush to clean the brows and lashes of powder. . . . There are eyebrow pencils nowadays of brown as well as black. . . . And, finally, the dot with the pencil in the outer corner of the eye to give the tilted-up look. . . . Get a good, fine, well-made powder and put it on with a puff or pad and it will *stick*.

And then there was eyelash-curling, meticulous lipstick application, and hairstyling, not to mention putting on earrings and pearl necklaces—all to be checked in the vanity mirror with the aid of a hand mirror, which was part of the complete dressing table.

For Men: Just Don't Call It Makeup

Men, too, checked their appearance in mirrors, as they had since reflective surfaces were invented. But they applied cosmetics in strict privacy, or they let their barbers do it after a shave, cut, and shampoo. "Suppose a man lacks color," the *Barbers' Journal* wrote in 1912, "the barber rubs his cheeks with rouge or a liquid preparation. . . . Suppose that the lips are not red enough or the skin on the lips is not soft, the barber rubs them with a lipstick." The barber would pluck and repaint male eyebrows, tint the mustache, or powder an alcoholic's red nose, then let the client judge the result in a mirror. "Having had his head and face beautified one would think that the man would be satisfied," the writer concluded, but he might go on to the wig shop. "Many men wear wigs, though women do not suspect it. Many men dye their hair, too."

In the twentieth century, however, more and more men were shaving at home. In 1895, King Gillette, searching for a disposable product that would mean repeat sales, came up with the double-edged safety razor for home use. Until then, men relied primarily on barbers to trim their beards. In the twentieth century, beards basically disappeared.

The new style was due in part to Gillette's relentless advertising. "The men who uphold the standards of American sport today are clean men—clean of action and clean of face," a 1910 Gillette ad stated, depicting beardless, smiling baseball stars like Honus Wagner. "Be master of your own time. . . . No stropping, no honing." The new germ theory portrayed beards as a breeding ground for an estimated 2 million "misanthropic microbes," as one scientist of the era estimated.

In the 1920s, men spent even more time at their mirrors, shaving daily and trying to slick their hair down like Rudolph Valentino or George Raft. "I've had a number of men come in recently and ask where they could get their lashes drooped or their brows weeded out and run into nice curves," a Cleveland barber noted in 1920. Carl Weeks, who produced the Armand line of cosmetics (named to sound French and chic), discovered that many men were buying his heavy cake makeup to cover beard stubble. In 1929, he introduced the Florian line for men with dashing names like Brisk, Dash, Vim, Keen, and Zest, in aggressive red-and-black containers to avoid appearing effeminate. Weeks insisted that his products were "for he-men with no women welcome no-how," but men still had to look into mirrors to apply Vim and Zest. By 1937, men were spending $200 million per year in barbershops—about the same amount women spent in beauty parlors—as well as buying shaving creams, blades, and other cosmetics.

Hollywood Vanities

The Depression ruined the overextended Carl Weeks, but for Elizabeth Arden, Helena Rubinstein, and Max Factor—all immigrants who helped create the cosmetics craze—the economic downturn of the 1930s only meant more opportunity. Canadian-born Florence Graham, the daughter of poor tenant farmers, changed her name to Elizabeth Arden when she opened an elite beauty shop on Fifth Avenue in 1909, parlaying it into a national cosmetics line. Rubinstein, whose father was a wholesale food broker in Krakow, sold cold cream in Australia, moved to London in 1908, opened a salon in Paris in 1912, and moved her salon to New York after the Great War commenced. The two competitors referred to one another as "that woman."

Fleeing the Russian pogroms in 1904, the wigmaker and cosmetician Max Factor moved four years later to Los Angeles. There he established a wig business and makeup studio specializing in beautifying movie stars. He introduced Society Makeup for the public in 1920, and by 1927 his cosmetics had achieved national distribution, thanks largely to his entrepreneurial children. Factor's Pan-Cake foundation, a modified water-soluble screen makeup introduced in 1938, sold extremely well, particularly because his ads stressed its use by famous actresses.

Max Factor, "makeup artist to the stars," exploited the obsession with glamorous Hollywood figures even as they exploited him through tie-in promotions in which the stars used his cosmetics in front of mirrors while touting their latest film. Elaborate home dressing-room tables were by now known as Hollywood vanities. Sex appeal sold makeup. In 1938, the cosmetics firm Volupté introduced two lipsticks. The glossy-red Hussy shade outsold the more demure Lady five-to-one.

Cosmetics advertisements in the women's and beauty magazines frequently featured mirrors, as in an innovative Elizabeth Arden ad for Skin Deep Milky Cleanser showing a woman's face reflected upside down in a mirror, as in a pond. In 1937, Arden paid for a mural painting called *A Pageant of Beauty*, in which women from Cleopatra to the modern era were shown gazing into mirrors.

A typical advertising come-on asked, "DO YOU ANXIOUSLY CONSULT STORE WINDOWS AND VANITY CASES AT EVERY OPPORTUNITY?" The mirrors, movie cameras, and bystanders that appeared in many such advertisements reminded women that they were constantly being scrutinized. "DO YOU YEARN FOR A CLEAR COMPLEXION?" a soap ad asked. In the illustration, a woman holding a hand mirror peers into her dressing-table mirror, which shows not her but a fantasy couple gazing at one another adoringly.*

The magic mirror was back—but in a twisted form. Instead of scrying the future, it showed people their worst fears or most desired fan-

---

*In 1933, the zany Marx Brothers mocked the American mirror craze in *Duck Soup*. While running away from Groucho, intruder Harpo smashes into a large wall mirror, breaking through into another room. To avoid being recognized, Harpo, who has dressed in the same nightgown and cap as Groucho and has put on a fake mustache, turns around and acts as if he is Groucho's mirror image. In a hilarious scene, Groucho squats, dances the Charleston, and does funny walks, all of which Harpo mirrors. They put on the same hat, too, but when they both bow, Harpo drops his hat. By this time, Groucho has bought into the ruse and just hands his double his hat.

tasies. In the Freudian 1930s, *Vogue's Book of Beauty* advised that a woman who failed to update her looks "destroys those potential personalities that psychologists tell us are lurking behind our ordinary selves." A cosmetics industry spokesman claimed that "many a neurotic case has been cured with the deft application of a lipstick." Women literally talked about "putting on a face" before going out.

Many Depression-era ads played on women's fear of aging. A 1932 ad for Dorothy Gray cosmetics featured a sorrowful woman gazing into her mirror, with the headline: "*'ARE YOU AFRAID OF TIME?'* YOUR MIRROR ASKS. . . ." Another pitch showed a worried woman's face in a mirror with the headline, "IF YOUR HUSBAND WERE AS FRANK AS YOUR MIRROR, WOULD HE SAY: 'DON'T GROW OLD, DEAR. . . .'" In a youth-obsessed culture, the elderly became pariahs, particularly when they tried to be stylish. "To me," wrote the actress Blanche Bates, "there is nothing more painful and pathetic than the sight of a grandmother—ancient enough to be one anyway—mincing along a public street with her cheekbones buried under a mass of crude vermilion, mascara dripping from her eyelashes, and her mouth a study in scarlet." Of course, such makeup was fine for the youthful actress.

Turn Around and There You Are

Regardless of their age, Americans were seeing themselves more and more often in mirrors. Reflective surfaces cast back images in subways, stores, airplanes, restaurants, dancehalls, and movie palaces. In high schools, universities, home extension services, and 4-H clubs, students of all ages got "beauty lessons" in front of more mirrors. As women entered the workforce, they sought mirrors frequently; one stenographer explained that she applied rouge and powder four times a day.

Architects put mirrors everywhere in the home. In 1928, Macy's featured a sample room with mirrors on floor, walls, and ceiling, so that visitors who entered could see themselves from ten angles. "Yesterday afternoon," the *New York Times* reported, "the crowds in front of the room had to be held back."

In a 1932 article, William Lawrence Bottomley recommended a mirror for "almost any room, provided it is treated not only as an expression of personal vanity, but as an important decorative element." A well-placed mirror could, for instance, redirect sunlight so that "one could almost believe that a mirror persuades the sun to enter."

Bottomley reserved his most enthusiastic use of mirrors for the bedroom and bath. Decorated mirror headboards, combined with mirrored ceilings, could make for "startling" results. Dressing tables, usually with a triple hinged mirror arrangement, were de rigueur, while tall chifforobes frequently featured full-length mirrors on their doors. An illustration of a lavish bathroom showed a tub surrounded on three sides by mirrors, "giving the room a rather gay and spacious appearance."

Around the world a mirror craze dazzled consumers. The French were even more lavish with mirrors and mirrored furniture than the Americans—not surprising, given the Gallic love affair with reflection since the days of Louis XIV—and they influenced other European styles. The Japanese, too, were obsessed with mirrors and cosmetics.

By the time World War II began, mirrors had become a commonplace part of modern life, and the vanity they encouraged was taken for granted. Soldiers were issued mirrors in their kits, but many also received gift cosmetic sets from family and friends. Pinups of Betty Grable and Rosalind Russell reinforced Hollywood values.

Even the women who took men's places in factories continued to take pride in their appearance; Rosie the Riveter wore lipstick. Cosmetics, considered "morale boosters," were briefly rationed in 1942, but social and business pressure reversed the ruling. A typical lipstick ad showed an attractive female pilot climbing into her cockpit and stressed "the precious right of women to be feminine and lovely—under any circumstances." The privilege of applying lipstick while examining oneself in a mirror was "one of the reasons why we are fighting."

In the postwar years, the use of cosmetics and mirrors spread further. "Women are now constantly buying additional shades of lipstick, even *before* they completely use most of them up," Max Factor Jr. observed. Avon representatives knocked on doors in new suburbs, with sales reaching $87 million by 1956. Movie tie-ins and fads continued. Charles Revson, the son of Jewish immigrants, introduced the Revlon lipstick called Fire and Ice in 1952, featuring a sultry woman in a slinky sequined gown. "For you who love to flirt with fire . . . who dare to skate on thin ice," the copy promised. Sex sold cosmetics and sent more anxious women—and increasing numbers of teenagers—to consult their mirrors.

Returning soldiers, conditioned by all those gift toiletry sets, bought $50 million of them (each with a mirror) in 1945. "Cosmetic sets have now become the popular gift for the man in the family," noted Herbert Rattner in a 1946 issue of *Hygeia*, illustrating his article with a cartoon of a man in polka-dot underwear powdering himself in front of a vanity mirror. While men didn't want to use effeminate face powder, they didn't mind dabbing on "after-shave talc." A 1955 *New Yorker* cartoon mocked such male vanity, depicting an actor in a dressing gown admiring himself in five surrounding mirrors. A man peeking around the doorway asks, "Tremaine, could I see you for a moment—alone?"

Mirrors, fashions, and cosmetics even survived hippies and the women's liberation movement of the 1960s and 1970s. The counterculture, with its bell-bottoms, tie-dyes, and long hair, had its own distinctive fashions. Mirrors did not go out of style. Punk stylists used them in later decades to admire their nose rings or spiked pink-and-purple hair.

The postwar boom also produced more glass. In the early 1950s, sleek glass-walled skyscrapers such as New York City's Lever House flashed and shimmered, providing even more windows for reflection. In 1959, the British manufacturer Pilkington revolutionized the manufacture of glass with the float-glass process, in which a continuous river of molten glass floats atop liquid tin, which has a lower melting point than glass. Gradually, the glass is cooled and annealed so that by the end of the line, it can be cut into huge sheets and stacked. Unlike traditional plate glass, the new variety did not need to be polished. Huge float-glass plants soon spread around the world.

Mirrored glass buildings made of float glass shot into the sky in the 1980s in the wake of the energy crisis, allowing in light while reflecting damaging ultraviolet rays. The skyscrapers acted like gigantic mirrors for clouds, sky, and other buildings, while office workers could look out. On the first floor, the one-way mirror effect could be amusing, as employees observed passers-by applying lipstick or combing their hair while staring intently at the unseen workers inside.

But the postwar preoccupation with celebrity and personal appearance was simply an extension of the obsessions of the 1920s and 1930s. Three portraits of teenage girls looking into mirrors capture different aspects of youthful initiation into a culture dominated by image and sex. In 1932, a British artist, Gerald Brockhurst, showed a naked sixteen-year-old facing herself in a three-part vanity mirror, gazing with solemn wonder, fear, and awe at her newly lush, sensual body. In 1946, the French painter Balthus portrayed a barely pubescent twelve-year-old striking a sexy pose while gazing raptly at herself in a hand mirror.

In 1954, Norman Rockwell's *Girl at the Mirror* captured an eleven-year-old rural American, wearing a white slip and sitting on a stool, elbows on her knees, face resting on her hands, a magazine open on her lap with a full-page spread of the actress Jane Russell. The girl, a picture of innocence with her hair parted in the middle and caught up in a braided bun in back, looks wistfully at herself, clearly wondering whether she will ever grow up to be such a sophisticated, sexy beauty.

Mirrors have always been ambivalent servants. Their magical surfaces reveal the truth, permitting people to see themselves as they are. Yet they have their sinister side, for the same reason. And sometimes they distort reality. The classic 1948 Orson Welles film *The Lady From Shanghai* ends in a dramatic scene in the funhouse of San Francisco's Playland amusement park, with distorting mirrors leading to a Magic Mirror Maze. Gorgeous, faithless Rita Hayworth and her disabled, bitter lawyer-husband shoot at one another's images, only to hit and shatter mirror after mirror, until finally they actually hit and kill one another. The character played by Welles stumbles out of a revolving door, where a sign says "Fun! Fun! Fun!" He concludes: "Everybody is somebody's fool. The only way to stay out of trouble is to grow old, so I guess I'll concentrate on that." And it might be best, we surmise, for him to avoid mirrors in the process.

Chapter 11

CELESTIAL RADIOS, DIVINE X-RAYS

A few archaic photons . . . left some remote quasar to travel through the void, without hitting anything, for almost as long as time itself had existed—a nice demonstration of how empty the universe is—until, some two or three times older than the earth, they ran into a mirror.

RICHARD PRESTON, *FIRST LIGHT*

ANOTHER TRANSFORMATION took place in the 1920s. Radios invaded American homes, binding the country together with invisible, long electromagnetic waves and contributing to the self-conscious culture that led people to the mirror. Cosmetics advertisements on the radio accounted for $3.2 million annually by 1930. But who would have thought that radio technology would lead to a different kind of mirror that would change our view of the universe?

In the fall of 1930, Karl Jansky, a twenty-five-year-old engineer working for Bell Telephone Laboratories on a former potato farm in Holmdel, New Jersey, assembled a strange contraption of brass pipe that looked like a mutated biplane wing supported by four Ford Model T tires. Driven by a 1/4 horsepower motor, the 95-foot-long antenna ponderously made a complete revolution every twenty minutes, scanning the sky for radio static at a 14.6-meter wavelength called "short-wave" in comparison with the "long" radio waves (over 200 meters between wave peaks) first used for radio transmission.

Ham radio operators of the 1920s had discovered that such short-wave transmissions worked amazingly well for overseas contact. The earth's upper atmosphere—the ionosphere—acts like a huge spherical mirror for such wavelengths, which bounce back to earth. In 1929, Bell began offering expensive shortwave transatlantic radiotelephone service, but mysterious interference frequently broke up the voice transmission. Jansky was assigned to investigate.

By January 1932, he had identified thunderstorms as the primary culprits, but there was in addition a mysterious low-level source "that changes direction continuously throughout the day, going completely around the compass in 24 hours." He thought it came from the sun but was puzzled that the "very steady hiss type static" arrived earlier as the days and weeks marched by. During a partial solar eclipse on August 31, 1932, with the moon blocking radiation from the sun, the hiss was as strong as ever, so the sun could not be the source.

The baffled Jansky discussed the mystery with Melvin Skellett, a fellow radio engineer at Bell Labs who also happened to be a graduate student in astronomy at Princeton. When Skellett learned that the hiss had moved exactly one day earlier over the space of a year, he immediately concluded that it was obeying sidereal, or star time, rising four minutes earlier every day with respect to the sun. In other words, Jansky's hiss was coming from outer space.

"The stuff, whatever it is, comes from something . . . outside the solar system," Jansky wrote his parents in December 1932. "It comes from a direction . . . *towards which the solar system is moving.*" Without attracting much attention, Jansky announced his conclusions at a small meeting of radio scientists. The *New York Times* picked up the story on its front page on May 5, 1933: "New radio waves traced to center of the Milky Way . . . no evidence of interstellar signaling."

More urgent news of Nazis and the deepening Depression pushed Jansky back into relative obscurity. Still, he yearned to pursue his discovery. Jansky knew that he was pioneering a new type of astronomy that would require a new kind of mirror, so he petitioned his boss for money to build a 100-foot-diameter dish, to no avail. Bell Labs kept him so busy with mundane projects that he never pursued his research much further. Suffering from Bright's disease, he died in 1950 at age forty-four.

Grote Reber's Radio Mirror

During the Depression years, a few intrigued scientists paid attention to Jansky's findings. Grote Reber was twenty-two when he read Jansky's articles in 1933. "It was obvious that Jansky had made a fundamental and very important discovery," he later wrote, but it was equally obvious that he needed better equipment to examine radio waves from space. "The only feasible antenna would be a parabolic reflector or mirror. By changing the simple focal device it would be possible to tune the mirror over a very wide frequency range."

So in four months of 1937, in his spare time (he worked for a radio manufacturer in Chicago), Reber built himself a 31-foot-diameter paraboloid dish out of wood, covering it with a skin of galvanized iron fastened with bolts. At the focal plane, 20 feet above the dish, he fastened what looked like a 50-gallon drum containing his detectors. Reber's dish could only swivel up and down, so that it was a transit radio telescope relying on the earth's rotation to bring objects into view.

It is important to understand how mirrors reflect "invisible light" on the long side of the narrow optical window of electromagnetic radiation. The light we see arrives in extremely small wavelengths of just billionths of a meter.* Astronomical mirrors accurate to within a fraction of a wavelength must therefore be incredibly well polished and accurate. Because radio waves are nearly 1 million times longer than visible light, they don't require such accurate mirror surfaces. In fact, the longer the wavelength, the less accurate the mirror can be, which is why some radio dishes look as if they were made out of chicken wire. As long as the spaces between the wires are substantially shorter than the wavelengths, most of the radiation is reflected.

Yet the shorter the radio wavelengths, the more finely meshed the wire must be, ultimately requiring solid surfaces such as Reber's. Because we listen to our radios, we tend to think of radio-wave receptors as "big ears," but they are also "big eyes" that "look" at the universe at a particular wavelength. You can tune into them and listen to the sta-

---

*You might want to review the chart of the electromagnetic spectrum in Chapter 8, showing the range from long radio waves up through the shortest wavelengths of x-rays and gamma rays.

tic, but it's better to record the waves more accurately with some visual means. (Since all forms of electromagnetic radiation travel at the speed of light, the frequency of waves gets higher as the wavelength gets shorter. References to "wavelength" and "frequency" are therefore interchangeable.)

For radio astronomers, one advantage of a paraboloid radio mirror is that it focuses the waves to one small circular plane, so that it is easy to change the detector, which selectively "looks" at a particular wavelength, translating the results into amplified electrical signals that can be graphed and mapped.

So, the good news about radio mirrors is that they are versatile, don't need to have perfectly polished surfaces, and can be used over a wide range of wavelengths by changing the detector. But there is also bad news. In order to "see" as well as the normal human eye, a radio mirror would have to be nearly a million times bigger.

Grote Reber knew that his 31-foot mirror wouldn't be able to gather many radio waves, but the shorter the wavelength, the greater the "invisible light grasp" he would achieve. In 1938, he started with 9.1 centimeters, pointing his telescope at various parts of the Milky Way, bright stars, the sun, and the planets. Although radio mirrors work equally well day or night (radio waves penetrate clouds and blue skies), he discovered that working at night avoided interference from the spark plugs of passing automobiles or amateur pilots who sometimes buzzed him.

So Reber stayed up from midnight to 6 A.M., then drove 30 miles to Chicago, where he designed radio receivers, came home, made supper, slept till midnight, then started all over again. But he found only irregular fluctuations rather than meaningful patterns. Reber then tried 33 centimeters. Nothing. Finally, with a more sensitive detector, he tuned in the radio emission from the Milky Way. With the addition of an automatic recorder (he had been graphing by hand), Reber conducted his first sky survey in 1941. With further improvements, using a longer wavelength of 1.87 meters, he eventually identified radio waves from the sun, Sagittarius (in the middle of the Milky Way), Cygnus, and Cassiopeia. In 1944, he submitted a paper along with the first map of the radio sky to the editor of the *Astrophysical Journal*, Otto Struve, who could find no reviewer willing to defend the paper. Convinced that

Reber's work deserved attention, and low on wartime submissions, Struve published it anyway.

Radar Comes of Age

As Grote Reber was scanning the sky from his backyard in Illinois, the British prepared for a war in which reflected radio waves would play a crucial role. In 1924, British physicist Edward Victor Appleton used radio echoes to determine the height of the ionosphere. The first practical radar (an acronym for "radio direction and range") was invented in 1935 by another British physicist, and by 1939 England had established a chain of radar towers along its southern and eastern coasts. The Germans, too, had developed Würzburg radar detectors, twenty-five-foot radio dishes to direct antiaircraft guns and detect enemy airplanes.

Many young British scientists found themselves caught up in a life-and-death struggle, with Winston Churchill pressuring them to develop new and more effective radar. Radar works by transmitting a powerful directed radio pulse and then detecting echoes at the same wavelength. Knowing the speed of light at which the beams travel back and forth, it is easy to determine how far away the object is and, by moving the beam back and forth, its direction. The first radar emitted 10-meter waves from aerials atop 240-foot towers, but it soon became apparent that in order to miniaturize radar that could be carried on airplanes, the scientists needed to explore much shorter wavelengths.

Bernard Lovell and his team reduced the wavelength to 1.5 meters, then 10 centimeters, 3 centimeters, and finally 1.25 centimeters by the end of the war. The most effective way to direct a powerful radio beam was to use a paraboloid shape in the same manner as a searchlight, sending the beams from the focal plane, with similar receivers to detect the returning radiation. Thus, airplanes and ships became "mirrors" that bounced back radar beams, with mirrors as well to send and receive. Tucked under the wings of bombers, these portable radar units made it possible to see through clouds and target cities such as Hamburg and Berlin. In modified form, in tandem with a powerful searchlight flipped on at the last minute, they could spot U-boats at night.

Airplanes and ground targets responded to radar by "jamming" it with a barrage of interfering waves. U.S. pilots going into battle dumped bales of thin aluminum strips (known as chaff) to create millions of artificial radio mirrors to confuse enemy radar. "The bombers are reproducing!" one alarmed Würzburg operator exclaimed.

In England, Stanley Hey was assigned to analyze the jamming problem. In February 1942, coastal radar sites reported severe interference and feared an imminent German Luftwaffe attack. Hey realized that it wasn't the Germans but a large, active sunspot that was jamming the radar with radio waves of "an amazing intensity." He wrote up a classified military report. Two years later, Hey scrambled to design a new radar system to detect incoming V-2 rockets. "Transient echoes at a height of around 100 kilometers gave rise to false warnings," Hey later recalled. They turned out to be radar reflections from meteor trails.

Meanwhile, Dutch scientists in occupied Holland couldn't take an active part in the war effort, so they theorized. Jan Oort, who had read Grote Reber's articles, suggested to his colleague Hendrik van de Hulst that there might be emission and absorption lines in radio wavelengths just as there were for optical spectroscopy. Van de Hulst predicted that atomic hydrogen—the simplest and most widespread element in the universe—ought to emit at the 21-centimeter wavelength. It was uncertain whether anyone would ever observe it, though. Once every 11 million years, a hydrogen atom's single electron switches its direction of spin and emits or absorbs a tiny bit of energy at 21 centimeters. Interstellar gas is so thin that there is only one atom per cubic centimeter. Yet a few years later, radio astronomers found the line. There is so much cold hydrogen in space that it emits strongly enough to provide a way to map many parts of the universe.

Discovering the Radio Universe

When World War II ended in 1945, the radar wizards turned their full-time attention to radio astronomy, particularly in England and Australia. In 1946, Stanley Hey published his previously classified papers on radio waves from solar flares and radar reflections from meteor trails. At first, most research focused on the sun, which is a relatively weak radio transmitter. Astronomers were astonished to find that while the temperature of the sun was 6,000 degrees Kelvin at optical wavelengths, it was 1 million degrees Kelvin at 1.5-meter wavelengths.* The longer radiation came from high above the surface of the sun, in the corona, which is (amazingly) much hotter than the surface.

Using surplus military radar receivers at the 1-meter wavelength, Hey studied Reber's "cosmic static," which revealed an intense source in the constellation Cygnus, subsequently called Cygnus A. But exactly *where* in Cygnus? The newly fledged radio astronomers could not easily pinpoint sources in the sky, because at a wavelength of 10 centimeters they would need a paraboloid mirror some 1,000 feet wide just to see with the same resolution that the human eye has for regular light waves.

So they began to use interferometry. To understand how it works, imagine constructing a paraboloid mirror 1 mile wide. Of course, that is impractical. Now imagine instead placing two smaller mirrors 1 mile apart. You won't collect as many radio waves, so you can't see as far or as clearly as you would with a full-aperture 1-mile-diameter mirror, but by combining the waves and looking at the interference fringes—and using a complex mathematical magic called Fourier synthesis—it turns out you can pinpoint positions fairly well.**

In 1949 in Australia, John Bolton used a "sea interferometer," a radio telescope set atop a high cliff over the ocean. The sea acted as a mirror to create a virtual second telescope. With it, Bolton located three discrete radio sources, one of which he identified with the Crab Nebula—also known as M1, since it was the first nebula listed in the eighteenth-century French comet-hunter Charles Messier's famous catalog. Astronomers had

---

*"K" stands for Lord Kelvin (William Thomson, 1824–1907), who pioneered the research of extremely cold temperatures. 0 K represents *absolute zero*, theoretically the coldest state, in which matter has no energy and nothing moves. To convert kelvins into centigrade temperatures, simply subtract 273.15.

**Interferometry isn't quite as simple as that. The astronomers must bring the interfering waves together with precise timing. For radio waves, that is much easier than for short optical waves. In 1920, Albert Michelson had managed to use optical interferometry to determine the size of the star Betelgeuse by attaching mirrors on arms outside the 100-inch Mount Wilson mirror, but no one else had been able to make optical interferometry work. Radio astronomers had a much easier time of it.

270 | MIRROR MIRROR

identified it as the remains of a supernova explosion recorded by the Chinese and other cultures in 1054 C.E. This was the first hint that radio waves might be associated with unusually violent events and that the after-effects could somehow last another 1,000 years.*

At Cambridge University in England, Martin Ryle and Graham Smith used two confiscated Würzburg radio mirrors set 918 feet apart as an interferometer to get a better location for Cygnus A. In addition, they found an even stronger radio source, Cassiopeia A, too far north to be seen from Australia. By 1950, they had identified some fifty "radio stars." It seemed inconceivable that such strong radio sources should come from outside the Milky Way galaxy, so Ryle and Smith thought that these must be a kind of relatively near dark star, since there were no obvious visible stars in those locations.

In 1950 the world's most powerful radio telescope, with the biggest mirror, had been built not to look into deep space but primarily as a radar device to study the earth's ionosphere. At Manchester University, Bernard Lovell discovered that electric trams near the university ruined his radar research, so he moved to Jodrell Bank, several acres of rural farmland owned by the university's botany department. There, he bounced radio waves off of meteor trails and tried without success to find radar reflections of ionization caused by cosmic ray showers.** "If we could improve the sensitivity of our equipment by several thousand times," Lovell thought, "we might observe one echo from a large cosmic-ray shower."

With that in mind, in the spring of 1947, he and his two assistants decided to build a large paraboloid dish that could be used simultaneously for both transmission and reception. The diameter, determined by the distance between a hopelessly mired truck and a hedgerow, would be 218 feet. They decided to make the outer rim 24 feet high, because that was as high as their ladder could reach. That dictated a relatively shallow dish with a focal plane 126 feet above its center.

---

*The Crab Nebula is 6,500 light-years away from earth, so it is actually inaccurate to say that the supernova explosion occurred in 1054 C.E. In fact, it took place 6,500 years before that, but the ancient Chinese recorded the event when the light reached the earth.

**Cosmic rays are charged particles of very high energy, such as protons.

In September 1947, after recruiting wives and children to tie a spider web of galvanized wire to heavy steel cable, they completed the radio mirror, with a tubular steel mast held by guy wires at the center. But when they switched on the power to beam radar waves, they still found no cosmic-ray echo. What they did find, however, was that as the Milky Way passed through the zenith—directly overhead—the noise level increased dramatically to a sharp peak followed by some squiggles, then another smaller peak. The giant mirror turned out to be useful after all, Lovell admitted, "but not for the primary purpose for which we had built it."

Late in 1949, Robert Hanbury Brown, another radar man, joined Lovell at Jodrell Bank because of the opportunity to work with the "pencil beam" of the 218-foot mirror. To make it even narrower, he scaled the 126-foot tower to put in a new primary focus feed for 1.89 meters instead of 4.2 meters. He also realized that he could bend the mast to widen the field of view beyond the zenith by carefully readjusting the eighteen guy wires holding it up. By the end of 1952, he and his colleagues had surveyed twenty-three sources within their field of view, finding many that had eluded the Cambridge interferometer.

Few optical astronomers paid much attention to the radio engineers' findings, yet the radar men needed help identifying these mysterious areas in the sky. Fortunately, two veteran astronomers at the newly opened Hale Observatory on Mount Palomar were eager to help. Walter Baade, in his late fifties, had lost none of his fondness for astronomical puzzle-solving. In the fall of 1951, he turned the 200-inch monster toward Cassiopeia A and Cygnus A. Baade's photographs identified the radio source in Cassiopeia as a few tatters of glowing gas.

Baade's colleague Rudolph Minkowski looked at the gas with a high-resolution, faint-object spectrograph that identified the spectral lines of hydrogen, oxygen, and sulfur. The Doppler shifts indicated that some of the gas was barely moving, while some was flying away at 3,700 miles per second. He concluded that Cassiopeia A was the remains of a supernova explosion some 10,000 light-years away and that the slow-moving gas was blown toward earth.

Cygnus A turned out to be even more astonishing. It was, as Baade wrote, "a queer object," which he interpreted as two distant galaxies in collision. Baade bet Minkowski a bottle of scotch that the spectrum would show hot gas produced by this cataclysm. He won the scotch, although he was perhaps wrong about the collision (the radio waves shooting out either side came from a black hole). But this *was* an incredibly strong source of radio waves at huge redshifts. While Cassiopeia A and the Crab Nebula were in our galaxy, Cygnus A was definitely far outside it, some 740 million light-years away.

Even with the 218-foot mirror at Jodrell Bank, Robert Hanbury Brown and his colleagues could not pinpoint the exact location of many sources within their field of view, so they built a small mobile antenna and, after "an awful lot of plodding about in muddy fields," connected it to the big mirror to create an interferometer. Using it, they were able to prove that most of the radio sources lying along the galactic plane—clearly within the Milky Way—were fairly large and relatively close. Like the Crab Nebula, they were true nebulosities, the wispy remains of exploded stars.

But they could not resolve the five sources above the galactic plane, so they extended the baseline of the interferometer. To their surprise, even when they mounted the mobile antenna at the Cat and Fiddle, the highest pub in England some 12 miles away from the big radio mirror, they still could not identify three of the sources. Finally, by 1961, with a baseline of 71 miles, almost all of the powerful radio sources were resolved as coming from narrow portions of the sky—all except 3C48, which was so small it still could not be located.*

Quasars and Lookback Time

Inspired by the success of his immobile 218-footer, Bernard Lovell spearheaded the construction of the world's largest radio mirror, a 250-foot fully steerable dish at Jodrell Bank, able to point anywhere in the sky. Disastrous cost overruns nearly landed Lovell in prison. The biggest cost increase came as a result of the decision to make the huge mirror surface out of solid steel sheets instead of wire mesh, in order to

---

*"3C" stands for the third Cambridge radio sky survey, completed in 1959; 3C48 was the forty-eighth on the list of 471 sources. The 1955 2C survey had been a disaster, coming up with 1,936 objects, most of which were spurious.

give it efficiency at the short 10-centimeter wavelength requested by members of the British military, who wanted to track Soviet guided missiles.

On October 1, 1957, with the telescope nearly finished, Lovell commented to a colleague that "only a miracle could raise us out of the bottomless pit of our troubles." Three days later, the Soviet launch of SPUTNIK 1, the world's first artificial earth-orbiting satellite, provided the miracle. No other facility on earth was capable of tracking the carrier rocket. Work that was supposed to require months was completed in forty-eight hours to allow the great mirror to point automatically to specified sky coordinates.

Ironically, Lovell and his team were thrown back to using the huge new radio mirror as a radar detector. First, they successfully bounced signals off the moon, then managed to track the launching rocket as it moved over the North Sea at 5 miles per second. Suddenly Lovell was a national hero, and he was knighted in 1961.

That same year, with seed funding from the Carnegie Foundation, Taffy Bowen supervised the construction of a steerable 210-foot radio mirror at Parkes, Australia. In 1962, Martin Ryle oversaw completion of the One Mile Radio Telescope, an interferometer consisting of three 60-foot paraboloid reflectors that could move up and down a mile-long railroad track.

The United States rushed to catch up. Its eventual dominance in radio astronomy came after a rocky start. The National Radio Astronomy Observatory (NRAO) was founded in 1954 at a remote rural site in Green Bank, West Virginia, chosen because of its sparse traffic and because the surrounding mountains blocked human radio interference. By 1959, an 85-foot radio telescope had been erected, to be followed by a 140-foot equatorially mounted radio dish, an engineering nightmare that took six years to complete. In desperation, the NRAO commissioned a cheaper, hastily constructed 300-foot transit telescope completed in 1962. It collapsed in 1988, fortunately with no one inside. In nearby Sugar Grove, West Virginia, the U.S. Navy commenced work in 1959 on a gigantic steerable 600-foot radio mirror that was abandoned in 1962 after $96 million had been squandered on it.

Also in 1962, radio engineer John Kraus built a clever, relatively inexpensive radio mirror at Ohio State University, using a long, tiltable flat reflector (analogous to a siderostat for optical telescopes) to bounce radio beams into a fixed section of a paraboloid reflector, 360 feet long and 70 feet high, which then focused the waves.

Finally, in 1963, the world's largest radio mirror, operated by Cornell University and funded by the Advance Research Projects Agency of the U.S. Department of Defense, was built in a natural bowl near Arecibo, Puerto Rico, the brainchild of William E. Gordon, an ionospheric physicist in Cornell's College of Engineering. One thousand feet in diameter, its wire-mesh mirror was immobile, but by moving the focus feed, held fifty stories aloft by cables hung from three gigantic concrete pillars (each as high as the Washington Monument), astronomers could observe a wide swath of sky.*

The Arecibo mirror surface was shaped spherically so that off-zenith objects would appear essentially the same as those directly overhead, but that meant that instead of collecting radio waves to a single focal plane, spherical aberration would bring them to a big fuzzy ball, which could be accommodated by a 96-foot-long focal feed. The huge bowl owed its funding indirectly to SPUTNIK and the Cold War, because Bill Gordon persuaded the military bigwigs that it could be used as a huge radar to detect hostile satellites or intercept Soviet communications bounced off the moon.

"It was astonishing now to find radio instruments surpassing optical in resolving power," one veteran radar man observed. In a mere fifteen years, radio astronomy had matured, and now optical astronomers scrambled for the latest radio observations. At Palomar, Allan Sandage, the anointed successor to Edwin Hubble, used the 200-inch mirror to delve ever farther into space, and in 1960 he photographed 3C48, which turned out to be a faint blue star. "I took a spectrum that night and it was the weirdest spectrum I'd ever seen," he marveled. The bright spectral emission lines made no sense, matching no known elements.

Sandage and Maarten Schmidt, a young Dutch astronomer, continued to pursue these strange objects, which were now dubbed quasi-stellar

---

*The gigantic bowl was actually the result of a theoretical error in determining the size needed for a detectable radar echo from the ionosphere, which turned out to be 100 feet rather than 1,000, but radio astronomers were delighted at the engineers' miscalculation.

radio sources, or quasars. Sandage photographed 3C273, a quasar with a thin jet protruding from it. Schmidt obtained multilined spectra of 3C273 that were a complete mystery to him. On February 5, 1963, Schmidt sat in his office, staring at the postage stamp–sized film with the spectral lines. Idly, he sketched the lines on a yellow notepad. Suddenly, it hit him that the pattern looked awfully familiar.

He shouted to Jesse Greenstein, a colleague who was walking past his door: "I think there's a 16 percent redshift in 3C273." What Schmidt had seen was that the pattern resembled hydrogen emission lines at very high temperatures, shifted a huge amount toward the red end of the spectrum. His mind whirling, Greenstein suddenly blurted out, "Thirty-seven percent! 3C48!" He had realized that this source's mysterious lines also fit the hydrogen-line pattern. These redshifts could only mean one thing. These quasars were incredibly small, bright, distant galaxies. 3C48 was approximately 4 billion light-years away.

These were not "radio stars" after all. Incredibly, there were radio galaxies out there, throwing off radio waves at nearly every wavelength. Theorists concluded that the radiation must be caused by high-speed electrons swarming through enormously powerful magnetic fields. Unlike normal galaxies such as the Milky Way or nearby Andromeda, which are relatively weak radio sources, these radio galaxies were mysterious powerhouses, and quasars were the most powerful, distant, and mysterious of all.

In the next ten years, 200 quasars would be identified, some receding at 90 percent the speed of light, at least 12 billion light-years away. Astronomers began to talk about "lookback" time, since they were now peering at beacons from the early days of the universe, when quasars, whatever they were, ruled.

Whispers from the Beginning of Time

The year after Maarten Schmidt realized that quasar light traveled billions of years to reach the earth, two Bell Lab researchers in Holmdel, New Jersey, looked even farther back in time, though they didn't at first realize it. Like Karl Jansky, they were just trying to figure out the source of pesky static.

When young radio researchers Arno Penzias and Robert Wilson joined Bell Labs in the early 1960s, they were assigned to a strange-looking rotating radio antenna, a 20-foot reflector-horn that looked like a giant scoop with an aluminum lining. This strange mirror pointed at right angles to the heavens, as if the open scoop at the end were meant to catch moonbeams. Its curved lower surface formed part of a parabola, so that when radio waves hit its jutting lower lip, they reflected down to the focus. The upper part of the horn just served as a shield. The horn provided much better protection from stray radiation (especially from the ground) than conventional radio dishes.

This mirror had been designed to detect radio waves that bounced off a 100-foot-diameter mirror-balloon of aluminized polyester put in orbit and inflated by the young National Aeronautics Space Administration (NASA) in 1960 for Project Echo.* It allowed voices translated into radio waves and transmitted skyward at a NASA station in Goldstone, California, to reflect from the giant mirror-balloon and be heard at Holmdel.

Then in cooperation with NASA, AT&T launched TELSTAR on July 10, 1962. Penzias and Wilson prepared the big horn antenna to listen in. The TELSTAR launch was a success, sending live TV pictures from Maine to France the same day. After its launch, Penzias and Wilson finally were freed to do real astronomy. Their plan was to look for an invisible gas "halo" outside the plane of the Milky Way by observing at 7.3-centimeter wavelength. Before they could locate the hypothetical halo, however, Penzias and Wilson had to eliminate an annoying low-level static that they assumed was coming from the antenna itself. They took the antenna apart, evicting two nesting pigeons, replacing parts, scrubbing everything. They covered the rivets that held the reflective aluminum plates with special conductive tape. Their sensitive receiver was chilled near absolute zero with liquid helium. Yet still the static persisted.

Then in February 1965, James Peebles, a young Princeton astrophysicist, gave a talk at Johns Hopkins in which he revealed that his group, under the leadership of Robert Dicke, was looking for radiation left over from the Big Bang. The ferociously hot event, blowing out

---

*NASA was founded in 1958 as a result of the panic over SPUTNIK.

high-energy radiation some 14 billion years ago, should theoretically still be visible, though highly redshifted and cooled. The next day, Arno Penzias happened to be talking on the phone with Bernard Burke, a Carnegie Institute astronomer. Penzias complained about the inexplicable hiss, and Burke, who had heard about Peeble's talk, said, "Call Bob Dicke."

When Penzias called, Dicke and his group were eating lunch in his office. After listening carefully to Penzias, Dicke hung up the phone and announced, "Boys, we've been scooped." The Penzias and Wilson static, indicating a temperature some 2.7 degrees above absolute zero, fit the theory. And it appeared everywhere, from all directions, just as a remnant of the Big Bang should. On May 21, 1965, the *New York Times* broke the news on the front page: "Scientists at the Bell Telephone Laboratories have observed what . . . may be remnants of an explosion that gave birth to the universe."

Penzias and Wilson had spent days cleaning pigeon droppings from their horn-reflector, and they were rewarded with the Nobel Prize. "They were looking for dung and found gold, which is just the opposite of the experience of most of us," an envious Bell Labs colleague commented.

Pulsating Messages from Little Green Men

In 1967, at Cambridge, Antony Hewish had set up a five-acre array of radio antennae to look for quasars by studying the scintillation of long radio waves. Just as stars twinkle because light is distorted by the earth's atmospheric turbulence, radio waves are refracted by ionized gas in the ionosphere and elsewhere in space. At 3.7 meters, the wavelength Hewish chose, interplanetary scintillation effects would be large for powerful, compact radio sources such as quasars.

Jocelyn Bell, Hewish's graduate student, operated the telescope and analyzed the data, which meant examining 96 feet of chart paper every day. Within a month of the start of regular observations in July 1967, Bell spotted regular, repeated fluctuations that she called "scruff." Hewish dismissed them as interference, perhaps from an electrified farm fence, but when they reappeared and persisted, moving in sidereal time, Hewish agreed that something was up. Using a recorder with a faster response time, he and Bell picked up a source that beeped with amazing precision every 1.3 seconds.

What could it be? Only half in jest, Hewish, Bell, and others on the team began to refer to the source as LGM (for "little green men"), but not wishing to appear fools, they carefully guarded their secret. "Here was I trying to get a Ph.D. out of a new technique," Bell recalled, "and some silly lot of little green men had to choose my aerial and my frequency to communicate with us."

By February 1968, when Hewish and Bell finally did make their findings public, they had discovered three other "rapidly pulsating radio sources," as they called them. Because there were four of them spread across the sky, because the signals were so powerful across such a broad band of frequencies, and because there was no Doppler shift as there would be from a planet revolving around a star, the Cambridge group in the end dismissed the idea of alien communication.* Not everyone gave up on the LGM, however.

In 1956, Frank Drake, a twenty-six-year-old graduate student at Harvard, had tuned a 60-foot radio mirror onto the stars of the Pleiades when an unusual, regular signal appeared—"too regular, in fact, to be of natural origin," Drake recalled. "I could barely breathe from excitement, and soon after my hair started to turn white." He thought he was detecting signals from extraterrestrial life-forms, but when he moved the telescope off the star cluster, the regular signal continued. "It had to be some form of terrestrial interference, probably military," he regretfully concluded.

Nonetheless, Drake remained obsessed with the idea of communication from other worlds. In 1959, using the new 85-foot radio telescope at Green Bank, West Virginia, Drake initiated the first deliberate search for such signals, named Project Ozma after the princess in an L. Frank Baum book. He tuned into the 21-centimeter hydrogen line, reasoning that aliens might use it, and looked at two nearby stars. Within five minutes of pointing the telescope at Epsilon Eridani, he found regular pulses appearing eight times a second. When Drake moved the mirror,

---

*Antony Hewish (not Jocelyn Bell, who later took her husband's last name of Burnell) was awarded the Nobel Prize for the discovery of pulsars.

the signal stopped. With mounting excitement, he turned it back to the star, but the signal never returned.

In the ensuing years, Drake pursued other, more conventional research in radio astronomy, but he is primarily known as the father of the Search for Extraterrestrial Intelligence (SETI). By 1968, he was living in Puerto Rico as the director of Arecibo, the largest radio mirror on earth. One February morning, a young Australian astronomer burst into his office with the latest issue of *Nature.* "Look at this!" he panted, stabbing his finger at the Hewish-Bell article. With quickening pulse, Drake read, thinking that here, at last, might be the long-awaited message from other life-forms.

Drake rushed out and bought a big TV antenna at the Sears store in Arecibo that would enable him to listen at the same long wavelength as Jocelyn Bell, and hooked it to the focusing feed arm high over the giant mirror. The pulsating source, which Drake named a *pulsar,* came in loud and clear, like a hyperventilating heartbeat. All over the world, other big radio mirrors—at Jodrell Bank, Parkes, Green Bank, and elsewhere—competed, and within a year a dozen pulsars had been located, firing as slowly as every two seconds and as fast as four times per second. Even Drake had to admit that there were just too many of them in too many different places to be little green men. But what were they?

Late in 1968, Green Bank astronomers, using the 300-foot radio mirror, identified a pulsar in the Crab Nebula, and then the 1,000-foot dish at Arecibo located a tiny source near the exact center of the nebula, fibrillating at the incredible rate of thirty-three times per second. Thus far, no optical astronomers had managed to locate a pulsar. In January 1969, Mike Disney and John Cocke, two novice astronomers, with the help of computer whiz Don Taylor, looked for it on Kitt Peak with an old 36-inch Newtonian telescope. They found the star in the Crab, pulsing visibly on their computer screen. A tape recorder had accidentally been left on, capturing the moment of discovery. "Good God, you know that looks like a bleeding pulse," the British Disney said, laughing incredulously. "It's growing, John." Soon thereafter astronomers at the Lick Observatory, using a stroboscopic camera attached to the focus of a 120-inch mirror, took two amazing photographs. In one, the Crab Pulsar was a bright dot close to another star. In the next, it had disappeared.

With this discovery, theoreticians solved the mystery of the pulsars, along with the power that had fueled the Crab Nebula for the last 1,000 years since a supernova explosion created it. If a star is sufficiently large, it will not simply shrink to a white dwarf, as our sun some day will. It will keep collapsing to form an incredibly dense *neutron star* with a gigantic magnetic field. The Crab Pulsar is only 12 miles across yet weighs 50 percent more than the sun, which is approximately 865,000 miles wide. Just as a figure skater spins faster with arms pulled in tight, the star's rotation speeds up as it collapses. As the pulsar whizzes around, electrons caught in the magnetic forces accelerate to near the speed of light, throwing off huge bursts of radiation near either magnetic pole. Pulsars are like lighthouse beacons sweeping the heavens. They spin extremely fast at first but very gradually slow down, which accounts for the high speed of the Crab Pulsar, a youthful 1,000-year-old.

Where Stars Are Born

All of the major radio discoveries thus far had come in centimeter or longer wavelengths, in the "radio window" of the earth's atmosphere. As the waves get shorter, into the millimeter and submillimeter range, and then into the infrared area just short of visible light, water vapor and oxygen absorb most—but not quite all—of the radiation. A few astronomers thought it was worth trying to look at these wavelengths between regular radio and optical wavelengths. In 1960, Frank Drake learned that Frank Low at Texas Instruments had developed a new kind of millimeter-wave detector, so he hired him and brought him to Green Bank, where the two men hoped that the winter cold would freeze most of the interfering water vapor. Only three winter nights, however, were sufficiently chilly for observations.

So in 1962, Low and Drake persuaded NRAO to fund a $1.5 million millimeter-wave telescope on Kitt Peak in Arizona, where the first federally owned optical observatory was taking shape, and where the high, dry air would allow the reflection of shorter wavelengths. The contract for the telescope was given to the Rohr Corporation, a California aerospace firm, which set about milling a one-piece aluminum 36-foot-diameter mirror. It had to be relatively precise and shiny because of the millimeter waves it would reflect, although it didn't need to be of optical quality.*

The Rohr mirror was a disaster. The seaside factory rose and fell subtly with the tides, which affected the milling machine. At one point, the automatic cutting tool went berserk and cut a hole in the reflector, which was then patched. In 1967, the flawed mirror was bolted onto a steel mount atop Kitt Peak. The computerized pointing system didn't work very well. With changing temperatures, the solid-aluminum mirror expanded and contracted at a different rate than its steel backing, so that the paraboloid surface flexed like a bimetallic strip, throwing the mirror badly out of focus.

It was easy to get time on the flawed telescope, because no one thought it would reveal anything of interest. In 1969, however, Arno Penzias and Robert Wilson got interested in shorter-wavelength astronomy. Their Bell colleague Charles Burris had built a sensitive millimeter-wave receiver, so Penzias, Wilson, and Keith Jefferts went to Kitt Peak to try it out.

By the late 1960s, radio astronomers had found spectroscopic lines for complex molecules in space at centimeter wavelengths. Surprisingly, ammonia, formaldehyde, methyl alcohol, and water vapor appeared infrequently with dark absorption lines. Penzias, Wilson, and Jefferts decided to look for carbon monoxide, which would be produced when ultraviolet light split up formaldehyde. They knew that when a carbon-monoxide molecule slows its rotation, it emits radiation at 2.73 millimeters. One day in May 1970, with a refined detector, they pointed the 36-foot mirror at the Orion Nebula. "I was idly looking at the oscilloscope display," Wilson says, "when I suddenly noticed points moving up." He asked the operator to move the mirror away from Orion, and the signal flattened out.

It turns out that carbon monoxide was "all over the place" and allowed the millimeter-wave astronomers to look through cold dust

---

*Because they can be used during the day, the shiny surfaces of millimeter telescopes can be dangerous if accidentally pointed toward the sun. On a January day atop Mount Graham in the early 1990s, for instance, one construction worker suddenly felt hot while working on the 10-meter submillimeter dish. "Your butt's on fire," another worker calmly observed. In other mishaps, such telescopes have melted their secondary mirrors.

clouds and see carbon monoxide's bright emission lines. Overnight, the ugly-duckling mirror became the most sought-after telescope in the United States. "For the first time, [we] were able to look not just at the hottest objects in the universe, but also at the coldest," radio astronomer Mark Gordon observes. "It is in these frigid regions, where temperatures seldom rise above 100 kelvin (–280 degrees Fahrenheit), that stars form." These dust-gas clouds were a fertile sea of chemicals— some 120 molecules have now been identified—where not only stars but perhaps life itself may gestate. Many scientists think that life was seeded when comet tails swept over the earth and drenched it in these molecules. At least half the energy in the universe is seen at submillimeter and millimeter wavelengths. An important new branch of astronomy had been born.

Radio Astronomy Comes of Age

In the ensuing years, radio mirrors built for varying wavelengths got more sophisticated and interferometry more complex, allowing more discoveries about the universe and its evolution. Under Frank Drake's leadership, the Arecibo dish was completely refurbished, the old wire mesh ripped from the reflecting bowl. The new surface, 40,000 panels of shiny aluminum with small perforations to let the rain through, allowed the mirror to reflect much shorter wavelengths. At the November 1974 dedication ceremony, Drake used the huge dish to send a three-minute message toward M13 in the constellation Hercules, 24,000 light-years away. If an alien on M13 gets the message and answers immediately, we may hear the response in the year 49974.

The SETI listening program continued around the world, with periodic claims for communication that could never be proved. When the cheaply built 300-foot radio mirror at Green Bank, West Virginia— which had been used for SETI programs—collapsed dramatically in 1988, paranoid Americans spread rumors that aliens had toppled it.

Radio astronomers also used large radio telescopes as giant radar transmitters and receivers, bouncing signals off relatively nearby planets to learn more about them. As a result, they were able to tell (before setting foot there) that the moon's surface was powdery or porous, that

Mercury did not keep one side facing the sun all the time, and that some Martian mountains were 6 miles high. By carefully timing the returning pulse, they were able to measure the precise distance to Venus and hence to the sun, a distance called the "Astronomical Unit," refined now to 149.6 million kilometers. Radio waves from Venus also indicated that thick clouds, mostly carbon dioxide, cover the planet and create a greenhouse effect, so that keeps the surface at 600 degrees Kelvin. And the giant Jupiter turned out to be a surprisingly strong source of radio waves, probably due to its magnetic field.

In 1974, on the Plains of San Augustin near Socorro, New Mexico, the NRAO began to build a huge interferometer of twenty-seven mobile paraboloid radio mirrors, each 82 feet wide and weighing 230 tons, arranged in a giant Y-shaped pattern. Upon completion in 1981, they could be spread to give the resolution of a single mirror 22 miles across with the sensitivity of a dish 426 feet wide. With the huge Arecibo mirror, they were featured in the film version of Carl Sagan's science-fiction novel *Contact*. The unimaginative NRAO bureaucrats named the Y configuration the Very Large Array (VLA). With refurbished electronics, better computer technology, and eight new dishes, the Expanded VLA will produce images with ten times greater detail when it is finished in 2010.

With the development of powerful computers and atomic clocks, it became possible to point mirrors from widely separated locales simultaneously at a distant quasar or other radio source in order to get pinpoint resolution by producing interferometric radio fringes later. The ten identical 82-foot-diameter Very Long Baseline Array radio mirrors are spread from St. Croix in the U.S. Virgin Islands, across the continental United States (including Socorro), and to the top of Mauna Kea in Hawaii.

Meanwhile, to replace the collapsed Green Bank telescope, a fully steerable 300-foot radio mirror was completed in 2000, featuring an "active" two-acre reflecting surface that can be tested by lasers reflected from corner reflectors and adjusted accordingly. Its mirror is shaped as part of an off-axis paraboloid so that the radio waves are reflected to a focus toward the side, which avoids blocking out any of the incoming waves. The new Green Bank Telescope is the world's largest precisely controllable mechanical device of any kind.

In the wake of the 1970 Penzias-Wilson-Jefferts discovery of the carbon monoxide line, millimeter and submillimeter mirrors bloomed in some of the driest regions on earth, but developments in the United States were delayed when Congress twice failed to fund a sophisticated 25-meter dish on the 14,000-foot peak of the extinct volcano at Mauna Kea. In desperation, Mark Gordon of the NRAO led a team to upgrade the 36-foot dish, replacing it in 1984 with a completely new mount and a superior 12-meter (39-foot) mirror.

The South Pole is an excellent place to do millimeter astronomy because it is high and incredibly dry, due to frozen water vapor, so the National Science Foundation funded a 1.7-meter submillimeter mirror there in 1994. Antarctica is such a hostile environment that so far it has told us as much about human frailty as the universe. One U.S. scientist went crazy there in 1998, taking off across the ice in a sailing sled loaded with Snickers bars, though he was saved. A thirty-two-year-old Australian astronomer died there under mysterious circumstances in 2000.

The Atacama Large Millimeter Array (ALMA) will dwarf previous efforts when it is completed some time after the year 2010 on a plain in northern Chile 16,500 feet above sea level, making it the highest observatory site on earth. Funded by U.S. and European organizations, a phalanx of sixty-four 12-meter mirrors, with receivers chilled by liquid helium to 4 degrees Kelvin, will be spread over a 10-mile area. This interferometer will be able to look deep into the heart of the dust clouds where stars are forming.

X-Ray Mirrors

At the opposite end of the electromagnetic spectrum from long radio waves, x-ray wavelengths are extremely short, around 10^{-7} millimeters, and incredibly energetic, which makes them difficult to focus. Instead of being reflected, x-rays plow straight into a regular mirror and are absorbed. So what sort of mirror might work?

This was the question that faced two Italian physicists in 1959. Riccardo Giacconi, who received a Nobel Prize in physics in 2002, was then twenty-eight and had just gone to work for American Science and Engineering (AS&E) in Cambridge, Massachusetts. Founded by stu-

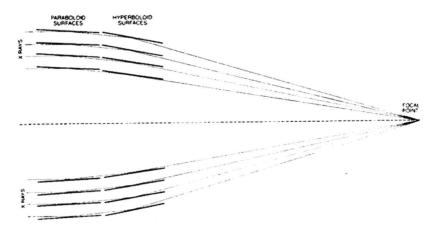

F I G U R E 11.1   A grazing incidence x-ray mirror system, with four concentric mirror surfaces.

dents of Bruno Rossi, an MIT professor who had worked on the Manhattan Project, AS&E was a private firm that worked closely with the U.S. Department of Defense to study the effects of nuclear weapons. Soon after joining AS&E, Giacconi was invited to a party at Rossi's house, where the older professor suggested x-ray astronomy as a potentially interesting field. But how to focus the x-rays?

A search of the sparse literature turned up an obscure 1952 article by Hans Wolter, a German physicist who had looked into building an x-ray microscope. Wolter showed that x-rays could be reflected at grazing angles of less than 1 degree from highly polished, dense mirror surfaces, like a bullet ricocheting off a wall. The proposal went nowhere, because it was too difficult to construct such tiny, precise mirrors for a microscope. But for a large-scale x-ray telescope, Giacconi thought it would work, and in 1960 he and Rossi published an article suggesting exactly how it could be done (see Figure 11.1).

Giacconi set up a laboratory in an old garage and, with NASA sponsorship, made a small x-ray mirror covering an area about half the size of a dime. He made it by machining and polishing the inner surface of an aluminum tube, then coating the interior with evaporated gold to provide high reflectivity. The x-rays were reflected twice by grazing incidence, first from a paraboloid surface, then a hyperboloid, to a detector at the focal point. Although far from perfect, the mirrors worked in a lab test. X-rays could indeed be focused.

Yet NASA administrators remained lukewarm toward x-ray mirrors, since simple detectors (like Geiger counters) seemed to be sufficient. In 1949, Herbert Friedman at the Naval Research Laboratory had flown an x-ray detector on a captured German V-2 rocket and proved that the sun did emit x-rays. In the next decade, Friedman's group studied solar x-rays through an entire sunspot cycle. "Few astronomers believed there was any prospect in the near future of extending x-ray astronomy beyond the solar system," Giacconi recalls. The x-ray luminosity of the sun is 1 million times smaller than the optical. To detect similar emissions from another star would take an incredibly sensitive detector, and it didn't seem worth the effort.

Giacconi thought otherwise. On June 18, 1962, with a detector launched aboard a rocket from White Sands in New Mexico, his group was looking for x-rays from the moon but instead found a surprising source in the constellation Scorpius, which they named Sco X-1. This star's x-ray luminosity was *1,000* times stronger than its visible light. "This was a truly amazing and new type of celestial object," Giacconi recalled. What could possibly cause it?

Giacconi yearned for an x-ray mirror to collect more x-rays, and he found an ally in John Lindsay at the Goddard Space Flight Center. Lindsay had worked with Herbert Friedman and remained a strong advocate of solar x-ray astronomy. Although Giacconi wanted to look beyond the solar system, he grabbed at whatever support he could get. The AS&E group made a grazing-incidence mirror, about the size of a tennis-ball can, that focused x-rays 31 inches behind it, and launched it on a rocket in 1963, obtaining x-ray pictures of the sun. They did it again in 1965 with an improved nickel reflective surface and were planning a much more ambitious effort when Lindsay suddenly died of a heart attack. Support from Goddard dissolved.

In that summer of 1965, a group of x-ray and gamma-ray* scientists met in Woods Hole on Cape Cod. Giacconi championed x-ray tele-

---

*Gamma rays are even more energetic than x-rays, with shorter wavelengths. They cannot be reflected even at grazing incidences, so there are no gamma-ray mirrors, although there are mirror telescopes on earth that look at *Cherenkov radiation*, brief blue flashes of light in the upper atmosphere caused by high-energy gamma rays and cosmic-ray particles. These optical shock waves, which last only 20 billionths of a second, were first observed in 1953 by British astronomers

scopes while Friedman argued that conventional detectors were sufficient. In a compromise, they decided that the first major NASA mission
should use Friedman's detectors, but the second would carry a large x-
ray mirror.

In October 1967, the AS&E team launched the first x-ray mirror
(about the size of a half-dollar) to be flown on a satellite. It successfully
observed solar flares. In 1970, Giacconi's group launched UHURU, the
first dedicated orbiting x-ray observatory, with a sophisticated detector
system that revealed sources pulsating every few seconds, while others
spurted erratic x-rays, sometimes in tenths of a second. UHURU also revealed a thin, superheated gas that doubled the known mass of some
galaxy clusters. The revolutionary discoveries underlined the need for
x-ray mirrors.

In 1970, NASA approved the Large Orbiting X-Ray Telescope
(LOXT), an ambitious satellite observatory that would carry two x-ray
telescopes. One would feature grazing incidence mirrors nearly 4 feet
wide. The other would be a design proposed in 1948 by the Stanford
physicist Paul Kirkpatrick and his graduate student Albert Baez, consisting of two flat mirrors bent in one plane to a paraboloid curve. X-
rays glancing off of the first mirror would be focused to a line, then to
a point by the second surface, set at right angles to it. But on January 2,
1973, NASA cancelled the LOXT. Giacconi recalls it as a "shattering
experience."

That same year, AS&E and Goddard both put small x-ray telescopes
aboard SKYLAB, a pioneering manned space station, to study the sun.
The superior AS&E version, made by Leon Van Speybroeck, a physicist who had joined the team a few years earlier, featured two near-
cylindrical mirrors about 1 foot wide, one nested inside the other, and
yielded stunning x-ray pictures of the sun.

In 1973, Giacconi moved his group to the nearby Harvard-Smithsonian Center for Astrophysics, where he began work on a NASA-

---

using a 30-centimeter parabolic mirror mounted in a garbage can, with a photo-
multiplier tube at its focus. In 1968, Irish astronomer Trevor Weekes pioneered
construction of a 10-meter aperture reflector made with 248 hexagonal mirrors
atop Mount Hopkins in Arizona, but it took another eighteen years before he finally detected gamma rays from the Crab Nebula.

funded x-ray telescope for the High Energy Astronomical Observatory (HEAO) program. As their project would follow Friedman's, it was called HEAO-2. For it, Van Speybroeck was to oversee the creation of four nested cylindrical mirrors, the biggest nearly 2 feet in diameter, with ten times the collecting area of the SKYLAB effort. NASA bureaucrats ruled that as a cost-cutting measure there would be no test prototype models—instead, it would be a "protoflight"—and pushed the harried x-ray astronomers to hurry it up.

Although the four detectors that would rotate in and out of the focal point were important, the mirror assembly was crucial, and nothing like it had ever been made before. Schott, the German company, first made fused silica "barrel staves," then melted them together in a heated centrifuge that left no seam. Then the Perkin-Elmer Corporation in Connecticut used diamond grinders to produce the approximate paraboloid and hyperboloid surfaces.

But how would they test the surfaces? The cylinders had to be thin enough to fit inside one another, but that meant that they sagged under their own weight in earth's gravity, even though that would not be a problem in outer space. Any rigid support system would also change the shape of the mirrors. Pondering the problem, Van Speybroeck realized that the closest he had come to weightlessness on earth was when he was floating in water. Glass wouldn't float in water, but it would in mercury, so the x-ray mirrors took a mercury bath during testing.

The next step was to polish the mirrors to a smoothness of 1/10-millionth of an inch. Because of the short x-ray wavelengths, even a small imperfection in the surface might cause major distortions. The impatient NASA managers ordered Perkin-Elmer to stop polishing a day early, causing a loss of efficiency at the "hard" end of the x-ray spectrum with the shortest wavelengths. With polishing stopped, a coating of chromium-nickel was evaporated over the interior mirror surfaces in a vacuum chamber. Finally, the four nested mirrors were held in place with an elaborate system of thirty-two counterweights on each mirror so that they hung as nearly weightless as possible while the assembly was bonded together and aligned.

The mirror was to be shipped from Connecticut to AS&E in Cambridge, where the detectors would be added, then to Marshall Space Flight Center in Huntsville, Alabama, for a final test. Van Speybroeck insisted on a dry run before he allowed a crane to pick up his precious x-ray mirrors inside the Perkin-Elmer warehouse. The crane dropped the dummy crate. After much rechecking, however, the crane successfully lifted the mirrors onto a truck. At Marshall, the mirrors were tested in a huge vacuum chamber, 20-by-40-feet long, with x-rays shooting into it from a 1,000-foot pipe. NASA once more stepped up the pressure, insisting on only one month of testing instead of the six months in the original plans. To test at different x-ray energies, angles, and mirror temperatures, the scientists worked in shifts around the clock.

Van Speybroeck, the mirror-maker, could not bring himself to attend the launch at Cape Canaveral. Riccardo Giacconi and Bruno Rossi were there for the launch just after midnight on November 13, 1978. Nearly two decades after he first conceived of an x-ray telescope, a tense Giacconi watched the rocket lift off successfully, and the x-ray observatory, named EINSTEIN, achieved orbit twenty-three minutes later.

By the time EINSTEIN's mirrors first opened onto the x-ray universe, Giacconi had already put Van Speybroeck to work on bigger mirrors that would eventually be launched in 1999 as the CHANDRA X-RAY OBSERVATORY. "Most of us think twelve minutes ahead, some a bit further, but Riccardo makes long-term plans come true," Van Speybroeck said. In 1963, Giacconi had proposed a 30-foot-long x-ray telescope with mirrors 4 feet in diameter. It took another thirty-six years for this vision to become reality. The four nested CHANDRA mirrors, almost 4 feet wide and coated with evaporated iridium, were the smoothest, cleanest mirrors ever made. In December 1999, the European Space Agency launched the XMM-NEWTON, with fifty-eight extremely thin cylindrical gold-coated mirrors, the largest a little over 2 feet wide. Though not as precise as the CHANDRA mirrors, they collect more x-rays for spectroscopy.

Unreflectable Black Holes

Outer space turned out to be full of x-rays, all originating from what Giacconi calls the "hot spots" of the universe such as supernova explosions. The x-rays are produced by incredibly strong magnetic fields, intense gravity, or extraordinarily high temperatures. The Crab Pulsar, a dense neutron star, throws its regular beacon into space not only in the radio and optical wavelengths but in x-rays.

The cylindrical EINSTEIN mirrors revealed that many more "normal" stars emit stronger x-rays than anyone had thought, allowing scientists to study the behavior of the turbulent outer layers of stars. But the most intriguing observations involved deep surveys in which the mirrors were pointed for an entire day at an "empty" part of the sky with no known radio, optical, or radio sources. These picked up many previously unknown quasars. What strange power made them spit out high-powered x-rays that could still be observed billions of light-years away?

Nothing. Or at least as close to an awful, active, sucking emptiness as we can imagine: a black hole. In 1784 the English geologist John Michell had theorized that there might be such bottomless pits in space, and in 1916 German astrophysicist Karl Schwarzschild had revived the concept. Now the x-ray astronomers were finding indirect evidence for them. When stars the size of our sun burn out, they expand to red giants, then collapse to form white dwarves that slowly cool. Larger stars collapse to form neutron stars. But monstrous stars, three times the mass of our sun, would theoretically collapse further still, so that nothing is left except a huge gravitational maw, a cosmic whirlpool that sucks in everything that comes near it, including light.

Thus, no mirror—whether radio, x-ray, or optical—will ever reflect a black hole directly. By definition it is invisible. But what if a nearby star provides "food" for the black hole, which slowly sucks material from it? As matter is ripped apart and pulled into the hole, it shrieks a death cry in x-rays and other radiation. That process is probably what powers quasars. Indeed, supermassive black holes probably lie at the heart of galaxies such as the Milky Way, accounting for the cosmic static that first puzzled Karl Jansky in 1932.

Chapter 12

BEYOND PALOMAR

Large mirrors are to astrophysics and astronomy what computational speed is to computing. . . . We want to peer right to the edge of the Big Bang. We want to directly detect earth-sized planets circling other stars. And the bigger the mirrors are, the deeper we can stare.

DAN GOLDIN, NASA ADMINISTRATOR, 2001

WHILE RADIO AND X-RAY mirrors were revolutionizing our view of the universe, the 200-inch Hale mirror on Mount Palomar remained the premier telescope for optical wavelengths. In 1976, the Soviets put up a 236-inch mirror in the Caucasus Mountains, but it never worked very well. "The era of constructing large optical telescopes on Earth is drawing to a close," science writer Timothy Ferris concluded in 1977, "and observers' hopes for the future involve orbiting large instruments in space."

Ferris was right about larger mirrors in space, but he was wrong about earthbound telescopes. A new generation of huge mirrors, along with innovative methods of making them work better, would lead astronomers into the twenty-first century.

Aden Meinel Thinks Big

If anyone could be called the father of larger mirrors, it is Aden Meinel. As a teenager in Pasadena in the 1930s, Meinel could hardly avoid telescopes. His girlfriend, Marjorie Pettit, was the daughter of Edison Pettit, a Mount Wilson astronomer. While attending junior college, Meinel hung out in the Caltech astrophysics building, then landed a volunteer job in the Mount Wilson optical shop.

By 1953, Aden Meinel had married Marjorie (who had a master's in astronomy and would collaborate on his work), begun a family, secured his Ph.D., and moved to Yerkes Observatory in Wisconsin. There he learned that the McDonald Observatory atop isolated Mount Locke in West Texas wanted something bigger than its 82-inch mirror, which had seen first light in 1939.

One evening as Meinel was about to enter the Yerkes Observatory, he looked at the circular drive around it and thought, "Look at all the photons that are being wasted on that grass!" Then he thought, "Why not pave a spherical bowl of similar diameter with fixed mirrors and move only the observer's cage?" Inspired, Meinel built a wooden model of a 500-inch segmented mirror, a bowl two-and-a-half times bigger than the Palomar mirror. He planned to place a large segmented Schmidt corrector plate at the center of curvature.

But Palomar director Ira Bowen expressed severe doubts about Meinel's plans. Bowen said that he was the only one who could keep spectrographs with four mirrored diffraction gratings lined up properly on the 200-inch telescope. It would be a lot harder to keep 150 spherical mirrors aligned. And how did Meinel intend to make the big segmented Schmidt plate? How would he deal with the chromatic aberration produced by refraction through the glass? As Meinel pondered these problems, he learned of plans for the huge Arecibo radio dish, which were quite similar to his 500-inch idea, but he wished he could solve his focal problems as easily in visible wavelengths.

Then, in 1957, Meinel was tapped by the National Science Foundation to do a site survey for a new national observatory, an optical counterpart to the Green Bank facility for radio astronomers. He eventually chose Kitt Peak in Arizona. "I proposed quite a few rather novel telescopes for Kitt Peak," Meinel recalled. But the solar astronomer who held the purse strings insisted on a traditional, safe telescope with an 84-inch Corning Pyrex mirror. Aside from its Ritchey-Chrétien design, it was essentially a Palomar clone.

No one thought of besting the 200-inch Palomar mirror. Lick Observatory acquired the 120-inch Corning disk made as a testing flat for the

Palomar mirror and, after polishing it for eight years, installed it in a telescope atop Mount Hamilton in 1959. McDonald Observatory opted for a 107-inch fused silica mirror in its new telescope, which saw first light in 1969.* It, too, mostly mimicked Palomar. "There was a real feeling," an astrophysicist observed, "that the 200-inch telescope ... had been produced by wizards and elves and set down on this earth." Astronomers relied on much-improved detectors such as photomultipliers to optimize the light collected in the mirrors they already had, but those who wanted to study newly discovered quasars were frustrated. "We were crying out for light," recalled one quasar specialist.

Multiple Mirror Mania

So things stood in the spring of 1970, when Fred Whipple, the adventurous head of the Smithsonian Astrophysical Observatory in Cambridge, Massachusetts, told Meinel that he had finally secured congressional funding for his dream telescope. Along with his staff, Whipple had been grappling with how to build an innovative large telescope. He was explaining that they had considered floating it on a rotating platform and building it from segments molded over a convex master, when Meinel interrupted him: "Well, I've got the mirrors."

Meinel now headed the University of Arizona Optical Sciences Center, funded largely by the U.S. Air Force. He also advised the Air Force on lightweight mirrors for spy satellites. Astronomers referred to such military involvement as working on the "dark side" or researching in "deep black," but aside from that, no one talked much about it because the information was classified.

One day at a luncheon meeting at the Pentagon, Meinel had said, "One way of solving the problem of very large apertures in orbit would be to cluster a number of smaller ones." He built a model with a configuration of six mirrors mounted together, and the military brass asked if

---

*Everyone at McDonald Observatory took great pride in the fused silica mirror until February 5, 1970, when an unbalanced astronomer bashed the middle with a hammer, then pumped seven bullets into it. Amazingly, the quality of the optical surface suffered little from the damage.

it would really work. "What would be really nice would be a test bed," Meinel said. He knew that the Air Force had funded some experimental 72-inch mirrors made of two thin sheets of fused silica separated by crossed glass supports. "How about a deal?" he asked. And that is how eight mirrors happened to be warehoused at the Optical Sciences Center when Whipple and Meinel had their historic phone conversation.

The resulting collaboration between the University of Arizona and the Smithsonian was called the Multiple Mirror Telescope (MMT), six mirrors mounted in a hexagonal array that turned together, with a small guide telescope in the center. The mirrors, which were barely curved to begin with, were "slumped" over a mold at Corning, where they had been made, to a focal ratio of f/2.7. One of them collapsed during the process, but seven survived and six were polished in the Optical Sciences Center in Tucson.

When the telescope finally saw first light in 1979, it exceeded expectations. The "seeing" atop Mount Hopkins, 40 miles south of Tucson, was superb because the peak stood alone. Thus, winds sweeping over the mountain were uniform rather than turbulent. The lightweight MMT mirrors lost heat quickly, so at night they stabilized quickly. Because of the relatively fast focal ratio (for the time) and the multiple-mirror configuration, they didn't need a huge dome. Instead, the mirrors were housed in what looked like a big rectangular warehouse, the entire building turning along with the mirrors. Rather than the traditional equatorial mount, which took up a lot of room, the MMT could use the simpler, compact computer-guided altitude-azimuth (alt-az) mount, in which two motions—a rotating base and an up-and-down pivot—were necessary.*

Even before the telescope was officially commissioned, it produced good science. "We felt it would be very nice to have some results on hand at the dedication," Smithsonian astronomer Nat Carleton recalled. So he hosted a Friday-night observing program. In March 1979, British astronomers Dennis Walsh and Bob Carswell asked if they could bring up a spectrograph and attach it to the MMT. They had found two adja-

---

*An alt-az mount requires a third rotation if image orientation is important, because the image of the sky revolves slowly otherwise. Big alt-az telescopes have an instrument rotator at the back of the primary mirror.

cent quasars but couldn't get good enough spectra on Kitt Peak telescopes to prove that instead of two objects they might really be seeing duplicate images of the same one. The MMT mirrors revealed clear, identical spectral lines, proving for the first time that the "gravitational lensing" hypothesized by Einstein, in which light from a farther-off source is bent and focused by a nearer galaxy, really does occur.

Laser Light

The only major problem with the MMT was that the elaborate laser coalignment system didn't work very well.

Lasers (an acronym of "light amplification by stimulated emission of radiation") were invented in 1960 when Theodore Maiman, a research physicist for Hughes Aircraft, excited chromium atoms in a 1.5-inch-long artificial ruby rod by wrapping a flashing bright light around it. By itself, this would have produced only a faint red glow. But Maiman silvered both ends of the rod, turning them into mirrors, so that the light bounced back and forth between them, creating a chain reaction in which electrons were bumped to an excited state and then emitted light. All of the light waves traveling back and forth between the mirrors were polarized as they flowed in the same direction. To allow the light to escape, Maiman partially silvered one end. It reflected most of the light back but allowed a small amount to escape in a thin red beam.

Within five years, lasers of all sorts were invented, producing lased electromagnetic radiation in a thousand different wavelengths, mostly from excited gases rather than solids like ruby. Each had a full mirror on one end and a partial mirror on the other, although silver was replaced by dielectric coatings to manipulate specific wavelengths. These strange mirrors consist of alternate thin layers of materials with different refractive indexes so that one layer bends the waves much more than the next. With just the right materials for the chosen wavelength, the radiating waves can be made to "interfere" with one another so that 98 percent are bounced back and only some are allowed through.

Lasers of various kinds soon found applications in industry, medicine, communication, business, military, and science. In July 1969, the first men on the moon left a panel of corner-cube reflectors that, true to their name, are shaped like the three-sided corners of a cube. Any light beam that hits such a mirror reflects from one side to another and another, then directly back to its source. A powerful ruby laser, guided by a telescope mirror at MacDonald Observatory in Texas, is able to reflect and return from these tiny moon mirrors.* By timing the pulse, the astronomers are able to determine a very precise distance to the lunar surface. Most of us see corner-cube reflectors every day, since bicycles and cars sport red molded plastic panels of them.

In 1976, NASA launched a 2-foot-diameter aluminum ball called the Laser Geodynamics Satellite (LAGEOS), with 426 corner-cube reflectors embedded evenly around the surface. By measuring the time it took a reflected laser beam to return, scientists could build up information about the motion of the earth's tectonic plates, the exact length of a day, and the strength of the earth's gravitational field.

Lasers also made Alexander Graham Bell's dream of a "photophone" possible through fiber optics. When laser light directed down a thin glass "wire" hits the side of the glass, it acts as if it has hit a perfect mirror in the process of total internal reflection, which will always occur if it hits the boundary at a suitably oblique angle.

By the MMT's introduction in 1979, lasers and fiber optics were hot scientific topics, but the engineers who planned the laser coalignment system hadn't counted on moths. Attracted by the lasers, they blocked the beams bouncing from mirror to prism. Although the system probably could have worked eventually, it was abandoned in favor of manual control.

Looking Through Dust with Infrared

The MMT had an alternate set of smaller secondary mirrors that could vibrate simultaneously many times per second in a precisely controlled jitter. This "chopping" action was necessary for observa-

---

*To direct a laser with a telescope mirror, simply reverse the lightpath from the focal point behind the hole in the primary Cassegrain mirror. The laser beam bounces from the secondary mirror back to the primary and then into space.

tions in the infrared, that area of invisible light just beyond the red end of the visible spectrum.

Infrared astronomers had multiple problems. The earth's atmosphere—particularly water vapor and carbon dioxide—blocks much of the infrared spectrum. Even worse, everything on earth emits infrared radiation as heat, so conventional telescopes, with their black baffles around mirrors to limit stray light, were themselves sources of confusing infrared radiation. One early infrared astronomer likened his effort to looking for a match in a blast furnace.

The MMT was one of the new breed designed (at least partly) for infrared observations. The smaller secondaries didn't pick up heat waves from beyond the edges of the primary mirrors. The chopping motion allowed observers to measure the infrared noise level of the telescope and adjacent sky so that it could be "subtracted" from the star being observed. The best observatories were the highest and driest, avoiding water vapor as much as possible.

Mirror chopping was invented by Frank Low, whose millimeter-wave detector (discussed in Chapter 11) also worked in the infrared. Low had come up with the idea of mounting six mirrors together to do infrared astronomy, based on the mistaken notion that small mirrors worked better, and had suggested the idea to Aden Meinel.*

Low was not the only infrared pioneer. Because infrared detectors could "see" heat at night, they could spot airplanes, missiles, and ground vehicles. In the late 1950s and early 1960s, the military sent classified infrared telescopes and cameras high into the sky on rockets or balloons to avoid water vapor. They also developed the Sidewinder heat-seeking missile.

Bob Leighton, a Caltech physics professor, learned about the military infrared detectors used on Sidewinders from Gerry Neugebauer, a newly minted Ph.D. serving as a commissioned officer at the nearby Jet Propulsion Laboratory (JPL). Leighton convinced Neugebauer that they should use the infrared detectors to conduct a sky survey at 2.2

---

*Low had been frustrated in his attempts to use the 200-inch Palomar mirror for infrared observations; it wasn't the mirror size that was the problem but the heat retention of the massive mirror, he later concluded.

microns,* one of the infrared windows allowed by the earth's atmosphere. But to do that, Leighton needed his own telescope.

Leighton hit upon an ingenious way to make a cheap paraboloid mirror. As a child he had been fascinated by the swirling patterns in his mother's stirred mop bucket. As a physics professor, he now knew that "if the speed is just right, the upper surface of the liquid will then have precisely the shape of a parabola." He decided to spin slow-setting epoxy at the proper speed in a mold turned on an air bearing (a thin film of compressed air that isolates from vibration). It worked, producing a deep f/1 62-inch paraboloid epoxy mirror that was then aluminized and set up on Mount Wilson in a shed with a roll-off roof. Meanwhile, Gerry Neugebauer built a detector cooled by liquid nitrogen to reduce thermal noise.

The mirror wasn't perfectly polished and resolved only about as well as the human eye, but it was a start. The scientists pointed it at a different altitude on every good observing night for the next five years, using the earth's rotation to scan a strip of sky. They looked at every area of the sky at least twice, and when they located sources, they would look at that region repeatedly to make sure it was correct and to check for variable stars.

In a preliminary 1965 publication, Neugebauer and Leighton revealed that "a number of strikingly red stars have already been found." Around 1,000 degrees Kelvin, these were extremely cool by solar standards. Faint when viewed through optical telescopes, they glowed brightly in the infrared. One night, Neugebauer and Leighton were watching the infrared signal being recorded, and they saw a "huge triple bump," as Leighton recalled. Yet on the optical channel there was nothing. They had located the first "invisible" infrared source, but others followed—"so cool," said Leighton, "that they were not even red"—and they were dubbed "dark brown" stars.

They could see innumerable sources glowing through the dust clouds in the center of the Milky Way. There were also stars lurking in nebulae. "What wasn't appreciated at the time," Leighton recalled, "was how many sources . . . were intrinsically quite bright, but were embed-

---

*A *micron* (short for micrometer) is one-millionth of a meter, or one-25 thousandths of an inch.

ded in nebulosity, possibly of their own making." By the time they published their final survey results in 1969, Leighton and his team had detected around 20,000 distinct infrared sources. For the first time, it occurred to astronomers that infrared radiation offered a way to see through dust clouds.*

Meanwhile, Frank Low and his associates explored the infrared skies at longer wavelengths of 10, 20, and even (rarely) 34 microns with a 60-inch aluminum mirror on Mount Lemmon, near Tucson, Arizona. With Low's special germanium detector cooled to within a few degrees of absolute zero by liquid helium, they concluded in 1973 that "some stars have very cold (<100 K) circumstellar dust shells or clouds."

In 1974, the Air Force published an infrared sky survey made with rocket-borne detectors looking at 4-, 11-, and 20-micron wavelengths, causing a flurry of ground-based observations, and NASA began planning for an infrared satellite observatory. The following year, NASA—inspired by infrared work Frank Low had done on a Lear jet—dedicated the Kuiper Airborne Observatory, a converted C-141 military cargo plane that could carry a 36-inch mirror to 40,000 feet, above most atmospheric water vapor. The telescope was stabilized by gyroscopes and isolated from aircraft vibrations by an air bearing and shock absorbers.

A search for the highest, driest earthbound site led to 14,000-foot Mauna Kea, an extinct Hawaiian volcano, where the British completed the 3.8-meter (150-inch) United Kingdom Infrared Telescope (UKIRT) in 1978. For its time, UKIRT featured a remarkably thin mirror, supported by eighty pneumatically controlled pads. Initially conceived as a cheap light bucket for spectroscopy, it surprised everyone with its superb performance. Soon after, it was joined by NASA's 3-meter infrared telescope and the Canada-France-Hawaii 3.6-meter, designed for both optical and infrared. Mauna Kea quickly became the most sought-after telescope site in the world.

---

*Leighton went on to make low-cost 10-meter aluminum mirrors for millimeter observations at Owens Valley and submillimeter on Mauna Kea. With associate Dave Woody, he carved them from corrugated aluminum, then glued on aluminum plates sucked down by a vacuum to produce what Leighton called "this big, strange 400-inch diameter structure . . . good to maybe one or two ten-thousandths of an inch."

Astronomers who fantasized about a Hawaiian paradise were shocked when they arrived at the summit. "Mauna Kea is an utterly desolate spot," wrote an infrared astronomer, a "mound of cinders." The thin atmosphere was conducive to observations but not to human breathing or thought.

Mirrors on the Dark Side

In the 1970s, opticians working at places like Kodak, Perkin-Elmer, and Itek worked in deep-black obscurity to produce lightweight mirrors like those Aden Meinel procured for the Multiple Mirror Telescope. Few civilian astronomers penetrated so deeply into the dark side, but all of them sensed that the military, with its huge budget and insatiable appetite for Cold War secrets, was far ahead of them in many ways.

Lincoln Laboratory, near MIT—one of several university-military research organizations created during the Cold War—planned one of the earliest and most bizarre mirror experiments in space. In 1958, a series of high-altitude nuclear bomb tests tore a hole in the ionosphere over the Pacific Ocean and disrupted military radio communications. As insurance against such disruptions in the future—what if the Soviets exploded a warhead high in the atmosphere to mess up communications?—Walter Morrow of Lincoln Labs planned to scatter 480 million hair-thin copper wires into an orbiting, reflective belt.

The 1961 launch of Project West Ford* from Vandenberg Air Force Base in California went into orbit but failed to deploy the needle-mirrors. Two years later, a second attempt succeeded in spewing a half-billion tiny radio mirrors into orbit. They worked after a fashion, but retransmitting communications satellites made the reflective space debris obsolete, and they gradually fell harmlessly into the earth's atmosphere.

In the meantime, the U.S. Air Force and Central Intelligence Agency spied on the Soviets, at first by using large cameras. After Francis Gary Powers was shot down in his U-2 spy plane over Russia in 1960, the

---

*It was originally called Project Needles but was changed to something less controversial.

super-secret National Reconnaissance Office (NRO) was created to launch spy satellites that ejected film capsules to be snagged in midair. To supplement them, surveillance satellites with infrared detectors flew at geosynchronous orbits—high enough to remain over the same place—with flat rotating mirrors to scan for the sudden heat coming from a ballistic missile.

By the late 1960s, a space armada of secret mirrors was flying overhead, but they became much bigger and more sophisticated in the following decade. The DSP–647s, a new generation of surveillance satellites, carried infrared Schmidt telescopes with 40-inch spherical mirrors. Meanwhile, the NRO sent giant radio ears into space. Dubbed RHYOLITE, these huge umbrellas unfurled in space to make radio mirrors from 70 to 120 feet in diameter, picking up long-distance telephone calls and radio waves, ranging from a Soviet commissar talking to his mistress to evidence of a ballistic missile test.

The KEYHOLE (KH) satellites were the real eyes in the sky, carrying lightweight mirrors polished for optical wavelengths in space telescopes that replaced the simple U-2 cameras. The size of a Greyhound bus, the first KH-9 (nicknamed BIG BIRD on the dark side) was blasted into space in 1971, carrying a hyperbolic primary mirror over 6 feet wide in a Ritchey-Chrétien Cassegrain telescope, along with other small mirrors and prisms to redirect the light to infrared sensors and photomultipliers. Big Bird flew for only fifty-two days, by which time it had run through its film supply.

The first KH-11 rocketed into orbit from Vandenberg in December 1976. The 92-inch mirror, nearly 8 feet wide (even larger in subsequent versions), reflected light to the secondary mirror and thence to charge-coupled devices (CCDs) that could amplify and transmit signals to the ground. Invented at Bell Labs, CCDs were far more sensitive to light than photographic film, yielding instantaneous images that didn't need to be parachuted to the ground. Now satellite mirrors could remain useful for years in space. In 1980, the big mirror of a KH-11 peered down on the U.S. embassy in Tehran, looking for the Americans held hostage there, and its transmitted pictures helped plan the route the rescue team intended to use had the mission been successful.

Adaptive Optics on the Dark Side

At the same time, the military desperately wanted to obtain decent pictures of orbiting Soviet satellites, using telescopes at the Air Force Maui Optical Site (AMOS) on Mount Haleakala, a 10,000-foot-high dormant Hawaiian volcano. That pictures came out fuzzy wouldn't have surprised Isaac Newton, who wrote in 1704: "For the Air through which we look upon the Stars, is in a perpetual Tremor."*

The U.S. Air Force and the Defense Advanced Research Projects Agency (DARPA) sponsored classified research on so-called rubber mirrors. The solution, known as adaptive optics, involved sensors that detected how the light's wavefront was distorted by the atmosphere, then corrected it with a small flexible mirror somewhere in the lightpath. Soviet satellites reflected enough sunlight so that they could be used for adaptive optics, which required a bright light source.

In 1973, John Hardy, an optician at Itek near Boston, oversaw the creation of the first successful deformable mirror, a thin wafer of aluminized glass just 1.5 inches in diameter, glued to the top of a half-inch-high stack of twenty-one piezoelectric ceramic rods. Each rod's length changed in response to local voltage, moving it up or down to reshape the mirror surface. In laboratory tests with a simple helium-neon laser, it worked, but it was a long way from there to a functioning adaptive optics system. Eventually, in 1981, Hardy ventured to Mount Haleakala, where he attached a much more complicated deformable mirror with 168 actuators (still only 3 inches wide) onto an Air Force telescope to look at stars and satellites.

Meanwhile, the Air Force was trying to develop weapon-beam lasers for the Airborne Laser Laboratory (ALL), housed in a modified Boeing 707. However, aircraft vibration and atmospheric turbulence made it nearly impossible to maintain a coherent beam. Using a deformable mirror to help direct its ALL laser, the Air Force did eventually shoot down a few Sidewinder missiles at very close range, but the carbon

---

*Spy satellites looking down on the earth are not troubled by the turbulent atmosphere because the turbulence is so close to the target. Similarly, you can read a newspaper through a piece of ground glass placed on it, but not if you move the glass farther away.

dioxide laser wasn't powerful enough, nor the mirrors sufficiently sophisticated, to be practical. Besides, by turning potential targets themselves into mirrors of shiny metal, enemies could foil most lasers by reflecting their light harmlessly. The ALL was grounded in the early 1980s.

Adaptive optics showed more potential on the Air Force's solar telescopes on Sacramento Peak in New Mexico, where young officer-astronomer Pete Worden, working with John Hardy, succeeded in using a deformable mirror to obtain clear images of sunspots in 1978.*

All of the innovative military mirrors remained in deep black as far as the civilian world was concerned. No one outside the NRO knows just how much money has been paid for how many or what type of military mirrors. Yet those secret mirrors led directly to the manufacture of one of the most famous—indeed, infamous—mirrors of all time.

The Space Telescope

Just after the end of World War II, physicist Lyman Spitzer Jr. and astronomer Leo Goldberg shared a park bench, excitedly discussing the possibilities of a large telescope in space, and in 1946 the thirty-one-year-old Spitzer wrote a paper, "Astronomical Advantages of an Extra-Terrestrial Observatory." He envisioned a 200- to 600-inch mirror for his orbiting telescope, which would, he predicted, "uncover new phenomena not yet imagined, and perhaps . . . modify profoundly our basic concepts of space and time."

Few astronomers read Spitzer's paper, part of a classified military-industry study of the fledgling RAND (standing for Research and Development) Project. Many of those who did hear about it scoffed. Spitzer went on to a distinguished career as a Princeton astrophysicist, pioneer-

---

*The military was interested in solar astronomy because of the dramatic effect the sun can have on ground and space operations and communications. Because of the intense heat of the sun, mirrors of solar telescopes were just slightly curved so that the heat was not too concentrated and the sun's image was larger. The resulting telescopes at Kitt Peak, Sacramento Peak, and elsewhere had relatively small mirrors but very long focal lengths, requiring tall towers and deep shafts under them.

ing ultraviolet astronomy, which meant getting telescope mirrors above as much of earth's ultraviolet-blocking atmosphere as possible, first with balloons, then rockets. But he never forgot the space telescope.

In 1972, Spitzer's Princeton team placed a 32-inch mirror aboard the satellite COPERNICUS to study the ultraviolet spectra from bright stars, which it could keep in view for several minutes at a time. It circled the earth for nine years. In 1978, the International Ultraviolet Explorer, marking the first collaboration between NASA and the European Space Agency, flew a 17.5-inch mirror made of lightweight beryllium into geosynchronous orbit, where it would send pictures back to earth for eighteen years.

In the meantime, Spitzer served on various committees to study the feasibility of a large space telescope, which shrank from a proposed 400-inch mirror in 1962 to a 120-inch reflector in 1965. Spitzer and his fellow enthusiasts continued to beat the drums for what Fred Whipple jokingly named the "Great Optical Device" (e.g., GOD) but that was now officially called the Large Space Telescope (LST). Political infighting in NASA and uncertain funding during the escalating Vietnam War plagued Spitzer's brainchild.

The LST finally appeared to be lurching toward reality when Congress suddenly denied funding for the telescope in 1974. Spitzer and his younger Princeton colleague John Bahcall mounted an intensive lobbying effort, and funds were finally restored in 1977, but the mirror was downsized to 94 inches, a good fit for the planned Space Shuttle. Not by coincidence, this was about the same size as the KH-11 spy satellite's mirror. As a Lockheed executive pointed out to visiting subcommittee members, the LST plans looked "very similar to the low earth orbit satellite we developed for the Air Force," and he encouraged NASA to "keep the cost down" by paying Lockheed and Perkin-Elmer to make it. The optics would be a bit different, because the KEYHOLE spy satellites looked down while the LST would look up, but the essential technology—particularly the mirror—was quite similar.

Based in Danbury, Connecticut, Perkin-Elmer had been founded in 1937. During World War II, the firm made tank periscopes and later became well established as a manufacturer of precision optics, particularly for the military. Now, in October 1977, the company got the contract to make the LST mirror (a Ritchey-Chrétien Cassegrain design). Eastman Kodak was commissioned to make a backup mirror, just in case.

Perkin-Elmer planned to use computer-assisted polishing on a special mount to simulate zero gravity. In the spring of 1979, Corning delivered a fused silica egg-crate mirror blank. By the end of the year, the Perkin-Elmer team had finished grinding the 12-inch-thick disk to its approximate hyperbolic shape, then proceeded with ever-finer grits. In August 1980, the opticians began fine polishing.

NASA put enormous pressure on Perkin-Elmer to produce a perfect mirror in a hurry. The LST would exemplify the risky "protoflight" policy in which no prototype system was built and tested on earth. The Perkin-Elmer opticians struggled to keep their composure and produce an exquisite surface.

By this time, opticians had developed a standard, though complicated, device called a refractive null corrector that tested mirrors by "correcting" the aspheric reflection so that it appeared to come from a perfect spherical curve, then interfered the light to produce—if all went well—straight alternating black and white lines. But the Perkin-Elmer optical metrology team felt that there were too many elements in this all-lens corrector to control precisely. Al Slomba, the team director, helped make a *reflective* null corrector, whereby the light bounced between two concave spherical mirrors and a very thin, easily made field lens. It worked fine in the location where the team made it before shipping it up to Danbury. There it was used to fabricate and test a 60-inch hyperbola. Then it was prepared for use on the 94-inch LST mirror.

Because the distance between the mirrors and the small lens had to be longer than for the test on the 60-inch mirror, an outside lab prepared a measuring rod made of invar—a nickel alloy that does not change with heat—calibrated to within a couple of microns, in order to adjust the null corrector's internal spacing of mirrors and lens. A laser beam would be sent through the lens, then bounced off the shiny rounded end of the rod to determine its precise placement. To make sure the light reflected from the exact middle of the rod, a small cap with a hole in the middle was placed over the end. The cap was painted with nonreflective black paint.

The technicians measuring the proper distance to the lens were frustrated because the adjusting screw wouldn't carry it to the proper position, so they had to add small metal shims to hold it in place. Once the reflective null test corrector was put together, it was used as the sole testing device for the LST mirror. The director of metrology on the Perkin-Elmer LST polishing team had great faith in it. Thus, when an "inverse" refractive version used in conjunction with it—to check its alignment—produced wavy rather than straight interference lines, he put it down to problems with the refractor's lenses, which indeed had a known error.

Then Slomba tested the central part of the 94-inch mirror, which involved carefully aligning a specially made test plate in the central hole in the mirror. Once this vertex plug—shaped precisely to be the center of the hyperbola—was in place, exactly even with the primary mirror, the refractive null tester should have produced straight-line interference fringes. Slomba found that there were indeed straight lines across the test plug, but the lines curved when the light reflected from the main mirror. He told Perkin-Elmer's director of metrology, his boss on the project, what he'd seen, but the director again dismissed the results because the refractive tester was just "cobbled together," as Slomba recalls, and the reflective null had been thoroughly checked out.

By the time the Perkin-Elmer opticians finally finished polishing the LST mirror in April 1981, Slomba's tests indicated that it was smooth to within 1/125th wavelength of red light. The mirror was aluminized, then given a protective transparent overcoat of magnesium fluoride. *Sky & Telescope* magazine called it "the crown jewel of astronomical optics—the most precise large mirror ever made."

Meanwhile, the Space Telescope Science Institute (STSI) was set up in Baltimore to create a comprehensive guide-star system for aiming the space telescope, as well as to coordinate scientific goals, allot observing time, and handle the immense amount of data that would stream down from space.

After several more years, during which the instruments, fine guidance systems, and other intricate parts were completed, the precious mirror, inside the Optical Telescope Assembly, was shipped across the country to Lockheed headquarters in Sunnyvale, California, where it was mated with the Support Systems Module and the European Space Agency's solar arrays in 1985. By that time, it had been christened the Hubble Space Telescope (HST) in honor of Edwin Hubble, and the planned launch date was pushed back (for the second time) to October 1986. Then the Space Shuttle *Challenger* exploded on January 28, 1986, killing the seven crewmembers, and all shuttle flights were cancelled indefinitely. The much-touted Hubble mirror remained in the dark, bundled inside the telescope assembly, still awaiting its liberation and first light in space.

Jerry Nelson's Ingenious Solution

In 1977, Jerry Nelson, who as a grad student had helped Bob Leighton make his spun-epoxy mirror, found himself on a committee at the University of California–Berkeley looking into building a bigger telescope than the old 120-inch (3-meter) device at Lick Observatory. The largest mirror being considered by the committee was 6 or 7 meters wide—quite daring, considering that the Palomar mirror was 5 meters.* "Being young and adventurous," Nelson recalls (he was thirty-three), "I thought that was boring. Given that Palomar was built in the 1930s, we should be able to advance significantly over that." Nelson informed the California committee, "We should aim for ten meters." That would be twice as large as Palomar in diameter, with four times the light-collecting area.

But how to make such a mirror? Even if someone could produce such a monstrous piece of glass, polishing, aluminizing, and supporting it to maintain its shape would be incredibly expensive. Also, if something went wrong, the entire project would be ruined. So Nelson considered other alternatives, such as the multiple-mirror approach and simple spherical segments, but in the end he opted for a segmented Ritchey-Chrétien system, which required a 10-meter hyperbolic surface and would allow a relatively large field of view. But the segments would require off-axis curvatures like the sides of a bowl—impossible to create with traditional grinding and polishing methods.

"Nobody in astronomy had a scheme for making these things," Nelson recalls. A colleague suggested he try mechanical engineers. In 1979,

---

*In the late 1970s, U.S. astronomers began to use the metrical system rather than feet and inches.

Nelson called on Jacob Lubliner, a professor of structural engineering at Berkeley. For months they talked, pondered, sketched, and wrote out equations together. Finally, they came up with stressed mirror polishing, a sophisticated variant of Bernhard Schmidt's method in which his corrector plate sprang into the proper shape after being sucked down with a vacuum. But the Lubliner-Nelson method was based on pure math rather than instinct.

"If you have a circular glass plate of constant thickness," Nelson explains, "you have to apply specific shear forces and bending moments around the perimeter, without pushing on the back of the mirror." In practice, this meant gluing twenty-four small metal bars around the edge of a piece of Zerodur (a low-expansion glass), then bolting a horizontal lever on the end of each bar and attaching lead weights. "It's sort of a teeter-totter fulcrum bar to allow you to push up or down on it."

At the end of 1979, Nelson and Lubliner took a 14-inch blank with appropriately placed levers and weights to Tinsley in Berkeley, where the chief optician polished it to a simple spherical curve. When the weights were removed, it sprang magically into an approximation of the shape they wanted. More testing convinced them that they could fine-tune the process until they reached the desired off-axis curve.

Finally Thinking Big

In January 1980, Nelson presented his stressed mirror polishing technique, along with plans for a 10-meter telescope, at a weeklong conference in Tucson hosted by the Kitt Peak National Observatory. Five years before, spurred in part by the new MMT, observatory director Leo Goldberg had initiated plans for PALANTIR (Program for a Large Aperture Novel Thousand-Inch Reflector), a 25-meter mirror for what came to be known as the Next Generation Telescope.

At the 1980 Tucson conference, many who would contribute to larger mirrors spoke. Aden and Marjorie Meinel offered plans for a 10-meter "optical table telescope" featuring a segmented primary mirror. Frank Drake argued for a 35-meter Arecibo-style fixed optical telescope. The Texans planned a thin 7-meter mirror. The visionary French astronomer Antoine Labeyrie touted inexpensive molded concrete arrays that looked like puffball mushrooms for doing interferometry with an unspecified number of 6-meter mirrors.*

A few astronomers contended that with the advent of sensitive CCD detectors there was no need for larger mirrors. Sandra "Sandy" Faber, one of the few female astronomers at the meeting, sarcastically commented that "this defense of existing apertures . . . implies that, by some amazing coincidence, there is nothing interesting in the heavens to look at that is fainter than what we can see right now in the year 1980." She went on to present an overwhelming scientific case for big mirrors.

There was also much debate over the role of space- versus ground-based astronomical mirrors. Most agreed with Leo Goldberg that they were complementary to one another. When the Large Space Telescope finally flew, its 2.4-meter mirror would be unobstructed by the earth's atmosphere, but much bigger mirrors were still needed on the ground for spectroscopic work. John Hardy gave an enthusiastic presentation about adaptive optics, describing a 4-inch deformable mirror with 300 actuators and giving as much information as he could without violating military secrets. During the question-and-answer period, it emerged that the actuators cost $10,000 apiece, which meant that the mirror Hardy had mentioned must have cost around $3 million. Where would they get that kind of money? And could adaptive optics ever really work? Still, there was the potential to transform large earthbound mirrors into unblurred eyes on the universe.

Jerry Nelson, who was planning to use much slower, cheaper actuators to keep his thin segments aligned, took Hardy seriously. So did Roger Angel, thirty-nine, an ever-curious British astronomer at the University of Arizona who had too many ideas to stay long in one astronomical field. From particle physics to x-ray pulsars to studying polarized light from quasars, Angel bounced from one challenging field to another.

---

*Labeyrie had invented *speckle interferometry*, the best way to compensate for atmospheric turbulence before adaptive optics techniques made it obsolete. A millisecond snapshot of the image formed by a telescope reveals a "speckled" photo, with light broken up by the turbulence. By analyzing multiple photographs, it was possible to reconstruct something of what the star really looked like.

Neville "Nick" Woolf, another transplanted Briton, had approached Angel in 1979. As the temporary acting director of the MMT during its construction, Woolf was pondering how to scale it up for a Next Generation proposal, and he figured Angel was the idea man he needed. Between them, they came up with the MT-2 (Multiple Telescope No. 2) proposal, which they presented at the 1980 conference. The MT-2 would use eight 5-meter mirrors held in a square arrangement on an MMT-style mount, yielding an effective aperture of 14 meters. To keep the weight down, they would make the mirrors only 4 inches thick.

During the conference, Leo Goldberg made a passionate plea: "The astronomical community must unite behind a single concept and proposal if early funding of a Next Generation Telescope is to materialize." With so much money involved in making the big jump to a post-Palomar mirror, it was vital that everyone come together to support the winning design. But it was not to be.*

Roger Angel's Honeycombs

After the conference ended, Roger Angel couldn't stop thinking about how to make big mirrors. He wasn't satisfied with the flimsy thin-mirror solution, but he was impressed with the performance of the egg-crate MMT mirrors. When he got some Pyrex custard cups from a Tucson bank in return for opening a new checking account, he ventured to a pottery shop and bought some firebricks, from which he constructed a backyard kiln. A few days later, he appeared at grad student John Hill's apartment, holding the two cups, melted together. "We can make telescope mirrors out of this!" he said.

---

*Yet another innovative large mirror idea spun off the 1980 Tucson meeting. Ermanno Borra, an Italian physicist at the University of Laval in Quebec City, wrote a 1982 paper proposing a zenith-pointing telescope with a 30-meter mirror made of a rotating pool of mercury, inspiring a generation of mercury mirrors now approaching 6 meters in diameter. They are relatively inexpensive, if potentially hazardous and nonsteerable. The enthusiastic Borra nonetheless pictures huge mercury mirrors spinning in space and on the ground, "relegating tiltable mirrors to specialized niches."

Over the next few years, Angel and Hill, with input from Woolf, worked together to make ever-larger lightweight mirrors. Long ago, George Ritchey had suggested making cellular mirrors, much like those now in the MMT. "I have from the first taken as models the honeycomb, the egg-shell and the spider's web," Ritchey wrote in 1928, and Angel now took Ritchey as his hero and guide, making thin-topped mirrors stiffened by a hexagonal honeycomb backing. At first, he made them by placing short vertical cylinders next to one another, then heating them while air blew up into the tubes so that they expanded and melded into a honeycomb pattern.*

By March 1985, Angel and crew were making a 1.8-meter-diameter mirror in an abandoned Tucson synagogue by attaching carefully shaped hexagonal ceramic fiber pieces inside a round mold. On top, they put chunks of Ohara borosilicate glass (similar to Pyrex but made in Japan), and heated it to 1,200 degrees Celsius while the entire mold-furnace spun fifteen times a minute. The glass melted to fill the mold, with just enough left over to form a thin paraboloid sheet on top, then continued to spin while the oven cooled. Dan Watson, who operated the computer that controlled the procedure, sat underneath the furnace and rotated with it. "He enjoyed his celebrity as the 'oven pilot,'" Angel recalls, "though it made him somewhat giddy."

The spinning furnace allowed Angel to make a deeply dished mirror with an extremely short f/1 focal ratio that eventually formed the heart of the Vatican Advanced Technology Telescope (VATT) atop Mount Graham.** Intrigued by Jerry Nelson's method of stressed mirror polishing, Angel pondered using the same principle, but instead of bending the mirror itself, he wanted to deform the polishing "lap" to change its shape appropriately as it moved across the mirror surface. The curvature of the deeply paraboloid VATT mirror was much steeper in the center, then flattened toward the outside. Perhaps if he put actuators on the lap to tighten or loosen a steel band. . . .

---

*Tucson-based Hextek continues to make medium-sized telescope mirrors with this method, inspired by Angel's early experimentation.

**The Roman Catholic Church had an Italian observatory 22 miles southeast of Rome. The city's bright lights polluted the skies; hence the need for the VATT.

At that point, in 1986, Angel was about to move the Mirror Lab to the only place on campus big enough to accommodate the large mirrors he planned to make—underneath the stands of the University of Arizona football stadium, where a more sophisticated spinning furnace did not require a cowboy oven pilot. But he realized he needed help with the stressed lap polisher, so he brought aboard Buddy Martin, a young radio astronomer with a math background. In 1988, Martin solved the problem by putting the steel bands in a large triangle and varying the tension appropriately. Although it took a while to perfect it and program computers, the revolutionary new polishing technique worked.

That same year, Roger Angel cast a 3.5-meter mirror, with plans to work his way up to 7.5 meters.

Downsizing and Disunity

By the mid-1980s, the grand plans for a 25-meter Next Generation Telescope had shrunk to 15 meters and been renamed the National New Technology Telescope (NNTT). The Kitt Peak National Observatory had been rechristened as the National Optical Astronomy Observatory (NOAO), which incorporated a telescope at Cerro Tololo, Chile, and solar telescopes at Kitt Peak and Sacramento Peak. With oil prices collapsing, the Texans bowed out of the competition,* leaving only Jerry Nelson's segmented mirrors versus Roger Angel's honeycomb concept. Angel and Woolf now proposed a multiple-mirror telescope composed of four 7.5-meter mirrors, and they won NOAO approval in a late-1984 shootout.

Nelson wasn't overly concerned, since he had a donor, a California multimillionaire named Howard Keck whose father had started Superior Oil. The newly named Keck telescope project was launched as a consortium between the University of California and Caltech. The

---

*The Texans eventually built the Hobby-Eberly telescope, featuring an 11-meter mirror composed of ninety-one easily made spherical segments—an inexpensive but imperfect way to get a lot of light, because it has to correct for spherical aberration and is not fully steerable. It saw first light in 1999.

thirty-six off-axis segments of Zerodur were made using the stressed mirror polishing technique.

When the round segments were cut into hexagons, some warping occurred. To touch up the mirror segments, Nelson sent them to Kodak in Rochester, New York, for ion figuring, a process in which precisely directed ion beams in a vacuum chamber modify the glass surface atom by atom in targeted areas. With active mirror supports on the telescope itself, Nelson hoped to correct any remaining flaws.

In the meantime, Jacques Beckers was frustrated. A Dutch solar astronomer who had switched to night work as the MMT director, Beckers had joined the fledgling NOAO in 1984 as head of the Advanced Development Program, where he was to foster the National New Technology Telescope, build a civilian adaptive optics program, develop optical interferometry, and encourage innovative infrared detectors. Beckers approved of Roger Angel's plans to increase the four NNTT mirrors to 8 meters each, but he soon realized that there was tension between the NOAO and the University of Arizona's Steward Observatory, staring at one another across Tucson's Cherry Avenue in uneasy alliance. The NOAO director didn't like the perception that it had too cozy a relationship with Angel's Mirror Lab. It seemed to some in the federal facility that Angel's outfit was self-promoting and slow, using NOAO funds to build too many intermediate mirrors, which Angel then felt free to dispense as he pleased.

Beckers's initial enthusiasm for his job dwindled as the NOAO director whittled away at his program, which was finally abandoned in late 1987. At the same time, the NNTT was descoped to a single 8-meter mirror, a copy of the one planned for Cerro Tololo in Chile. These telescopes were eventually renamed Gemini.

Beckers departed the following year for Garching, Germany, where he joined the European Southern Observatory project. Unlike the fractious, individualistic Americans, European optical astronomers had learned the necessity of banding together for major projects. They planned to build the aptly (if boringly) named Very Large Telescope (VLT), four separate but adjacent telescopes, each holding an 8.2-meter mirror made of a 7-inch-thick "meniscus" of Zerodur blanks. As with the Keck, they would keep the mirror in shape with actuators. Beckers helped plan for interferometry between the telescopes, which would be extremely challenging, because the light waves from each had to meet as if they had arrived on earth at precisely the same instant. This would require many extraordinarily precise mobile mirrors to reflect waves back and forth to delay their arrival at the interferometer in order to produce useful fringes.

But those fringes were a long way off in 1988. First, Schott, a German specialty glass firm, had to produce the mirror blanks, which they did (much like Roger Angel) by spinning the molten glass in a paraboloid mold. Zero-expansion Zerodur wasn't nearly as forgiving as borosilicate glass because it required a crystallization process *after* the spinning. With such a large piece of glass, it was vital that the crystals form simultaneously, or the glass would fracture. After three spin-castings ended in shattered fragments, Schott finally got it right, with the first mirror blank delivered to the French polisher REOSC in 1993.

By that time, the relationship between NOAO and the University of Arizona had completely deteriorated. Roger Angel's Mirror Lab had made three 3.5-meter mirrors, one of which went to a Kirtland Air Force Base telescope in New Mexico for adaptive optics research. As an intermediate step before attempting an 8 meter–class mirror, Angel had announced that he would make a 6.5-meter mirror that would by happenstance just fit into the old MMT telescope to replace all four of its mirrors. He also intended to supply two 6.5-meter mirrors for the Magellan Project, a university collaboration at Las Campanas Observatory in the Chilean Andes. Finally, Angel revealed plans in 1992 to cast two 8.4-meter mirrors, a direct descendent of the MMT, with two big mirrors on one mount, now called the Large Binocular Telescope. He had pushed the size just beyond Japanese plans for an 8.3-meter meniscus mirror so that his would be the largest monolithic mirror in the world.

Everyone in the astronomical community assumed that the Mirror Lab would also make the NOAO Gemini mirrors, but the rules for placing closed bids posed a problem for the University of Arizona outfit. There were to be separate bids for making the mirror blank and polishing, and all bidders had to guarantee their results. Unlike big corporations like Corning or Schott, which could absorb losses on special projects by selling cookware or other products, the University of Arizona had no deep pockets, and the Mirror Lab had always made mirrors on a "best effort" basis, where a broken blank or other unforeseen disaster would be forgiven. Forced to guarantee delivery, the Mirror Lab's bid for making the blanks came in too high, and the contract was given to Corning in September 1992.

The announcement stunned many U.S. astronomers, whose loud protests, having failed to sway the Gemini board, led to the creation of a National Science Foundation investigative committee chaired by astronomer Jim Houck. The hearings, held in Tucson in January 1993, were a nightmare for the Gemini staff, who were heckled and interrupted by antagonistic astronomers. The Houck committee's conclusion favored a Roger Angel honeycomb mirror, but to no avail. After Corning threatened lawsuits and Gemini held more design review hearings, the original decision stood. Angel's miffed Mirror Lab refused to bid on the polishing, which went to the French firm REOSC.

Corning makes its Ultra-Low Expansion (ULE) glass in boules, small slabs made from spraying fused silica with titanium doping. These are then stacked, fused in an oven, cut to hexagons, set next to one another, and fused in another huge furnace to make a single slab that can be cut to the proper thickness and then slumped over a convex form made of fire bricks.

In the end, all of the new generation telescopes in the 8- to 10-meter range turned out to be successes, whether made of slumped ULE, crystallized Zerodur, or spun-cast honeycomb borosillicate. The first 10-meter segmented Keck telescope beat the others by years when it saw first light on Mauna Kea in 1992. After the traditional shakedown period, it worked magnificently. Howard Keck was so pleased that he funded most of a nearby twin, Keck II, which began operations four years later, with plans to connect them as an optical interferometer.

Adaptive Optics Comes Out of Its Dark Closet

In 1982, John Hardy gave a talk entitled "Active Optics—Don't Build a Telescope Without It!" in which he suggested using actuators to adjust the shape of large segmented mirrors as well as using a small deformable mirror to correct for atmospheric turbulence somewhere in the lightpath. Although few thought his proposals were workable then, ten years later his advice seemed prophetic. By 1992, adaptive optics had spawned an industry, and deformable mirrors, though by no means cheap, were being placed in the light path of many telescopes and were part of most new telescope plans.

A lot had happened in those ten years. Hardy's deformable mirror, mounted on a 63-inch U.S. Air Force telescope at Mount Haleakala, first functioned to unscramble atmospheric turbulence in 1982.

A team from MIT's Lincoln Lab, including Bert Willard, soon joined the Itek opticians on Mount Haleakala. "The idea was to develop high-energy lasers that could hit incoming missiles," Willard explains. Using an Itek deformable mirror mounted on a 24-inch Air Force telescope, Willard first attempted to correct for turbulence using a bright beacon on an airplane or test missile to provide enough light to adjust the rubber mirrors, but his team puzzled over a way to perform adaptive optics on enemy missiles without such artificial lights.*

Meanwhile Julius Feinleib, who owned Adaptive Optics Associates in Cambridge, Massachusetts, visited Mount Haleakala, where he watched lasers piercing the night sky in lidar (*light detection and ranging*) experiments—the optical equivalent of radar, in which laser beams are bounced back from distant objects. He had an inspiration. Why not use lasers to create artificial bright "stars" that could be used for adaptive optics? Then they could be aimed anywhere.

The idea worked, first with a laser that created so-called Rayleigh backscatter from air molecules in the lower atmosphere. At the Kirtland Air Force Base in the New Mexico desert, Bob Fugate was already experimenting with laser weapons at the Starfire Optical Range. In 1983, his team, using a small 15-inch telescope, successfully demonstrated that a Rayleigh beacon could serve as an artificial star, and Fugate then convinced the Air Force to commission a 60-inch mirror for a telescope to conduct adaptive optics experiments.

---

*At one point in 1985, a Space Shuttle carried a retroreflector to serve as a beacon for one of Willard's laser beams. The astronaut entered "10,000" into the shuttle's computer to tell it to orient properly for the 10,000-foot-high peak of Mount Haleakala, but the computer was set for miles, so the shuttle turned upside down to face a spot 10,000 *miles* from earth. Once they corrected the error, the experiment worked. In reflecting the laser from the shuttle's retroreflector, they had to take into account the finite speed of light, aiming the laser beam just in front of the shuttle's actual position.

Bert Willard's group in Hawaii switched to Rayleigh beacons with some success in the late 1980s. By then a Princeton University professor, called to a secret think tank meeting by the military, had suggested an improved laser beacon that could create an artificial star 60 miles above the earth by exciting sodium atoms—a major improvement because it allowed correction of more atmospheric turbulence. A Lincoln Lab group working at the White Sands Missile Range in New Mexico proved in early 1985 that a sodium laser beacon did indeed work.

In the meantime, civilian astronomers were independently developing similar schemes. In a letter published in *Astronomy and Astrophysics* in the summer of 1985, French innovator Antoine Labeyrie and a colleague introduced the laser-beacon concept for adaptive optics, mentioning the possibilities of both Rayleigh and sodium beacons. For astronomers, an artificial beacon meant that they would not always need to have a bright star near the faint object they wanted to observe.

The military was alarmed. When they discovered that University of Illinois astronomer Laird Thompson was about to publish an article on sodium laser beacons, they sent Major Pete Worden to convince Thompson not to publish. Worden, who had pioneered solar adaptive optics, now worked at the Pentagon as special assistant to the director of the Strategic Defense Initiative (the so-called Star Wars initiative). He came back from his Illinois mission and said, "They're going to publish anyway. Why don't we fund them?"

Worden began to lobby for declassification. As the Star Wars program wound down along with the Cold War and the departure of President Ronald Reagan,* Worden was able to cut loose some of the expensive military hardware and deformable mirrors, giving them to the astronomers. In 1991, Worden, now a colonel, took over as head of technology for Star Wars and promptly declassified most of the adaptive optics work.

---

*Few experts thought the Star Wars program would work because it required far more powerful lasers, bigger mirrors, and more sophisticated deformable mirrors than any available. When asked whether an effective laser missile defense could ever be built, Bert Willard responded, "I'll believe it when I see it." Nonetheless, the U.S. military is still trying to get improved laser beams to work.

The Perfectly Wrong Mirror

On April 24, 1990, the Space Shuttle *Discovery* finally blasted off from Cape Canaveral, carrying the long-awaited, much-hyped Hubble Space Telescope (HST) into a flawless orbit. A Marshall Space Flight Center spokesman called it "the finest optical telescope ever built. Its mirrors are virtually perfect."

But then the first pictures were beamed back to earth on May 20. At the Jet Propulsion Laboratory (JPL), John Trauger, an optical physicist who had helped make the small mirrors for the Wide Field/Planetary Camera (WFPC, pronounced "Wiffpick"), looked at his camera's first pictures along with Aden and Marjorie Meinel, who now worked at JPL. "They looked at each other," Trauger recalls, "and said that it looked like spherical aberration."

Across the continent in Baltimore, at the Space Telescope Science Institute, British astronomer Chris Burrows puzzled over one of the star images, which looked like a faint sun seen though fog. After a thorough analysis, Burrows concluded that the Meinels were right—the images indicated severe spherical aberration, which would have resulted from polishing the mirror too much toward the edges. Burrows thought it might be fixed by moving the actuators on the back of the primary mirror.

No one wanted to believe anything was wrong with the primary, particularly not NASA officials or Perkin-Elmer's new owner, Hughes Aircraft. Maybe the secondary mirror was misaligned, they said, or maybe something was wrong with the WFPC. Perhaps there was so-called gyro drift due to poor gyroscope pointing, or the telescope was still degassing, dispersing air into space. Sandy Faber, a member of the WFPC team, tried to ask an optical engineer at Hughes about the actuators, but he rudely brushed her off, warning her, "You better not let anybody hear that you're talking about them [the actuators], you'll really get in trouble."

When the Faint Object Camera, another of the five instruments on the HST, produced blurry pictures of the same type on June 17, it became clear that the trouble was with the Hubble mirror, and the actuators, designed only to fine-tune the mirror, could not fix the problem. On June 27, 1990, NASA held a press conference to tell the world that its outrageously expensive telescope was flawed. A media feeding frenzy ensued. "PIX NIXED AS HUBBLE SEES DOUBLE," the *New York Post* tittered. *Newsweek* put the telescope on its cover with the headline, "STAR CROSSED: NASA's $1.5 BILLION BLUNDER." Cartoonist Gary Larson's *Far Side* featured a blurry flying saucer holding blurry aliens, with the caption, "Another photograph from the Hubble telescope."

Investigators found the reflective null corrector at Perkin-Elmer and discovered that its lens was 1.3 millimeters (1/20th of an inch) out of position. They also discovered that the little cap over the end of the invar measuring rod was missing a flake of black paint. The laser beam intended to reflect from the end of the rod had instead bounced from the shiny portion of the cap where the paint was missing, which resulted in the misplacement of the lens in the testing device. The investigators also discovered that two different tests with refractive null correctors had indicated a problem but had been dismissed as flawed.

Why? Remember that NASA was applying immense pressure on Perkin-Elmer to finish the mirror. The mirror was holding up everything else. In a May 1991 memo, a Perkin-Elmer review group had suggested using a simple Hartmann test to double-check the primary, "to uncover some gross error such as an incorrect null corrector."

The test was never done. Kodak's backup mirror, completed in 1980, had never been compared with the Perkin-Elmer mirror with the same testing devices. Neither was there ever a ground test of the complete telescope optics, with light reflected from the primary to the secondary and thence to a focus. All of the Perkin-Elmer dark-side mirrors for spy satellites had been tested end-to-end, but NASA, frantic to prevent further cost overruns and delays, insisted on its protoflight concept—with disastrous results. The mirror did indeed feature the most perfectly polished surface ever produced, but its shape was perfectly wrong.

Morale among Hubble scientists and NASA bureaucrats plummeted. Sandy Faber stated flatly at a WFPC meeting, "Our scientific program is now fully compromised—devastated."

Around this time, WFPC team member John Trauger, about to give a talk to a California optical society, found himself standing next to Aden Meinel again. "By the way," Meinel told him, "here's how you fix it," then explained that in the WFPC camera, a dime-shaped mirror intercepting the image of the Hubble primary could be polished to reproduce the error, but in the opposite direction, so that it would correct the spherical aberration. Trauger was already working on an improved WFPC-2, due to replace the first version, so it was a relatively simple matter to make the new little mirror.

A similar idea occurred to Jim Crocker, an American engineer from the Space Telescope Science Institute, while he was taking a shower in a German hotel room, which featured a movable showerhead mounted on adjustable rods. Having strained for a way to correct the aberrated light for the other Hubble instruments, he suddenly pictured a small corrective mirror on the end of the showerhead. This was the genesis of the Corrective Optics Space Telescope Axial Replacement (COSTAR), a Rube Goldberg affair with mechanical arms that swung mirrors into the lightpath to correct the light and send it to the three remaining in-struments—the fourth, a high-speed photometer, would be sacrificed to make way for COSTAR. The heart of COSTAR consisted of "12 DC motors, 10 mirrors, 4 arms, and countless wires and sensors, all in the space of a shoebox," as one of its creators observed.

As plans were made for the all-important Space Shuttle rescue-repair mission, a new NASA administrator took over the beleaguered agency in 1992. Two years before, the editor of an astronomical journal had lamented that NASA bureaucrats tended to appear as "arrogant, boring parodies of Dr. Strangelove or as talking potatoes." NASA needed "a real communicator." In Dan Goldin, NASA had all that and more. As an engineer, Goldin had worked in deep black on reconnaissance satel-lites. He also had supervised Perkin-Elmer opticians making the superb x-ray grazing incidence mirrors for the Chandra telescope.

At NASA, the aggressive, sometimes abrasive Goldin announced that there would be no more billion-dollar programs. His mantra was "faster, better, cheaper." NASA would send up multiple smaller mis-sions, each with a specific goal. At the same time, Goldin emphasized that they must think big, push the envelope, take risks.

Except with the Hubble Space Telescope. If they failed to fix the flawed mirror, NASA's future was in jeopardy. At JPL, John Trauger showed Goldin the WFPC-2 with which, he said, they would fix the Hubble. "No," Goldin said, "what you're going to do is save the agency."

Fortunately, the $692 million shuttle mission went off without a hitch. In five days of space walks in December 1993, the seven astronauts proved that human hands could be indispensable for space mirrors, as they replaced gyros and solar panels, and installed the all-important WFPC-2 and COSTAR. In celebratory mood, the astronauts played the popular song lyrics, "I can see clearly now," and the corrected Hubble mirror did indeed yield spectacular images over the next decade. In 1994 the HST found evidence for black holes and planet-forming disks around nearby stars and provided riveting pictures of Comet Shoemaker-Levy 9 as its fragments bombarded Jupiter.

The following year Hubble captured a photo of the soaring star factories in the Eagle Nebula that, with color enhancement, became one of the most popular astronomical pictures of all time, dubbed the Pillars of Creation. For ten days in December 1995, astronomers held the Hubble mirror steady, peering toward a blank spot in space where little was known to exist, and let the scarce photons rain in. The resulting picture, known as the Hubble Deep Field, showed swarms of galaxies billions of light years away, clumped, oddly shaped, as glorious and surprising as the swimming protozoa revealed in the first microscopes.

The Hubble mirror also enabled astronomers to find Cepheid variables at a greater distance than before, allowing them to adjust the Hubble constant at which the universe is expanding, leading to an estimated age for the universe of 11 billion years or so. This was confusing, since other Hubble observations found quasars that appeared to be 14 billion years old. The cosmological debate over the size and age of the universe heated up.

Lyman Spitzer, who had championed the space telescope for half a century and had served on the panel planning the Hubble fix, was finally vindicated by the time he died at eighty-two during the evening of March 31, 1997, after a full day of work on a manuscript about interstellar matter, based on data from the Hubble Space Telescope.

Keeping Infrared Mirrors Cold in Space

As impressive as it was, the Hubble mirror was optimized only for optical and ultraviolet light. Held at a constant, moderate temperature, it couldn't be used for fine infrared observations. As the twentieth century drew to a close, astronomers realized that infrared wavelengths held the key to two vital areas of research. The light streaming from the farthest reaches of time, relatively near the Big Bang, may have been emitted in the ultraviolet or x-ray, but by the time it reaches earth, it has been red-shifted and stretched to longer infrared wavelengths. Much nearer by, astronomers wanted to find evidence for earthlike planets revolving around other stars. To look for oxygen, water vapor, and other signs of life required large mirrors optimized for the infrared. That meant *cold* mirrors, because warm mirrors emit their own infrared radiation.

Putting frigid mirrors into space proved to be complicated, time-consuming, and expensive. In 1983, after years of delay and near-cancellation, NASA (with help from the Dutch and British) flew the In-frared Astronomical Satellite (IRAS), with a small 22-inch mirror surrounded by liquid helium. The mirror and telescope assembly were made of beryllium, a lightweight, strong metal that quickly loses heat. Opticians at Perkin-Elmer had to overcome the temperamental metal's "hysteresis"—its failure to resume its original shape after heating and cooling—but they eventually got the mirror to cooperate by cycling it repeatedly through large temperature changes.

The IRAS mirror had to be so small because most of the room in the satellite was taken up by liquid helium, required to keep the mirror at 10 degrees Kelvin and the detectors just 2 degrees above absolute zero. That way, the satellite could look at far infrared wavelengths out to 100 microns. IRAS died when the last of the helium evaporated, ten months after launch, but by that time its sky survey had painted a new infrared picture of the heavens. "You could see the entire Milky Way blazing out," Frank Low recalls. Some galaxies appeared fifty times brighter in the infrared than in visible light. IRAS also discovered several comets.

Perhaps the most exciting discovery involved Vega, a bright young star only twenty-six light-years away. Astronomers wanted to use Vega, a traditional "standard candle" at visible wavelengths, for calibration at the longer wavelengths used by IRAS. But the star surprised them by demonstrating an infrared excess, eventually interpreted as a ring of debris. "The Vega ring," wrote Gerry Neugebauer in 1984, "is the first compelling example of solid matter in orbit around a star other than the sun, and it may well represent an early stage in the condensation of a planetary system."

The thrilled infrared astronomers immediately accelerated plans for SIRTF, the Space Infrared Telescope Facility, which would feature a larger mirror and would be a true observatory instead of a survey instrument.* But they still faced the seemingly insurmountable problem of keeping the mirror cold, which implied a huge helium supply and limited scientific lifetime. SIRTF team members—Low, Neugebauer, and others—struggled with these issues. By the time the Hubble was launched in 1990, SIRTF plans called for a gigantic Titan missile to launch a massive amount of helium surrounding a 36-inch mirror, at a projected total cost of $2.5 billion.

Meanwhile, Tim Hawarden, a South African working on Mauna Kea at the UKIRT infrared telescope, was involved with plans for a European Space Agency satellite, the Infrared Space Observatory (ISO), with a 24-inch mirror surrounded by helium. "It struck me as horrendously complicated," he recalls. Since space is relatively cold anyway, why not set up a sunscreen and let the telescope cool by natural radiation? His calculations indicated that it would work. In January 1990, Hawarden submitted his plan—the Passively Cooled Orbiting Infrared Observatory Telescope (POIROT)—to the European Space Agency, which turned it down, along with twenty-one other proposals. Harley Thronson, a University of Wyoming astronomer who met Hawarden while on sabbatical in Edinburgh, loved the idea and Americanized it as Edison (legendary inventor Thomas Edison had performed early infrared observations during an 1886 total solar eclipse in Wyoming).

"Tim and I did modeling on money we scraped together from leftover grants and the kindness of strangers," Thronson recalls. He paid for the engineering study and published the paper on his own credit card. In May 1991, Thronson and Hawarden formally presented Edison to a NASA scientific advisory group as an alternative to SIRTF. Using the same resources, they promised to launch a 2.5-meter mirror and bring it down to 37 degrees Kelvin without helium. It would weigh less

---

*Actually, SIRTF originally stood for *Shuttle* Infrared Telescope Facility (IRAS was a scaled-down version), but it turned out that the shuttle was a terrible environment for infrared observation because it was surrounded by a halo of dandruff, paint flecks, and dehydrated sewage that some astronomers called the "shuttle shit cloud," every particle of which glowed brightly in the infrared.

than SIRTF and would have an indefinite lifetime. Its only disadvantage would be that without getting even colder, it wouldn't be as sensitive as SIRTF at far infrared wavelengths longer than 40 microns. But with such a large aperture, it would revolutionize infrared astronomy.

Members of the SIRTF team were aghast, defensive, and afraid that this upstart plan might completely derail their expensive mission, which was already in jeopardy because of the Hubble fiasco. A JPL scientist presented hastily compiled data purporting to show that Edison would take twelve years to cool in space. Thronson was infuriated. It was still winter in Wyoming, and Thronson convinced a neighbor to stop heating his guesthouse trailer, which was about half the mass of Edison. It took only a few days for the trailer interior to match the frigid outdoor conditions. "These JPL guys lived in Southern California. What did they know about the cold?"

After several versions of Edison failed to secure funding, Thronson burned out on the project.* In 1996, he joined NASA, where, ironically, he oversaw SIRTF, due for launch in 2003, before being appointed science chief of the NASA Exploration Team (NEXT), which is exploring the possibility of a new space station between the earth and moon to serve as a human outpost and possible construction/launch site for future observatories.

Hawarden and Thronson did have an impact on SIRTF plans, although few on the infrared scientific team will admit it. The beryllium mirror is still only 33.5 inches, accompanied by 95 gallons of liquid helium, but it takes advantage of passive radiant cooling to extend the life of the mission considerably. If all goes well, it may last five years or longer.

Gossamer Mirrors in Space

But the real POIROT/Edison legacy will be the Next Generation Space Telescope (NGST, recently named the James Webb Space Telescope),

---

*Thronson helped secure funding for the Stratospheric Observatory for Infrared Astronomy (SOFIA), a Boeing 747 that will fly a 2.7-meter mirror for infrared observations high in the atmosphere to replace the smaller Kuiper Airborne Observatory, grounded in 1995. SOFIA will finally fly in 2004.

the successor to Hubble, which will feature a 6-meter (20-foot) light-weight mirror cooled by passive radiation, just as Tim Hawarden first proposed in 1982.

That is one of the few certainties about the NGST. Like most major astronomical space plans, its size, shape, and launch date keep changing. In 1994 and 1995, Alan Dressler of the Pasadena-based Carnegie Observatories (formerly called the Hale Observatories) chaired an eighteen-person committee that published its recommendations that the NGST pursue two goals: "(1) *the detailed study of the birth and evolution of normal galaxies such as the Milky Way,* and (2) *the detection of Earth-like planets around other stars and the search for evidence of life on them.*" With a budget of $500 million, they suggested that the NGST mirror be 4 meters or larger, optimized for near infrared wavelengths.

In a January 1996 speech to the American Astronomical Society, Dan Goldin recounted how he had talked to astronomer Geoff Marcy the night before, who had told him that he had found indirect evidence for two Jupiter-sized planets orbiting nearby stars, and Goldin enthusiastically committed NASA to the search for other earthlike planets. For that, he said, they would need a huge NGST mirror. "I see Alan Dressler here. All he wants is a four meter optic." Why not double that goal to *eight* meters?

With this ambitious goal in mind, Lockheed Martin, Northrop Grumman, and Roger Angel's team at the Mirror Lab scrambled for the prize. Northrop Grumman's beryllium design was chosen in late 2002, though the mirror size had shrunk back to a more manageable 6 meters. The NGST is scheduled to blast off in 2009, but no one knows when it will really fly or precisely what it will look like. The mirror will have to deploy, unfolding like a flower into a precise configuration controlled by actuators, and it will rest in space 1.5 million kilometers from earth, far from any possible service by astronauts.

The big space project slated to follow the NGST is the Terrestrial Planet Finder (TPF), which will strive to identify earthlike planets orbiting other stars, an incredibly challenging venture because the planets will be tiny glowworms in the searchlight of their suns. To get the resolution to spot such planets will require free-flying telescopes in space so precisely aligned that they can perform nulling interferometry in which the infrared waves from the star would interfere destructively to cancel themselves out, allowing the long-sought planet to stand out. An alternative plan for the TPF features a coronograph in which a small ball would be placed in front of a single mirror to block the starlight and permit the planet to be seen, but it would require an absolutely perfect mirror. JPL's John Trauger believes he can produce it with a new deformable mirror that can correct irregularities so tiny that it has no application on earth.

Looking out even farther, NASA's Gossamer Optics project encourages wild futuristic plans for placing huge, fragile, accurate mirrors in space.* They will have to be unfolded or constructed in orbit. Most Gossamer plans use existing technology pushed to the limit, such as thin-shell graphite fiber composite or nickel mirrors, stretched reflective mylar, or inflatable membranes. Mark Dragovan of JPL has two proposals. One is a variant of the Kirkpatrick-Baez crossed-mirror scheme, in which two curves at right angles to one another create a paraboloid focus. The other involves blowing bubbles of fast-curing epoxy in space, using specially shaped frames, and then aluminizing them. On the far visionary fringe, Antoine Labeyrie promotes the ultimate lightweight mirror. It would consist only of metallic molecules trapped by laser beams acting as "optical tweezers" to hold the diaphanous reflective curtain in the proper curve.

Viable plans for solar sails—huge expanses of thin reflective sheets that would allow spacecraft to be propelled by light, either from the sun or powerful lasers—probably remain far in the future. This intriguing idea, first proposed by the Soviet Konstantin Tsiolkovsky in 1921, awaits much lighter weight material, although Louis Friedman, the executive director of the privately funded Planetary Society (based in

---

*Some space mirrors may be better left as fantasy, such as American Michael Lawson's Space Marketing, Inc., which wants to place inflatable aluminized Mylar billboards in space, the last frontier for advertising Pepsi and Coke. A stalled Russian project called Znamya (banner) has plans to unfurl giant mirrors in space to control the world's weather, directing sunlight to Siberia, major cities, and disaster areas and, critics fear, upsetting the earth's fragile ecosystem and causing light pollution.

Pasadena), is determined to demonstrate its viability soon on a limited earth-orbiting basis, with help from Russian rockets.*

Overwhelmingly Large Plans on Earth

Meanwhile, huge but considerably heavier mirrors were in the works back on earth. By the beginning of the twenty-first century, sixteen post-Palomar telescopes with 6.5- to 10-meter mirrors had seen first light or were nearing completion. The two Kecks on Mauna Kea and the four separate mirrors of the Very Large Telescope (VLT) in Chile were beginning to make interferometry work, and Roger Angel's two Large Binocular Telescope (LBT) mirrors awaited installation.

Some of the new mirrors were made of honeycomb borosilicate, others segmented, and still others meniscus. "Looking back," Alan Dressler observed, "people were partisan that theirs would work while others wouldn't, but the lesson is that they all work." And they worked not only in the formerly dominant United States and Europe but also in many other countries from South Africa to Japan.

With the full promise of adaptive optics still years away, astronomers geared up for another big leap in mirror size, gambling that multiconjugate adaptive optics—a system originally proposed by Jacques Beckers in which multiple deformable mirrors correct for different atmospheric layers—would unscramble most of the light thus captured, making the projects worthwhile.

It appears that Jerry Nelson's team, which has secured funding, may again beat everyone else with the California Extremely Large Telescope (CELT), a scaled-up 30-meter version of the Keck with 1,080 segments. The 30-meter Giant Segmented-Mirror Telescope planned as a national telescope by NOAO/Gemini soon merged with CELT. Roger Angel has proposed the 20/20, featuring two 21-meter mirrors, each composed of seven segments, mounted on a circular track, so that they can always be maneuvered to receive light at precisely the same time.

---

*In December 2001, the visionary Dan Goldin was replaced as NASA administrator by businessman Sean O'Keefe, and it is unclear which programs he will back, especially in the wake of the 2003 *Columbia* Shuttle disaster.

Thus, as with the LBT, interferometry will be much easier, without the need for complicated delay lines involving dozens of extra mirrors reflecting and relaying the light.

Paul Hickson of the University of Columbia, whose 6-meter paraboloid mirror made of spinning mercury is nearing completion in Canada, is pushing his cheaper alternative as the Large Aperture Mirror Array (LAMA), consisting of eighteen 12-meter liquid-mercury telescopes placed near one another in New Mexico or Chile. The mercury mirrors are not only inexpensive but can be cleaned as easily as a swimming pool. Aside from the possible health hazards, they have another drawback—they point straight up, although they would feed optics that could look at a somewhat broader patch of sky.

Solar astronomers are also looking toward the future. Incredibly, the forty-year-old McMath-Pierce device on Kitt Peak, with its 1.6-meter mirror, remains the world's largest telescope to study the sun. Equally remarkable is what we still don't understand about our nearest and dearest star. Over the years, various ambitious plans for bigger solar instruments have been scuttled or scaled down. Now the Advanced Technology Solar Telescope (ATST), the brainchild of Jacques Beckers, may actually come to be, with twenty-two institutions, headed by the National Solar Observatory, collaborating on it. Its 4-meter mirror will be mammoth as far as solar telescopes go, posing a major challenge. Most solar telescopes have required very long focal ratios (around f/65) to prevent frying the secondary mirror, but for the ATST such a focal length would require a shaft 800 feet long. Instead, the ATST will resemble a much shorter night-time telescope with an f/3 focal length, requiring a well-dished mirror. The mirrors will be cooled, and some sort of "heat stop" will somehow have to siphon the heat away, though it is not clear how.

The most exciting plans for new telescopes, jumping beyond 30 meters, come from Europe. Under the leadership of Danish astronomer Torben Andersen, the Swedes have worked up a 50-meter telescope design and, joining with Finland, Ireland, and Spain, are pushing their renamed Euro50 Telescope, with off-axis segments to be manufactured with an innovative polishing technique pioneered by David Walker at the University College London Optical Science Laboratory.

The most outrageous proposal, from Italian astronomer Roberto Gilmozzi, is the Overwhelmingly Large Telescope (OWL) with a 100-

meter segmented mirror. "Everyone thought I was completely crazy when I started talking about this," Gilmozzi recalls, so he couldn't discuss it at European Southern Observatory headquarters in Garching, Germany. Instead, over numerous beers in Munich's *biergartens*, he and his associates plotted how to build a mirror the size of a football field.

They plan to mass-produce identical, easily made spherical segments. If Schott and REOSC can produce and polish one per day—which they say they can—it would take about five and a half years to make 2,000 segments. Then they would reflect the light from a huge flat secondary, which would forgive some of the jitters inevitably produced by wind, thence to 8-meter corrective mirrors that would take out the spherical aberration. After moving to Paranal, Chile, as director of the Very Large Telescope, Gilmozzi continues to champion plans for the monstrous mirror, which may just be feasible, though it must overcome wind effects and will require sophisticated adaptive optics.

The Fate of the Universe

The new generation of mirrors will enable us to look ever farther back into time and space. In 1998, two teams of astronomers, looking at newly identified standard candles called Type IA supernovae, peered as far back as possible with large mirrors such as the Keck, the Very Large Telescope, and Hubble. They concluded that stars and galaxies are not only fleeing from another in the wake of the Big Bang but are *accelerating*. This resolved the apparent contradiction about the age of the universe, which is, at around 14 billion years, older than previous redshift figures from the Hubble indicated.

The startling discovery of a runaway universe calls for an inexplicable force that astrophysicists have dubbed dark energy, and they have fudged by creating a so-called cosmological constant to make all the math work—an irony in itself, because Einstein originally posited such a constant to accord with a theoretically static universe, and then later called it his "greatest blunder."

It appears that the more we learn about this incredible world and universe we inhabit, the less we understand. What *is* the dark matter and dark energy that makes up most of the universe? We really have no idea.

But we *do* know that the reflective metals used for our mirrors were produced in the incredible heat and pressure of star interiors, then blown into space by supernova explosions. Carefully shaped as reflectors, this metal in turn allows us not only to search for distant galaxies but also to keep a lookout for asteroids or comets such as that which probably wiped out the dinosaurs. Unless we somehow manage to deflect near-earth objects, the odds are good that a celestial bullet will someday render human beings extinct.

Until that day, however, we will remain a curious species, using mirrors not only to look out into the cosmos but also to peer into our own eyes, searching for answers that may be as elusive as the explanation for dark energy. The human soul holds its own share of dark matters.

Chapter 13

FINAL REFLECTIONS:
ILLUSIONS AND REALITIES

What is your cosmos but an instrument containing small bits of colored glass which, by an arrangement of mirrors, appear in a variety of symmetrical forms when rotated?

VLADIMIR NABOKOV, *BEND SINISTER*

"Are you coming or going?"
"Yes! Yes!"

CONVERSATION OVERHEARD
IN MIRROR MAZE

As I grope my way through the mirror maze in Lucerne's Glacier Garden, I keep bumping into a mirror where it appears that I am walking down a long corridor framed with Moorish columns. At the far end of the corridor, I see a young man approaching. He keeps disappearing as both of us turn corners or pursue blind alleys. Finally, I bump into another mirror, turn right, then left, and there he is again, now quite close. He reaches out and, as he touches my face tentatively, he asks, *"Sind sie echt?"*

In German, he is asking me if I am real. I laugh and say *"Ja,"* and he laughs too. But for a confusing moment of disequilibrium, he wasn't joking. That's what mirrors can do to us human beings. They can jolt us out of reality or fantasy with equal ease.

My younger brother once worked in a factory where he was the only white laborer. One day he caught a glimpse of himself in a mirror down a hallway and thought, "What is that white boy doing here?" Then he realized he was looking at himself. Most of us have had experiences like that, moments of shock when we see a stranger before recognizing ourselves. When it once happened to Sigmund Freud on a train, it caused him "profound displeasure." In researching this book, I had more than my share of such moments, though I generally found them intriguing.

In this final chapter, I decided to come out from behind the impersonal authorial mirror, which I've held up to reflect this history of humans and reflective surfaces, and thus to become somewhat more *echt* and to share some of the extraordinary experiences my mirror quest brought me.

Distortions, Magic, Psychics, and Feng Shui

The Prague mirror maze atop Petrin Hill, known as the Bludiste, was built in 1891. "I see my back at the end of one aisle," I wrote in my journal there, "my side in another, and I can look sideways at myself writing in this notebook. It is strange to turn everywhere and see yourself. There is a big fly in here with me, whacking against the glass, buzzing madly."

At the end of the maze is a superb gallery of distorting funhouse mirrors installed in 1911, which turned out to be the real attraction. Children shrieked and giggled while adults flirted and laughed. One mirror makes you look like a dwarf with cute little knees but a very long body. Another stretches your head into grotesque shapes. As in the Lucerne mirror maze, the laughter here rippled over barely concealed anxiety. *Do I really look like that? Could this be a part of who I am in my dreams?* Later, at the London Natural History Museum, I saw a woman in a low-cut blouse look at herself in a distorting mirror, where she saw her breasts absurdly stretched downward. She gasped and backed away, exclaiming, "I've seen the future!"

In Paris, I attended one of the last traditional performances in the Palais des Mirages, a large hexagonal mirror room installed in 1882 at the waxworks Museé Grevin.* More than a hundred people came to

---

*In order to push more paying customers through the mirror room, the management decided in 2001 to halt the shows, simply letting people wander through instead.

see themselves and the fantastic light show reflected back and forth to infinity. While eerie music played, the cornices between the mirrors featured Arabs, elephants, Shivas, and Buddhas, until the lights went off. Children gasped. When the lights came up again, the cornices had rotated to reveal a primeval forest scene, followed later by ethnic women of different cultures.

Adrian Fisher, the modern mirror-maze master, creates optical and other labyrinths throughout Europe, Asia, and the United States, working from his home in England. Fisher, who relishes both precise geometry and fairytales, plans his mazes with tiny mirrors in which he can study sight lines, using colored balls on the end of sticks. "The whole thing about a mirror maze is to create a kind of surreal experience, entering what you know is a small space that feels five times larger," he says. "You move a fraction and something else moves as well. It's quite an overwhelming, abundant experience, teasing you at every turn."

Surreal would describe the effects of mirrors that I found in the three French museums of magic I visited in Cap d'Agde, Blois, and Paris. They allowed cubes to appear suspended in midair, made a narrow panel look like a deep box with lights recessing within it, and permitted a woman's disembodied talking head (Le Femme Fleur) to appear in the midst of a bouquet of flowers in a box. Half-silvered one-way mirrors looked like regular reflectors until a light on the other side rendered them transparent, transforming my own face into a devil's or a pig's.*

In the Blois museum dedicated to the great nineteenth-century magician Jean Eugène Robert-Houdin, I walked through a narrow passageway that made it seem as though I were negotiating a bridge over an abyss because of slanted mirrors on either side and on top. In La Salle des Illusions, I saw my head upside down, sideways, stretched, transmogrified into a monster, and set atop numerous chess pieces in an Alice in Wonderland–type setting.

Humans love illusion and magic as long as we feel sufficiently safe. That's why the classic Pepper's Ghost illusions at Disneyland's Haunted

---

*On my long train rides through Europe, I noticed the same effect in reverse. When the train entered a tunnel, the otherwise transparent windows turned into mirrors.

Houses are so popular. At Eurodisney outside Paris, I watched the translucent ghosts waltz and dematerialize, just one scene on a semi-scary ride filled with witches and goblins. At the end, the car jerked past a series of mirrors. In the last one, a skeleton appeared to be grinning at me and hanging onto the top of my car, but when I looked up, there was nothing there.

During that Haunted House ride, I saw a woman's head talking inside a crystal ball, which reminded me of the scrying tradition in magical reflective surfaces. Just as a scrying craze ended the nineteenth century, magic mirrors have made a comeback as part of the New Age silliness that ushered out the twentieth century. For instance, in 1995 Donald Tyson, a Nova Scotian who describes himself as a "practitioner of ceremonial high magic," published *How to Make and Use a Magic Mirror: Psychic Windows into New Worlds.* In it he tells readers how to construct a black mirror and, by dedicating it to the moon, charge it with "lunar magnetism." As with most pseudosciences, heightened expectation is essential to success. "Call to the spirit by name, at the same time holding in your mind a clear image of the spirit. . . . Begin to talk to it as though it were already present. . . . As you do this, scry into the depths of the glass and strive to bring the face of the spirit upward."

Feng shui, the now-popular ancient Chinese folklore, uses mirrors to redirect *chi* (invisible energy) or to deflect evil spirits. "I'm wondering if it is bad feng shui for our house and our family to have cars pointing themselves at the house," a worried homeowner inquires of Kirsten Lagatree, feng shui advice columnist for the *Los Angeles Times.* Yes, she answers, it is *terrible* feng shui. "Hanging small convex mirrors over the garage . . . might be helpful." Lagatree advises another reader to hang a mirror on a stair landing "to reflect and disperse the incoming chi before it can rush out again." Feng shui mirrors are also useful on desks so that no demons or bosses can sneak up on you.

Kristine Nyhout of Toronto explains that since feng shui dictates that "no one in a home should have their heads cut off by a mirror," she and her tall husband moved all of the mirrors up, so that Kristine can only see her eyebrows and forehead. "It's a bit difficult to do my makeup," she writes, but this small sacrifice is worth the avoided hazard.

The hardheaded Malaysian *Business Times* explains that "mirrors placed behind or at the side of a cash register are believed to enhance the profit of the proprietor," then offers more intimate advice not to situate mirrors facing a bed. "A mirror has mystical force and can disturb the soul of a sleeping person." A mirror beside the bed, however, can ward off evil spirits.

Sun-Struck in the Pyrenees

High in the French Pyrenees, I found huge solar mirrors that could not only fend off but also incinerate any evil spirits. They are the legacy of a long tradition of French solar research inspired by the legend of Archimedes. In 1747, for instance, Georges-Louis Lecler, comte de Buffon, a famed naturalist, used ninety-eight mirrors set in a paraboloid frame to ignite a tar-smeared plank at 126 feet. In 1872, August Mouchet completed his solar engine, an immense conical mirror that boiled water for a steam engine.

In 1946, Felix Trombe, a chemist seeking to study the effect of high temperatures on rocks and minerals, used a surplus military searchlight to concentrate sunlight, then convinced the French government to fund a large circular solar mirror at Mont-Louis, a seventeenth-century Pyrenees military citadel. The sun in those mountains is particularly strong, since its rays penetrate only a thin atmosphere, usually unobstructed by clouds. In 1968, Trombe built an even more ambitious solar furnace in nearby Odeillo.

The front side of the eight-story Odeillo building is a giant paraboloid mirror, fed by sixty-three flat sun-tracking heliostat mirrors on a sloping hillside in front of it. When I approached it in late June 2000, I was struck by the contrast between this technological marvel and the quiet little medieval town that hosts it. Cows grazed nearby while the ancient church in the middle of the hamlet hosted a giant bonfire for Saint Jean's Day, a semipagan Catalan celebration of the summer solstice. Yet the sun, which can fry your skin before you know it in Odeillo, is vital to both scientist and peasant.

As he gazed out his window at the snow-covered mountains, Gabrielle Olalde, the director of the huge Odeillo mirror, commented that he lived in paradise, with an average 250 days of perfect weather. On his desk is an inch-thick piece of iron with a hole in it, from which frozen iron tears flow—evidence of the 3,000 degrees centigrade reached at the solar mirror's focus.

A few miles away at Themis, an array of 200 heliostats was installed in 1981 to direct sunlight to a paraboloid mirror atop a tower, which concentrated it to boil water and produce electricity. As with the similar Power Tower in Barstow, California, it proved too expensive and was abandoned within a few years.* Now the Themis facility is used only by astrophysicists who study brief flashes of Cherenkov radiation caused by gamma rays hitting the upper atmosphere. Meanwhile, French engineer Denis Eudoline has restored the old Mont-Louis mirror, using it to fire ceramics, make jewelry, and promote solar energy in underdeveloped countries.

The Odeillo facility is still used for scientific research,** primarily to simulate extreme conditions such as those that the NASA/European Space Agency Solar Probe will encounter when it is launched toward the sun in a few years. "Thanks to the control of sunlight by men on earth," Olalde told me, "we can go to explore the sun. I like the idea of this cycle."

Down the hill, the French military has its own solar furnace, initially used to study the effects of nuclear bombs on people and their protective clothing. The U.S. military has a solar furnace at White Sands Missile Range in New Mexico, where young pigs in military uniforms were once reputedly tested and incinerated.

Factories, Explosions, Blind Spots

Saint-Gobain made the original mirrors for the Odeillo facility. In Paris, I visited the sleek mirrored skyscraper headquarters (known as Les Miroirs) of this venerable firm, founded by the Sun King, Louis XIV. Now a multinational corporation, Saint-Gobain makes only a small

---

*The original Barstow Power Tower ceased operation in 1988, but in 1996 it was upgraded as Solar Two, using a salt solution to store energy for use on demand, though it still costs far more than fossil fuels.

**There are similar solar mirrors in Almeria, Spain, Cologne, Germany, Tashkent, Russia, and elsewhere in Brazil, Israel, and Australia.

fraction of its profits from mirrors. Then, in the northern French town of Aurys, I toured the only Saint-Gobain mirror factory in France. Thirty nozzles spray a silver nitrate solution over glass pieces ten feet wide as they travel down a conveyor belt. After the silver deposit forms, the new mirrors are rinsed with distilled water, followed by a secret copper-free protective spray and a pass through an infrared oven to bake it all on.

Even though the process is automated, it is not without its hazards. The silver nitrate can explode; every mirror-silverer is familiar with this hazard. British experimental psychologist Richard Gregory recalls that his father, who served as an astronomer in Egypt just after World War I, had trouble with thieves. The astronomers solved the problem by leaving small pools of silver nitrate (used to resilver the telescope mirrors) on the floor. "In the middle of the night," Gregory says, "stealthy footsteps became small explosions—followed by yells of terror and a rapid exodus."

In Weiherhammer, Germany, I visited a Flachglas float-glass plant, now owned by Pilkington, the British firm that invented and introduced the revolutionary process in 1959. There are hundreds of similar float-glass plants around the world, mostly owned by the Big Four in flat glass: Pilkington (Libbey-Owens-Ford in the United States is a Pilkington subsidiary), Saint-Gobain, Belgium-based Glaverbel, and Japan's gigantic Asahi, which owns AFG (formerly American Flat Glass) and a chunk of Glaverbel. I was awed by the mountains of pure white sand that, mixed with soda, limestone, and other ingredients, goes into a gigantic vat from which a broad ribbon of new red-hot glass flows continuously, twenty-four hours a day, for years at a time. The two Flachglas lines produce 1,350 tons of glass a day.

By the end of the 1,600-foot line, the glass has cooled and is cut and stacked. If something goes wrong, the line cannot stop. The glass must all be smashed as it comes off the line and returned to the oven as cullet. I asked plant manager Joachim Bretschneider what would happen if a person fell into the vat. "It might cause a little local brown color in the glass."

A great deal of this glass is used for sideview and rearview automobile mirrors. When I rented a car in France, I noticed that the driver's-side mirror was rounded with a small convexity on the outer edge. It took me only a few minutes to adjust to the slight distortion, and I was delighted that it avoided the notorious blind spot that all American cars have, since our laws insist that driver's-side mirrors must be perfectly flat. Only passenger-side mirrors can be convex, and they must feature the wording, "Objects in Mirror Are Closer Than They Appear."

Back in Vermont I bought a convex mirror for my driver's side. If this book accomplishes nothing else, I hope it provokes a change in this misguided American law. We can now purchase cars with all manner of high-tech mirrors that automatically dim harsh lights, but we still have a blind spot.

Mirror Artists

By themselves, mirrors are empty slates, as the suicidal poet Sylvia Plath observed: "I am silver and exact. I have no preconceptions. / Whatever I see I swallow immediately / Just as it is unmisted by love or dislike." But I met mirror-makers who are artists, projecting pride and even love onto their products. Take Guido Barbini on the island of Murano. His family tree includes Gerolamo Barbini, one of the Murano mirror-makers lured to France in 1665, only to return three years later. With his older brother Cesare, Guido Barbini carries on the family tradition, silvering small pieces of glass by hand. These are then attached to the borders of the main mirror (made of already-silvered float glass). "The place is quite messy and nothing special to see," I wrote in my journal, "until you go through a wooden door to the showroom & Boom! exquisite mirrors everywhere, ornate frames, engraved glass."

Far to the north in Newcastle-upon-Tyne in northeastern England, I met David Sinden, now age seventy, who looks and sounds uncannily like Sean Connery, though his Northumbrian accent isn't quite Scottish. At fourteen, he made a telescope, propped it on a cushion, and aimed it out the window. "Suddenly Saturn swung into view, this little planet with rings around it, dazzlingly clear and bright and sharp, and my hair stood on end. I was hooked."

Sinden later worked in a chemical factory, but in his spare time he made more than 100 telescope mirrors for friends and amateur astronomers before landing his dream job at Grubb Parsons. "I was blazing with enthusiasm," Sinden recalls, and at twenty-nine he was appointed chief optician. During his twenty years at Grubb Parsons, Sinden oversaw the creation of some 80,000 custom-made mirrors and lenses. "Every one would have a little love," Sinden says, "along with sweat, toil, grit, and grease."

Since the demise of Grubb Parsons in the mid-1980s, Sinden has carried on the tradition with his own small company, making specialized optics. "We've made some things which only a crackpot would work on, but I've done it because of the sheer creative pleasure." Among other things, Sinden has made superb camera obscuras mounted high above Cadiz, Spain; Lisbon, Portugal; and Havana, Cuba, offering incredible live views of those cities. He refused to make the all-aluminum mirror for the revamped Leviathan at Birr Castle, since it is inferior to glass.

For Sinden, making a beautiful mirror, accurate to within one-millionth of an inch, is an end in itself. "Although I have a great passion for astronomy, the real truth is that I don't care if my mirror goes into a telescope or not. You could hang it on the wall as a work of art as far as I'm concerned." Sinden likens himself to Tennyson's Ulysses, quoting from memory: "How dull it is to pause, to make an end, / To rust unburnish'd, not to shine in use!" He wants to continue to make mirrors shine, too. "My express intention is to live to 100 and make mirrors until I am 99."

David Sinden is representative of many of the passionate, meticulous, obsessed mirror-makers I met in my travels. At Stellafane, atop Breezy Hill near Springfield, Vermont, amateur astronomers gather every summer to show off an extraordinary range of homemade mirrors and telescopes and to stay up all night peering at the sky, weather permitting. They can also attend workshops on "pushing glass" to learn to make their own mirrors.

At the 2000 Stellafane, I met Bert Willard, the author of a wonderful biography of Russell Porter. Willard began to attend meetings of the Springfield Telescope Makers as a thirteen-year-old in 1953. He went on to a career at MIT's Lincoln Laboratory, where he pioneered adaptive optics, but he loves Stellafane above all else. So does his contemporary Paul Valleli, who became a TN (telescope nut) after landing a job at the Harvard Observatory as a teenager, then going on to work at Itek and other optical firms, where he made many of the military mirrors flying over our heads as well as those aboard the Voyager and other NASA space travelers.

Bob Thicksten, who has lived atop Mount Palomar for nearly a quarter-century, polished his first mirror when he was sixteen. As the site manager, he is still in awe of the Hale Observatory, with its great 200-inch mirror. "Many people have referred to it as a cathedral," he told me, "and I get that feeling." So did I, particularly when I was allowed to ride up the curved track elevator and look down into the depths of that lake-like mirror. Even though Thicksten is responsible for the famous 200-inch, however, he still refers to the 12.5-inch mirror he made as "my precious," a phrase he adopted from Gollum in *Lord of the Rings*.

Farther north, I met David Hilyard, the University of Santa Cruz optician for the Lick Observatories atop Mount Hamilton. As a teenager, Hilyard got a job sweeping floors at Laser Optics in Danbury, Connecticut, and found his calling. "A good optician is meticulous but calm," he told me. "It takes ten years or so before you really know what you're doing." Hilyard has a callous on his thumb from using it to polish small zones of a mirror in its final figuring. "About ten minutes before you make your last mistake is when you should call a mirror done."

Most mirror-makers are male, yet one of the original 1920 members of Russell Porter's Springfield Telescope Makers was Gladys Piper, and I met several mirror-making women at Stellafane, though they were a distinct minority.

I also met two famous women who use mirrors to find comets and asteroids. In 1982, after her children were grown, Carolyn Shoemaker began to help her husband, Gene, look for near-earth objects on the battered 18-inch Schmidt telescope (the "Baby Schmidt") at the Palomar Observatory, where writer-astronomer David Levy eventually joined them. Carolyn used a "stereo microscope," a modified form of Wheatstone's reflecting stereopticon, to look at two negatives of the night sky taken about an hour apart. "I discovered that I have good stereo vision," she told me. "When you put the two films on the stereo, everything lies down nice and flat except comets and asteroids, which look as if they are floating or sinking." She became a kind of latter-day Caroline Herschel, discovering more than thirty comets. Now widowed, she and Levy continue the search once a month at his Arizona home observatory.

By the time Carolyn Shoemaker commenced her hunt for comets, Eleanor "Glo" Helin had already been searching for near-earth objects with the Baby Schmidt for nearly a decade. In 1976, she discovered Aten, the first asteroid with an orbit smaller than the earth's—and a prime candidate to hit the earth at some point. Her hunt began as a collaboration with Gene Shoemaker, and she has continued the search on her own after they had a falling-out.

Helin also peered at mirrors in a stereo microscope to spot comets and asteroids, but she now uses two bigger telescope mirrors to search the sky robotically—with an Air Force telescope on Mount Haleakala in Hawaii and the newly revamped 48-inch Schmidt atop Mount Palomar—in the Near Earth Asteroid Tracking (NEAT) program she runs from her office at the Jet Propulsion Laboratory in Pasadena.

John Dobson, Mirror Evangelist

The most extraordinary mirror-maker I met is a former Hindu monk named John Dobson. His maternal grandfather founded Peking University, and his father taught zoology there, which accounts for Dobson's birth in China in 1915. The Dobsons reluctantly left China's social and political unrest, arriving in San Francisco in 1927. Distressed by the contradiction between the Golden Rule and the threat of hell, Dobson became a "belligerent atheist" until he attended a life-changing lecture by Swami Ashakananda in 1937. Eventually, after securing a degree in chemistry and working on the atom-bomb project, Dobson joined the San Francisco Vedanta monastery in 1944.

The strict life of prayer and celibacy did not dampen Dobson's curiosity. "I wanted to see what the universe looks like." He made a tiny, ineffective refracting telescope from lenses he got at a junk store; then one of his fellow monks told him that he could grind his own mirror for a reflecting telescope. Dobson recalled seeing a slab of porthole glass on a friend's kitchen table, so he requisitioned it, then found another at a marine salvage store. With instructions from a library book, he ground these together with carborundum grit, the top piece turning into a concave mirror and the bottom into a grinding tool.

After polishing the spherical mirror, he parabolized it by taking it a tiny bit deeper in the middle with jeweler's rouge, coated it with a solution of silver nitrate, ammonia, and glucose, put it into a cardboard construction tube, and equipped it with a flat front-surface secondary scrounged from a junk store, along with an eyepiece from old binoculars. Then Dobson pointed his new telescope toward the third-quarter moon. "I couldn't believe what I saw. It looked like you were coming in for a *landing*," Dobson recalls. He thought, "Everybody's got to see this." He had found his life's mission.

Shortly thereafter, in 1958 Dobson was transferred to Sacramento, where the monks were building a retreat. He began making 5.5-inch mirrors from the flat bottoms of gallon jugs, screening sand for ever-finer grits. "I had to do all of this on the QT," Dobson says, "because this wasn't part of my duties at the monastery." At night he would sneak out and grind his mirrors in a sympathetic neighbor's pump house. Buck Turgis, a visiting San Francisco monk, was flabbergasted by viewing the rings of Saturn and said, "J. D., you've got to make bigger mirrors."

Turgis began to send clandestine shipments of 12-inch surplus port-hole glass from the marine salvage store, hidden in monastery orders for 100-pound bags of fertilizer. "They couldn't understand how 100 pounds got so heavy," Dobson recalls gleefully. He and Turgis developed a secret code in which telescopes were referred to as "geraniums." If Dobson wrote that he had a 12-inch potted geranium in bloom, it meant that the mirror was successfully silvered and in its tube.

Dobson put wheels on his telescopes and trundled them through the streets of Sacramento. Kids would ask, "What's that?" and were hooked once they looked through the eyepiece. Dobson would then ask the mother if she would mind storing it in the garage for a month or so. The obsessed monk couldn't keep his mirror-making a secret forever. The local swami didn't mind so much, but the Vedanta abbot in San Francisco warned him that he must choose between his vows and his avocation. Finally, in 1967 Dobson was thrown out of the monastery.

He began a nomadic existence, staying with different families who had his telescopes. "I lived on dog biscuits and cider for a while." By the time he moved to San Francisco, there were fifteen telescopes sprinkled in Sacramento homes, each with a 12-inch mirror made of port-hole glass.

Dobson founded the Sidewalk Astronomers, which has since grown into an international organization. He also began to teach classes in mirror-making and cosmology at museums and community centers. He and his followers set up telescopes—with names like the Psychedelic Giraffe, Tumbleweed, and Stellatrope—on every clear night at the corner of Jackson and Broderick Streets in San Francisco, and then Dobson began to travel to dark-sky observing sites at the Grand Canyon, Yosemite, and other national parks. His 24-inch mirror had a 13-foot tube that required a ladder to get to the eyepiece.

Dobson has never charged anything for letting people use his mirrors. "People ask why we are doing it," he says. "Because nobody else will. I go out of my way to help other people see where the hell they got born. We are the dust of exploded stars." Someone once called the cops to report a group of strangers huddled around a huge gun on the corner.

I met John Dobson, then eighty-five and world-famous, in Los Angeles in January 2001, where he was staying in the Vedanta monastery. A slightly built, lithe man with long gray hair held in a ponytail, Dobson lives a peripatetic existence, roaming the world, teaching mirror-making workshops, and showing as many people as possible the wonders of the night sky.

I served as Dobson's assistant one evening when we set up two telescopes on the front lawn of the Griffith Observatory. People indoors waited in long lines to look through the expensive old 12-inch refractor, but we gave them better views of Jupiter and Saturn through Tumbleweed, with its 9.5-inch mirror, and a nameless 6-inch reflector. People were genuinely astonished at seeing Saturn. "Is it real?" they would ask. One man thought it was a cartoon I had painted. "It was wonderful to see the children's solemn faces register wonder and delight," I wrote that night in my journal. "Most of the people were Japanese, Korean, German, Italian, or Mexican. But they didn't need English—the planets cut across all cultures."

Dobson also teaches cosmology classes in which he presents a fascinating combination of Albert Einstein and Swami Vivekananda. He scoffs at the Big Bang, insisting: "Nobody ever saw a singularity [the hypothetical point from which the universe exploded]. It's all guesswork." He prefers his own theory in which fleeing hydrogen atoms somehow recycle from the edge of the universe back to the center.

Most scientists believe that the evidence for the Big Bang is over-whelming, but no one can deny that Dobson's evangelical mirror-making has changed the face of amateur astronomy. He invented the Dobson-ian mount, an inexpensive alt-azimuth arrangement made of plywood that turns on an old LP record and bits of Teflon. Before he came along, most Stellafane mirrors were made of thick glass 6 inches in diameter. Thanks to Dobson, they are now much thinner and larger, usually rest-ing in sturdy Dobsonian mounts. For more than thirty-five years, he has taken his mirrors to the streets, introducing tens of thousands of aver-age people to the wonders of the heavens. Late in 2002, I checked to see how the octogenarian was doing. He had just returned from a trip to Russia, Ukraine, England, and Ireland and was preparing for his 2003 journey across America.

Kaleidoscope Renaissance

Cozy Baker is to the modern kaleidoscope what John Dobson is to the amateur telescope. In 1981, her twenty-three-year-old son, Randall, a budding artist, was killed by a drunk driver. Determined not to suc-cumb to her grief, Baker wrote *Love Beyond Life*, a book dedicated to his memory, advising people to find beauty in the midst of tragedy. But she was having a hard time following her own advice until she saw a kaleidoscope in a Nashville crafts shop and was enchanted by its ever-shifting, gorgeous views. "As a little girl, I loved fireworks, church stained glass, rainbows, sunrises and sunsets," she says. "Now here they all were in one mirrored tube of magic."

Baker found a November 1982 *Smithsonian* article profiling seven fine-arts kaleidoscope-makers and tracked them down. Baker discovered that there were no books on kaleidoscopes, so she decided to provide one. In 1983, she embarked on a national kaleidoscope tour, staying in crafts-men's homes from Cape Cod to California, from Florida to Vermont.

Baker wrote her book, *Through the Kaleidoscope*—the first of sev-eral she would pen on the subject—and began to build a collection, in-cluding a $3,800 Van Dyke Series II, one of a limited number created by Connecticut glass artist Bill O'Connor, who made exquisite liquid-filled ampules for his object cells.

Baker's sprawling Bethesda home has been converted into a private museum, with more than 1,000 kaleidoscopes gracing every surface, as I discovered when I spent a visually exhausting day there, looking through everything from a miniature scope held on a ring finger to a 12-foot polyangular affair with shifting mirrors and blowing silk scarves. I operated the big Marbleater, a Rube Goldberg kaleidoscope with a conveyor belt of fifty-five hand-blown marbles. I squeezed a rubber atomizer to blow colored feathers in one object chamber, watched multiplied Monet flowers in another, and viewed the world in reflected reality through several teleidoscopes (they have only lenses at the end). Baker also has several projection scopes that throw images onto a screen, an outdoor mailbox/kaleidoscope, and a KaleidAquarium in which live fish swim in the object chamber.

The mirrors of the kaleidoscopes in Baker's house are held in containers of wood, acrylic, ceramic, gold, silver, fabric, stone, bronze, cardboard, steel, plastic, glass, and just about anything else you can imagine, including emu eggs. When one of her visiting grandchildren picked up a Spam container and shook it, her son Brant said, "Mom, it wouldn't surprise me if you told me *that* was a kaleidoscope." It was.

In 1986, Baker founded the Brewster Society (named for kaleidoscope inventor David Brewster), which hosts an annual meeting of kaleidoscope artists and aficionados and publishes a quarterly newsletter. One of its first members was Charles Karadimos, whose Damascus, Maryland, studio isn't too far from Cozy Baker's kaleido-house. By the time Baker found him in 1983, Karadimos, who began his artistic life as a stained-glass worker, had been making kaleidoscopes for a few years. At first he used three-mirror triangles, which reflect infinitely to fill the entire viewing space. Eventually he switched to two-mirror scopes, which produce the classic rose-window effect. Karadimos learned to heat and twist strips of scrap stained glass into delicate shapes. "Whatever I made was a personal impression of what I was doing at that moment. If I was listening to classical music, the colors were related to it."

Karadimos, who has made more than 10,000 kaleidoscopes, is a purist who uses only lamp-worked colored glass, never premanufactured buttons or beads. He uses only dry cells rather than floating colorful objects in viscous liquids. "I like to control the image. If it has liquid, it continues to move after you stop turning it. Also, I like the sound of glass falling and clicking. I can turn a piece and it can change radically or subtly, depending on the speed at which I turn it."

Karadimos gave me a quick lesson in how to make kaleidoscopes. Inspired, I subsequently bought some front-surface mirrors and, in my basement workshop, cut two strips and held them together with duct tape (*not* a Karadimos method), carefully adjusting the angle to 15 degrees, then holding it there with a third side of black velvet-covered glass. This would give me twenty-four narrow pie pieces producing a twelve-pointed mandala. I put these into a cardboard tube (a foot-long, 3-inch-diameter affair) covered with red foil, then attached my object case and took a first look into an enchanting new self-created world.

"Well, Charles," I e-mailed Karadimos on January 6, 2000, "I made my first kaleidoscope, and I must say the image is incredible, though the outer package leaves a bit to be desired." I called it my Christmas Kaleidoscope because of the dominant greens and judicious amounts of red. "I twisted and pulled bits of colored glass over a propane torch but violated your purity rule by putting in a piece of lace, a few beads, some bent wire from a paperclip, and because I didn't have much red glass, a bit of bright red plastic wrap from some gouda cheese. Oh, yes, and a round orange ring from the Dollar Store."

As Cozy Baker once observed, it doesn't much matter *what* you put in a kaleidoscope—"even eggshells and cigarette butts look beautiful." San Antonio artist Carmen Colley has used local items such as rattlesnake rattles, moth antennae, and cicada wings.

Sometimes the New Age angel-speak of modern kaleidoscope lovers can be a bit much—"May you treasure life as a rare crystal. / Let joy polish it till beauty is born," begins a typically saccharine poem in the *Brewster Society News Scope*—but these mirrored visions really can be soothing to those in distress or despair. That's why Cozy Baker donated them to doctors going to Guatemala and why some hospice nurses carry them for their terminally ill patients. Sherry Moser, a former pediatric oncology nurse whom I visited in her studio in rural Cleveland, Georgia, used to give kaleidoscopes to children with intravenous drips. "Now I feel that I'm helping people in a different way," she says, by making kaleidoscopes with lamp-worked pastel glass floating serenely in glycerin.

In New York's Catskill Mountains, I lay on my back inside a modified farm silo and gazed up at a huge kaleidoscopic sphere (apparently

50 feet across but actually less than 5 feet), where rear-projections of bees, butterflies, flowers, and stars swirled in a ten-minute show. Here at Catskill Corners on Route 28, Charles Karadimos designed the 38-foot aluminum mirrors for the world's largest kaleidoscope. The spherical effect is produced by three mirrors forming a tapered triangle so that the narrow end is farthest from the viewer.

I was even more impressed with Michigan artist Don Doak's huge star dodecahedron, a multicolored twelve-sided illusion encompassed by swirling yellow lines that looks like a celestial fantasy hanging in space, with planets swirling around it and viewers' faces reflected from the walls of this mini-universe. The actual opening is 20 feet wide, but the space inside appears to be 64 feet across (and deep). Doak, who came to kaleidoscopes after working as a photojournalist, sculptor, and craftsman, has pioneered exquisite polyhedra through a tapered-away-from-you three-mirror system such as that used in the silo, but much more complicated. By varying the mathematically determined angles at which the mirrors are cut at the front, back, and sides (they must be accurate to within thousandths of a degree), he can produce variations on a sphere, an icosahedron (twenty sides), or a dodecahedron.

Thanks in part to Cozy Baker's encouragement and promotion, there are dozens of such extraordinary kaleidoscope artists. Many of them are playing with innovative mirror systems. The craft and its appreciation are spreading. In recent years, the Japanese have taken passionately to kaleidoscopes, opening a museum and producing their own artists.

In 1851, Scottish Reverend Legh Richmond called kaleidoscopes "a kind of visible music" that soothed his soul. "As by magical influence, confusion and irregularity seemed to become the prolific parents of symmetry and beauty." Yet Richmond also realized that kaleidoscopes were emblematic of inevitable change. "The phantom which delighted me but a moment before was gone—forever gone—irrecoverably lost!"

Fearful Symmetry, Mirror Universes

The appreciation of mirror symmetry appears to cross all cultures, so that the sight of the tree reflected in the water, with which I began this book, is universally satisfying. The inkblot is a good example of bilateral or mirror symmetry. If you drew a line down the middle and put only half of it on a mirror, it would reproduce the entire blot. Humans and many other earthly life-forms have apparent vertical bilateral symmetry—imperfect because of moles and various other anomalies, as well as internal organs like the heart—primarily because of gravity. Our feet have to be different from our heads, just as a tree's roots differ from its leaves, but our left sides mirror our right fairly accurately.

For a long time, scientists thought that the entire universe must be symmetrical. While some molecules are built in a "left-handed" configuration (though this designation is obviously arbitrary), scientists believed that there must always be an equivalent right-hand version. They called this the "conservation of parity." As one scientist put it, "[We thought that] Nature's hardware shop always stocked an equal number of right- and left-handed corkscrews."

In 1956, physicists Chen Ning Yang and Tsung Dao Lee suggested an experiment to test parity, carried out by Madame Chien-Shiung Wu the following year. More electrons came out of the "south" end of spinning radioactive Cobalt-60 atoms than the other end. Not only that, but along with the electron an elusive little neutrino was emitted, always spinning in only one direction. Universal parity was dead.

All amino acids in living beings are "left-handed." Just before she stepped through the looking glass, Alice wondered aloud to Kitty whether cats drank mirror-milk in the reverse world. "Perhaps Looking-glass milk isn't good to drink," she said. Now we know it probably isn't, since many molecules with exactly the same makeup other than their handedness have radically different effects. All humans might be mirror-lactose intolerant. But no one has ever drunk such mirror-milk, which raises another interesting question. Where are all the right-hand organic molecules?

And why isn't the universe completely regular, rather than clumped into assorted galaxies, stars, and planets? If the universe began as a uniform singularity, then the Big Bang should have resulted in a perfectly uniform expanding sphere. Actually, the universe should have quickly destroyed itself, converting back to pure energy, since it would have theoretically produced an equal amount of matter and antimatter. Many theoretical physics believe that for every piece of matter there may be a piece of antimatter mirroring it somewhere in the universe.

Charles Howard Hinton, an American mathematician, first imagined something along these lines in 1888. "We must conceive that in our world there were to be for each man somewhere a counterman . . . exactly like the image of the man in a mirror," Hinton wrote. "And then when the man and his counterfeit meet, a sudden whirl, a blaze, a little steam, and the two human beings . . . leave nothing but a residuum of formless particles."

In their 1956 paper, Yang and Lee suggested that symmetry might be restored if there were a parallel universe somewhere with neutrinos rotating in the opposite direction. This notion of a "mirror universe" has become quite popular with imaginative physicists such as Rabindra Mohapatra at the University of Maryland and Robert Foot at the University of Melbourne. What if, in the first millisecond after the Big Bang, an antimatter universe formed a separate mirror cosmos?

Reputable scientists are taking this notion quite seriously. In 2002, Robert Foot published *Shadowlands: Quest for Mirror Matter in the Universe*, in which he argues that invisible mirror matter—a mysterious substance that somehow differs from antimatter—accounts for the so-called dark matter that astronomers have been unable to locate. In Foot's scenario, this mirror universe coexists in invisible union with our own. "If mirror matter exists," he says, "then there should exist also mirror stars, mirror planets, even mirror life."

Even Martin Rees, the Astronomer Royal for England, takes mirror universes seriously. He believes there may be *multiple* mirror universes being created all the time as stars collapse to create black hole singularities. "Our universe may be just one element—one atom, as it were—in an infinite ensemble: a cosmic archipelago."

Searching for Gravity Waves

The only way mirror matter could be detected in our world would be through gravity, since it presumably has mass. Although gravity holds us to the ground, keeps the earth in orbit, and accounts for the Milky Way's shape, its force is so weak that no one has yet detected gravity waves, although Einstein's General Theory of Relativity predicts their existence. Scientists are using—you guessed it—incredibly precise mirrors to search for gravity waves in the Laser Interferometer Gravitational-Wave Observatory (LIGO) program.

There are two identical LIGO sites in Hanford, Washington, and Livingston, Louisiana, L-shaped installations with vacuum tubes running 2.5 miles at right angles (and crossing in an X within the main building). At each end of the arms, there is a superbly reflective, slightly concave mirror made of fused silica, with dielectric coatings tuned to reflect 99.995 percent of the infrared laser light bounced off it. The idea is to send the laser beam through a beam-splitter down both arms simultaneously, then reflect the light back and forth 100 times and look for interference patterns that indicate one arm is just slightly shorter than the other—compressed by an unusually strong gravity wave. It is similar to the 1887 Michelson-Morley interferometer that searched for the mythical luminiferous ether.

So far, the U.S. experiments, as well as similar efforts in Germany, Japan, and Italy, have been plagued with startup problems. It is almost impossible to isolate the mirrors from nearby traffic or other vibrations. That's why there are several LIGO sites, looking for simultaneous results. Even if they work, though, the odds of detecting sufficiently strong gravity waves are slim. Hypothetically, such waves would be produced every few months somewhere within a 650 million light-year radius, when a dense neutron star falls with a horrible rending of the space-time fabric into a black hole, or some other similar cataclysm.

High-Tech Mirrors and Mother Nature's Example

The quest for energy from atomic fusion, ever-smaller computer chips, and ever-faster, more efficient communications all involve unusual mirrors. At Lawrence Livermore National Laboratory in Livermore, California, scientists and engineers are building the National Ignition Facility, a $4 billion, stadium-sized affair focused on a b-b–sized bit of deuterium and tritium. The idea is to zap these hydrogen isotopes simultaneously with 192 powerful lasers, setting off an implosion reaction that will fuse them into helium—the same nuclear fusion that fuels stars. Scientists at Livermore have been working on this idea for two decades and hope to reach the Holy Grail of Fusion in 2008. In the meantime, they have employed some of the country's finest opticians to make mirrors to direct and focus the lasers, including deformable mirrors for adaptive optics.

Lawrence Livermore opticians—and others at nearby Sandia National Laboratories—are working on a set of four specialized mirrors that are likely to revolutionize the computer industry in the next few years, making possible a radical reduction in the size of chips. Currently, lithographs of the intricate circuit patterns from a master "mask" are photoreduced and etched onto wafers, using lenses that refract and focus light waves. The shorter the wavelength, the greater the possible reduction, but as waves get into the ultraviolet range above visible light, the lenses begin to absorb the radiation.

Enter Don Sweeney and Norm Thomas of Lawrence Livermore, as well as others who have figured out a way to reflect rather than refract extreme ultraviolet waves of 13 nanometers (visible light is in the 500-nanometer range). They have done it with eighty-one alternate coatings of molybdenum and silicon so that the ultraviolet waves constructively interfere, thus reflecting 70 percent of the light. Although Lawrence Livermore applies the coatings, the four specialized Zerodur mirrors—two convex, two concave—were subcontracted in competing bids to Carl Zeiss in Germany and Tinsley Laboratories in California.

The Tinsley mirrors, produced by computer-controlled polishing, are apparently slightly superior. They bounce the ultraviolet light up and down, focusing and narrowing it to produce a tiny image on the wafer with virtually no aberration, using off-axis mirror segments to keep the light path completely unobstructed. "With this technology," Sweeney told me, "computers will be 100 times more powerful in ten years' time."

Meanwhile, at Bell Labs (owned by Lucent Technologies) in New Jersey, scientists have made tiny mirrors the size of a pinhead that may revolutionize the telecommunications industry. While multiple wavelengths of light reflect off the interior sides of fiber-optic cables incredibly quickly—duh, at the speed of light—they must be converted into electrons at every routing junction, then get reconverted to light, and so on. It's as if one flew on a jet but had to walk between terminals before taking another jet. Using micro-electromechanical systems (MEMS) to make minuscule mirrors that can swivel on two axes, Bell Lab scientists are working to reroute fiber-optic light by simple reflection.

They have worldwide competition from an array of startup companies and major corporations, all racing to find the best way to reflect light at these fiber-optic junctions. Agilent, a Hewlett-Packard spin-off company, has introduced a thermally controlled bubble to act as an optical switching device. Light messages go straight through intersections with no bubble, but they reflect down a different fiber if a bubble-mirror burps up.

Nature provides the template for the most promising method of manipulating fiber-optic light, through the process of iridescence, about which Isaac Newton speculated. As Newton noted, we see an object as a particular color such as blue because it *reflects* the blue wavelength and absorbs all others. But some objects, such as hummingbird feathers, abalone shells, fish scales, snake skins, bristle worms, and morpho butterflies, somehow reflect a shimmering array of light that changes as its angle to the viewer changes. A morpho's spread wings in the tropical sunlight appear incandescently blue, but when the wings change their angle, they can turn to a dull brown.

The secret to the morpho's iridescence lies in the microscopic scales on its wings, whose patterns allow most wavelengths of light to enter but that creates interference patterns that block blue light arriving at certain angles. The wings act something like a three-dimensional crystal with tiny holes the exact size of a particular wavelength—all produced by evolution to help the morpho attract a mate or confuse a predator. Even though they are as full of holes as a sieve, the butterfly's scales act like magical mirrors.

At MIT, physicist John Joannopoulos theorized that similar man-made arrays could be used to manipulate light, and now so-called photonic band-gap crystals are the hottest mirror prospects for solving the photonic switching problem. By creating channels in the crystal, Joannopoulos figures he can direct the lightpath of a particular wavelength. "Suppose you are a photon," he says. "You enter through a defect in the crystal. You look around and see a perfect crystalline environment. You cannot penetrate it because you have a forbidden wavelength. So you follow the defect, no matter how it twists and turns."

At Bell Labs, Pierre Wiltzius is working on self-assembling photonic crystals, made by putting microscopic particles in a colloidal suspension, then removing the fluid and allowing the particles to settle into patterns. His photonic gap mirrors, if they work, will be much smaller than MEMS mirrors and will not require their moving parts. Wiltzius is just one of the scientists racing to create workable photonic crystals. With the military funding much of the research, universities and private companies in twenty-nine countries are working hard to replicate the secrets of the morpho butterfly.

The Mirror Reversal Puzzle

Even if such high-tech mirrors render computers more powerful and make fiber-optic communications much faster, they will not change human nature. We will still get up every morning to face ourselves in our everyday bathroom mirrors.

But is that really me in the mirror? Not quite. I don't see myself as I really am, but with my left and right side reversed. My hair part doesn't really look like that to other people—it is actually on the other side. *Why? Why do mirrors reverse left and right, but not top and bottom? Why don't I see my feet looking back at me when I look into the mirror?*

This puzzle has plagued scientists and philosophers since ancient times. As we saw in Chapter 3, Plato thought that eyes emit rays that somehow coalesce on the surface of a mirror.* "Right appears left and left right, because the visual rays come into contact with the rays emitted by the object in a manner contrary to the usual mode of meeting." Three centuries later, Lucretius pondered the problem. The face in the mirror "turns inside out, as would happen with a plaster mask" if it were limber enough (like a modern-day rubber Halloween mask). Thus "it would happen that the right eye became the left." Lucretius explained that this reversal could be carried "from mirror to mirror, [so

---

*It is astounding that many supposedly educated people still believe that we see by emitting visual rays rather than receiving light waves into our eyes. In a 2002 study published in *American Psychologist*, more than half of the subjects believed in visual emissions.

that] what was left in the object becomes first the right and at the next reflection is true left again."

Some 1,800 years later, Immanuel Kant was still worrying over this philosophical riddle. For him, mirror reversal proved a profound truth. "Space and time [are] mere forms of our sensuous intuition," he thought. Everything depends on perception. What could be more similar to your right hand than its mate reflected in a mirror? "And yet I cannot put such a hand as is seen in the glass in the place of its original." The same glove wouldn't fit both hands. From this Kant concluded grandly, if incomprehensibly, "Space is the form of the external intuition of this sensibility."

Two modern thinkers have offered satisfactory answers to the puzzle, though neither of them will grant any validity to the other. In *The New Ambidextrous Universe*, Martin Gardner asks us to imagine standing on a mirror, where we would see ourselves upside down. "In a strict mathematical sense the mirror has not reversed left and right at all; it has reversed front and back"—or up and down, for those with mirrored floors. Gardner then asks us to imagine ourselves facing a mirror, with our left side to the west. "Move your west hand. The hand on the west side of the mirror moves. . . . It is the front-back axis, the axis that runs north and south, perpendicular to the mirror, that has been reversed. You are facing north. Your twin faces south."

In *Mirrors in Mind*, Richard Gregory offers an ingenious explanation that amounts to the same thing, though he pooh-poohs Gardner's theory. Gregory approaches the problem through mirror writing, which switches letters left-to-right as you see on the clever cover of this book. But what if you wrote the word *Mirror* on an overhead transparency? If you then turned it around to face the mirror, as you would with an opaque piece of paper or a book, then the mirror would reverse the word. But if you hold the transparency up to the mirror so that you can still read the word normally, it will appear unreversed in the mirror as well. Thus, Gregory observes, "When a book is rotated around its *vertical* axis to face the mirror, its *left and right* switch over. It is this that produces mirror reversal. It is really object reversal."

Frankly, I have never been able to get too worked up over the left-right mirror-reversal conundrum. It seems obvious to me that there is a point-to-point correspondence between a real object and its appearance in a flat mirror, and everything obeys the law of reflection.*

True Mirrors

What isn't so obvious to me is how some mirrors can *prevent* left-right reversal. I recently visited Callan Castle, the 1904 Atlanta mansion built for Coca-Cola magnate Asa Candler. There is a circular parlor there with a curved fireplace and a large concave mirror fitting into the wall above it. As I entered the room and waved with my right hand, the figure in the mirror waved back from the opposite side with *his* right hand. As I walked slowly into the room, my mirror image grew larger, disappeared, then reappeared as in a normal mirror and shrank as I continued to walk toward it.

It was an unnerving experience, not because of the funhouse mirror aspect but because when I was unreversed I looked *normal* and *odd* at the same time. Something wasn't quite right. I realized that it's because my face does not really have true symmetry. My smile is a bit lop-sided, there's a mole on my right cheek, and my hair is parted on the left. Yet I am used to seeing myself only in reverse in a mirror, where my smile goes the other way, the mole has shifted to my left cheek, and I have a right part.

In 1930, William E. Benton patented his Duality Mirror, a thin, silvered metal surface held perpendicular to the middle of facial portraits so that viewers could look at one side, then the other, to see how the face would look with two perfectly symmetrical right or left halves. Benton demonstrated his device on an old photo of Edgar Allen Poe, with astonishing results. With a face composed of two mirrored halves

---

*Another counterintuitive characteristic of mirrors is their inability to reflect more of your body, no matter how far you back away from them. Try it yourself by taping pieces of paper just above and below your head in a mirror. Now back up. Your head will continue to occupy the same space! If you measure it, you will find that the mirror head is exactly half the size of your real head. Why? Think about the law of reflection (equal angles of incidence and reflection) and you'll figure it out.

of the right side of his face, Poe looked like an intense, handsome man with deep-set eyes.* In the picture with two left halves, his face looked puffy, with a pursed mouth and worried eyes.

Not surprisingly, Benton's gimmick flopped. People were not keen on his interpretation that "there is a little of Dr. Jekyll and Mr. Hyde in each of us" and that the left side of the face reveals a subconscious, cruel, sensuous nature. People did not want to learn about their asymmetrical faces.

Entrepreneurial siblings Catherine and John Walter are trying to change that with their True Mirror, which does the same thing as the old concave Candler mirror I saw in Atlanta, but without distortion. They do it with two flat mirrors held at a 90-degree angle. Because these front-surface mirrors are precisely abutted, the middle joint is invisible.

It's not a new idea. Hero of Alexander described it in 1 C.E. In 1887, British priest John Joseph Hooker took out a patent for "useful Mirrors for Obtaining True or Positive Reflections," and Hooker was followed by many other similar patent-seekers, as John Walter found to his chagrin. He thought he had originated the idea in 1982 when he accidentally saw his twenty-four-year-old face re-reversed in a medicine cabinet mirror opened at right angles to a bathroom mirror. By that time, Walter was already attuned to facial asymmetry, or at least hair-part asymmetry, and had developed a theory about it based on personal experience.

It seems that as a young physics/math major, Walter was an insecure nerd. When he looked in his mirror, he couldn't see anything wrong, though. The problem? He was seeing himself reversed, looking as if he parted his hair on the left, when in fact it was parted on the right. So he dragged his comb the other way, and voila, overnight he became Mr. Popularity.

Walter's resulting Hair Part Theory observes that most men part their hair on the left, which is supposedly perceived as masculine and assertive. Men who part their hair on the right tend to be regarded as more sensitive, effeminate, and nerdy. Women, by contrast, tradition-

---

*Identifying left and right is confusing. This is the *right* side to Poe but the *left* side as we look at his photo.

ally part their hair on the right, and if they are left-parters (like Margaret Thatcher and Hillary Clinton) they may be perceived as powerful and masculine. Because many women tend to switch their hair arrangements and parts frequently, though, the effect isn't as strong for them.

Walter began to see evidence everywhere. In *Superman: The Movie* (1978), Christopher Reeve parted his hair on the right when he portrayed wimpy Clark Kent, on the left when he took to the skies. Early in 1979, Walter wrote to President Jimmy Carter, perceived as being weak and overly sensitive, advising him to change his right to a left hair part. That April, Carter did indeed switch, but by then it was too late.

For ten years, John Walter tried to perfect his True Mirror using regular reflectors before realizing that he needed aluminized front-surface mirrors to render the middle line invisible.* Since he incorporated the business in 1994, media attention and sales have steadily increased. Today, with younger sister Catherine as a partner, John Walter carries on his crusade to change perceptions and hair parts at 43 East First Street on New York's Lower East Side, where I met them and examined my True Self.

Typical reactions of first-viewers: "I look cross-eyed." "One of my arms is longer than the other." "I didn't know my lips were lopsided." "It's not me. I can't see the world seeing me like this." Manhattan psychiatrist Gerald Epstein, who has a True Mirror in his office, says some patients lose their balance and physically stumble when looking into it. That's why the Walters siblings offer a two-month money-back guarantee and urge people to give it at least a month to get used to themselves. "Some people love it. Some people hate it. Some people have even run away screaming," John Walter says.

Catherine Walter, an anthropologist who once spent several weeks living with a mirrorless Mayan family, thinks that regular mirrors are a

---

*Andrew Hicks, an assistant professor of mathematics at Drexel University in Philadelphia, has recently created a seamless, less bulky "true mirror," although he didn't at first realize it. He and associate Ron Perline were trying to invent a rearview mirror with an undistorted wide field of view to eliminate the blind spot. Using an arcane mathematical formula to control a computerized grinder, they produced a sort of saddle-shaped aluminum mirror that worked, reflecting a test checkerboard pattern with fidelity. Only when Hicks viewed writing in the mirror did he realize that it was unreversed.

mixed blessing. "Because we have mirrors everywhere," she told me, "we get much of our sense of self from them." She feels that the primary virtue of the True Mirror is to let us connect with our authentic selves by gazing deeply into our own eyes—right eye looking into left eye and vice versa.

Plenty of adults come to appreciate the True Mirror, as comments in the company guest book indicate. "It's very confusing—but I like it. It seems softer. I look more alive." "Perfect for me to do a self-portrait (I'm an artist) and see myself truly." "It is like looking at someone who looks familiar, but who I've never seen before."

Who Do You See in the Mirror?

And so we come back to the question the earliest hominid faced in the still water after the rainstorm. This is the question of identity, of essence, of soul that concerned the ancient Egyptians, Chinese, and Aztecs. It is even at the root of the questions asked by astronomers who have turned their big mirrors out toward space. *Who are we—and what is our place in the universe?*

This brings us into the realm of psychology, where we must pick carefully among unproven theories and scientifically valid concepts. Mirror pseudoscience is not restricted to psychics and feng shui practitioners. There are psychologists with advanced degrees who believe that people with so-called multiple personality disorder (MPD) see their different "alters" in a mirror. MPD is almost certainly an artifact of suggestive therapy and misguided books that encourage people to role-play different internal prostitutes, waifs, bullies, demons, angels, or animals.* Even Richard Gregory, a British experimental psychologist who wrote the otherwise excellent 1997 book, *Mirrors in Mind*, included an illustration captioned "How multiple personalities appear in a mirror," showing various alters in a hand mirror.

------

*MPD has been officially renamed "dissociative identity disorder," which is an equally dangerous diagnosis. It is possible (but unlikely) that MPD is a real but rare condition, but it has certainly been wildly overdiagnosed (and created iatrogenically) in recent years.

This induced disorder is no harmless fantasy, because MPD thera-pists encourage their patients to believe that they were the victims of horrendous, completely forgotten childhood sexual abuse that caused them to "split" into various personalities. As I detailed in my 1995 book, *Victims of Memory*, there is no scientific (or common-sense) evi-dence that people can forget years of traumatic events without organic brain damage. Yet psychiatrist Marlene Steinberg continues to promul-gate this theory in her 2000 book, *The Stranger in the Mirror*, replete with checklists most of us can at times identify with, such as "I feel that I need to find my true self" or "I have had the feeling that I was a stranger to myself or have not recognized myself in a mirror." Elizabeth Loftus, an expert on memory distortion, warns that this is a "most dan-gerous book."

There *are* bona-fide psychological conditions involving mirrors, as psychiatrist Katharine Phillips documents in her 1996 book, *The Bro-ken Mirror*. She quotes Sarah, a third-year medical student with "body dysmorphic disorder," a form of obsessive-compulsive disor-der in which she cannot stop trying to "fix" her hair for hours. "I try not to ever look in the mirror when I'm at work, because when I do I can get stuck there," Sarah says. "The mirror acts like a switch. When I look in, the obsession turns on, and it can get pretty out of con-trol," so bad that she keeps her shiny toaster hidden in a cupboard at home.

Sarah's problem is an exaggeration of our culture's obsession with appearance and image, which puts particular psychological pressure on women. Anorexics can look into a mirror and see too much body fat. Male bodybuilders often suffer from "bigorexia"—their mirror image looks too weak. Twenty years ago, I interviewed a beautiful young woman with bulimia, an eating disorder featuring compulsive binge eating followed by intentional vomiting. "When I look in the mirror," she told me, "I can't figure out what I really look like. It changes all the time."

She is not alone. In 1964, Chicago psychologists Arthur Traub and J. Orbach created the "adjustable body-distorting mirror," a Plexiglas re-flective surface about 4 feet high that could be adjusted to bend as a convex or concave mirror in various ways. Subjects stood 7 feet from the mirror and saw themselves at first as "tall, with pin head, large elongated body and legs tapering to tiny feet," then as "short with enormous horned head and tapering legs," and so on. When subjects were asked to adjust the mirror so that they looked normal, they had an unexpectedly difficult time. "Many subjects declare . . . that they have forgotten precisely what they look like," wrote Traub and Orbach.

Many schizophrenics react oddly to mirrors, sometimes staring at them for hours. Curiously, there are no blind schizophrenics, and in the single known case where a long-term schizophrenic went blind, she went into remission within a few days. Human vision is mediated by the visual cortex, which resides at the rear of our skulls, in conjunction with many other areas in the brain such as the temporal lobes (on the side above your ears), the limbic system, the parietal lobes, and the prefrontal cortex. Some cells in the temporal lobes respond specifically not only to faces but also to *parts* of the face, such as the mouth and hair, with particular attention to the eyes. There are cells, for instance, sensitive to the direction of gaze.

Such visual information appears to be collated and made self-referential in the right prefrontal cortex at the front of the skull. Therefore, it is intriguing that many schizophrenics appear to have abnormal activity in the right prefrontal cortex, as do autistic children, many of whom dislike looking at themselves in mirrors. Similarly, many Alzheimer's patients fail to recognize themselves in mirrors; they either have extended conversations with the person in the mirror or get angry because their double keeps mimicking them and won't go away. Because those with Alzheimer's tend to avoid confronting the stranger in the mirror, placing mirrors in front of exit doors seems to discourage them from wandering.

Many people with a mirror problem have some sort of damage to the prefrontal cortex and temporal lobe, as do people who suffer from prosopagnosia, the inability to recognize familiar faces, though prosopagnosics may identify voices. In extreme cases, they cannot recognize themselves in a mirror, sometimes walking into mirrors as a result. Such people develop elaborate strategies to get through life, as sufferer Bill Choisser writes in *Face Blind*: "That must be me in the mirror because I am the only one in the bathroom."

Australian neuropsychologist Nora Breen specializes in people suffering from Mirrored-Self Misidentification, which she distinguishes from prosopagnosia, because some of them recognize other people's faces. Fred, an otherwise healthy eighty-seven-year-old retiree, complained about a stranger who kept following him around in his home, car, shopping centers, and on an airplane. Fred tried talking to him, but he remained mute, so he shrugged it off. But one night while lying in bed with his wife, Fred was distressed to see the stranger beside her in a full-length wall mirror. Tom, seventy-seven, another Breen patient, could identify objects held behind his shoulder in the mirror, but when asked to grab them, he would scratch at the mirror surface or try to reach behind it.

California neurologist Vilayanur Ramachandran uses mirrors to make missing limbs reappear in his "virtual reality box." A vertical mirror divides the center of a topless box with two holes in the front through which the patient puts his real and "phantom" arm, so that it appears that he still has two good hands. Several amputees who experienced exquisite pain from mentally clenched phantom hands found temporary relief when they looked into the box and saw their illusory missing hand conducting an orchestra. "I can actually feel I'm moving my arm, Doctor," a patient reported.

A 1968 experiment shows how most people can hallucinate when staring at themselves in a mirror. Psychiatrist Luis Schwarz and psychologist Stanton Fjeld placed subjects 2 feet from a 16-inch square mirror, illuminated only by a tiny bulb 3 feet behind them, and then tape-recorded their impressions over the next half-hour in what amounted to perfect scrying conditions. "It is a transparent face, jelly . . . like a cloud changing its form completely . . . the nose is large and the ears smaller and smaller . . . now I am bald," a man identified as "neurotic" reported. But a "normal" male said, "My eyes are whole caves with dancing skeletons," while another observed, "I see several faces. . . . They change one from another. . . . Their hair cut is changing . . . are holy men . . . a Japanese . . . a Negro." Yet another normal male saw himself gradually fading: "The image is darker and darker . . . disappears . . . the mirror . . . and I see a deep black."

Some mirrors are intended to manipulate. A recent article in *Chain Store Age* offers the case study of Sally, who tries on a pair of Capri pants in a store's dressing room. "Alas, the sole mirror and bad lighting make Sally look pale and fat," and she leaves without buying anything. The moral? Store managers should "spend money on mirrors (and lots of them)," along with good lighting. Although no department stores will admit it, rumor has it that they sometimes use slightly convex mirrors that make people look slimmer.

Bruce Newman and Susan Larson of Assist Technologies in Lake Carmel, New York, sell the 4-inch-square, hinged PC Mirror that is attached to the side of my computer. Intended for use in workplace cubicles, the mirror can be useful in various ways, such as warning of a boss's approach or permitting a quick appearance check before a meeting. But its primary purpose is to help telephone sales people to "smile while you dial," on the assumption that the smile can be "heard" in a more pleasant tone of voice. Company-sponsored surveys claim an average 8 percent increase in sales after PC Mirror installations.

Apes, Elephants, and Dolphins Face the Mirror

One morning in 1964, when twenty-two-year-old grad student Gordon Gallup looked at his own reflection while shaving, he considered that it would be interesting to see whether other species of animals could recognize themselves in mirrors. Five years later, as an assistant professor of psychology at Tulane, he got the chance.

Gallup put four preadolescent chimps—two female, two male—into separate cages and placed a full-length mirror outside each cage for ten days. At first, they reacted as they would to a stranger—bobbing, vocalizing, threatening, or adopting submissive postures. On the third day, however, he noticed a dramatic change. In the mirror, the chimps began to examine the inside of their mouths, to groom the hair on their foreheads, to pick their noses, to examine their genitals—taking advantage of the mirror to see otherwise inaccessible areas. They made faces, blew bubbles, and manipulated food wads with their lips. It was obvious to Gallup that they knew they were looking at *themselves* in the mirror, but he needed to be able to prove this subjective impression to skeptical colleagues.

Gallup devised a test. He anesthetized the four chimps and marked an eyebrow ridge and top half of the opposite ear with an odorless red dye. He did the same thing to a male and female chimp with no previous mirror experience. When the chimps awoke, they were monitored to make sure they weren't touching their red marks. Then mirrors were introduced. The four experienced chimps immediately took notice, touching the red marks on themselves repeatedly, and then looking at that finger. One of them even smelled the finger. The two other chimps showed no mark-directed responses.

In subsequent experiments, Gallup did the same test on macaques and rhesus monkeys, habituating them to mirrors for two weeks before trying his mark test. They failed. In a two-page paper published in *Science* on January 2, 1970, Gallup summarized his experiments, concluding, "Recognition of one's own reflection would seem to require a rather advanced form of intellect." He added that this ability might imply a "concept of self" that set humans and the great apes apart from other species. That fly I met in the Prague mirror maze might wear itself out, buzzing and banging against its image, but it will never figure out who that other hardheaded fly is.*

Gallup and others have subjected all kinds of animals to the mark test with mirrors. Orangutans passed with no problem. To everyone's surprise, however, no gorillas touched the red marks. Only Koko, the famous gorilla who has learned sign language, clearly identified herself in a mirror, according to her owner. Bonobos, the endangered, peaceful apes of the Congo, knew themselves in mirrors. But all monkeys flunked. Gallup left a mirror with a pair of rhesus monkeys for eighteen years, and they still didn't get it.**

Elephants apparently failed the mark test in the late 1980s, but ten years later, in Nevada, animal behaviorist Patricia Simonet tested two performing Asian elephants—Bertha, a veteran in her forties, and eight-year-old Angel, although Angel got little chance to see herself because the dominant Bertha hogged the mirror. Within twenty minutes of the

---

*Even the clever, tool-using ravens studied by naturalist Bernd Heinrich do not know themselves in mirrors. When he unveiled a mirror, he writes in *Mind of the Raven*, "all the birds went bonkers," retreating to their loft. While some canaries cozy up to their reflections, male cardinals are notorious for attacking a perceived interloper.

**Even though they do not recognize themselves in mirrors, many animals, including monkeys, pigeons, parrots, chicken, and fish, can use mirrors to find hidden objects or to solve a puzzle.

mirror's introduction, Bertha stopped flapping her ears and trumpeting at the other elephant in the mirror and apparently began examining herself. During the mark test, white face paint was applied to a brow and temple, behind one front leg, and on one hip, all visible to Bertha only in the mirror. With her trunk, she touched the marks fifteen times during the two hours of testing.

Dolphins also know themselves in mirrors, according to a 1999 study by Lori Marino, an Emory University psychology professor, and Diana Reiss, director of marine mammal research at New York Aquarium. They marked Pressley and Tab, two teenaged bottlenose dolphins, with nontactile black magic marker and used a variety of control circumstances. In order to qualify for self-recognition, the dolphins had to (1) spend more time at the mirror when marked, (2) display no "social" behavior as if toward another dolphin, and (3) swim immediately to the mirror and expose the mark.

Pressley and Tab passed the test, obviously contorting their bodies in front of the mirror in order to observe marks under their chins or sides. When they marked Pressley's tongue, he opened and closed his mouth in front of the mirror as he never had before. Gordon Gallup, Marino's former mentor, is still skeptical, because dolphins have no hands or trunks with which to touch their marks. "It is entirely possible," he told me, "that the dolphins have learned that they have control over the behavior of the 'other' dolphin in the mirror, and therefore when they see the image with a mark they change their orientation to the mirror so that they can see it better."

No one denies, however, that dolphins are smart. Like humans, they have large brains, but they have meager frontal lobes, which are crucial to humans. Marino believes this may be a case of evolutionary convergence, in which different species arrive at the same survival strategy by independent paths, like the flying ability of bats and birds. The convergence is probably not toward the specific trait of mirror self-recognition, she says, but to a "certain level of complexity in how they process information."*

---

*Should we treat animals that recognize themselves in mirrors in an especially humane manner? So Steven Wise argues in his 2002 book, *Drawing the Line: Science and the Case for Animal Rights*, but he also makes the case for many other animals who don't pass the mark test, including honeybees.

When Babies Become Aware of Themselves

In a 1972 issue of *Developmental Psychobiology,* Beulah Amsterdam published the first mirror-recognition study for human babies, "Mirror Self-Image Reactions Before Age Two." She described how she had tested eighty-eight children between the ages of three months and twenty-four months by putting a spot of rouge on one side of the nose and seeing if they touched it while looking in the mirror. Early on, babies seemed to recognize their mothers in the mirror, but not themselves. By six months, infants were smiling and playing with themselves in the mirror, but they treated the reflection as another child. At one year, they began to search behind the mirror for their mysterious playmate.

Finally, Amsterdam concluded that from twenty to twenty-four months, "65 percent of the subjects demonstrated recognition of their mirror images." Subsequent research has substantiated her findings, indicating that most children's brains first register that they are observing themselves sometime during the latter part of their second year, when they become coy, embarrassed, clownish, or self-admiring in front of the mirror.

What exactly does mirror self-recognition imply? Gordon Gallup believes self-recognition means self-awareness. "You become the object of your own attention. You are aware of being aware. And that, in turn, allows you to make inferences about comparable states of awareness in others." Gallup doesn't deny that other animals such as dogs or even fleas may have alternate forms of self-concept, but the brain's capacity to allow us to know we are looking at ourselves appears to place us— along with higher apes and perhaps elephants and dolphins—in a unique category, and this simple ability to recognize ourselves in a mirror seems to be essential to the human enterprise.

Can it be a coincidence that toddlers develop language and begin to say "I," "me," and "mine" about the same time they learn mirror self-recognition? Or that the frontal lobes develop dramatically in the second year of life? Or that they reach Piaget's level of understanding "object permanence" (remembering and seeking out hidden objects) and begin to engage in pretend play? Or that they begin to act like strong, self-willed individuals in the terrible twos? Or that they begin soon afterward to develop empathy for others and moral standards? Or that their autobiographical memories supersede the period of "infantile amnesia" around the age of three?

In the late 1800s, Charles Horton Cooley, a Michigan sociologist, theorized that the human sense of self is created in infants through social interactions. Cooley—who was himself a shy semi-invalid—called this the "looking-glass self" because he believed that our self-concept is a reflection of what we perceive others think of us.* His disciple, George Mead, concluded, "It is impossible to conceive of a self arising outside of social experience." Gallup, suspecting that Cooley and Mead were onto something, gave the mark test to chimpanzees that had been raised in complete isolation, after habituating them to mirrors. As he predicted, they failed to identify themselves.

Similarly, the famed Wild Boy of Aveyron, captured in the French woods in 1799, reached behind a mirror to find the boy he thought was hiding there. The Wild Boy never learned mirror recognition or how to speak. Perhaps such abilities must be developed during the crucial developmental period when the brain is growing and establishing new branches, connections, and synapses.

Of course, mirrors are not necessary for self-awareness. Blind people know perfectly well who they are, for instance. Thus Sidney Bradford, blind before his first birthday, was an intelligent, self-assured fifty-two-year-old when his sight was restored by a cornea transplant in 1958. He was fascinated by mirrors, often preferring to see the world in their reflection rather than directly. But Bradford couldn't get used to his own face in the mirror and shaved by touch in the dark as he always had.**

---

*In 1949, Jacques Lacan, a French neo-Freudian, incorrectly theorized that infants go through a "mirror stage" between 6 and 18 months of age in which they discover their mirror image and believe it is themselves, thus dooming them to a life of alienation from their true selves. As Lacan put it: "This jubilant assumption of his specular image by the child ... [exhibits] the symbolic matrix in which the I is precipitated in a primordial form, before it is objectified in the dialectic of identification with the other." Is that clear?

**The story ends tragically. With his sight restored, Bradford became self-conscious, lost his self-confidence, and died within two years.

The ability to recognize oneself in a mirror correlates with (but does not cause) essential human traits such as logic, creativity, curiosity, the appreciation of beauty, and empathy, leading directly to tool use, scientific experiments, storytelling, poetry, art, theater, lawmaking, philosophy, religion, and a sense of humor. In other words, as humans evolved, the ability to *think*—to ponder themselves in mirrors, among other things—helped them to survive. Gordon Gallup quips, "I am, therefore I think," an inversion of Descartes' most famous statement. "It is our ability to conceive of ourselves in the first place that makes thinking and consciousness possible, not vice versa," Gallup concludes.

"Without self-awareness," Emory University primatologist Frans de Waal observes in his 1996 book, *Good Natured*, "we might as well be folkloric creatures without souls, such as vampires, who cast no reflections. Most important, we would be incapable of cognitive empathy, as this requires a distinction between self and other and the realization that others have selves like us."

As one would expect, other species that display mirror self-recognition also show the capacity to empathize, which is the very essence of the Golden Rule—to treat others as you would be treated. Dolphins, for instance, are famed for helping injured people. Yet the ability to put oneself in someone else's shoes also permits deception and cruelty. What would sadists know about exquisite torture unless they could imagine what it felt like? As Jane Goodall discovered with her beloved chimps, they could murder as well as comfort one another.

Sex also seems to be connected with mirror self-recognition, as we have seen throughout this history of human interaction with mirrors. Bonobos and dolphins are highly sexed animals always ready for intercourse. Pan and Delphi, two half-brother dolphins studied by Marino and Reiss, always enjoyed sexual play with one another, but when mirrors were available, their libido went wild, so that in one half-hour session they attempted to penetrate one another forty-three times. In all cases, they assumed positions so that they could watch themselves in the mirror, breaking off if their bodies drifted out of sight, then resuming sex play in front of the mirror.

Self-awareness may lead to more satisfying sex, but it also makes humans, and perhaps some other animals, aware of their mortality. Humans want to believe in a humanistic deity—a mirror image of sorts—

who will guarantee us immortality in heaven. Fear of death, Gordon Gallup believes, may account for the religious impulse, but I think there's more to it. Our search for meaning and our innate reverence for this world in which we live are also probably related to self-awareness and mirrors.

The Mirrorless Biami Garden of Eden

Throughout the developed world, mirrors are commonplace. "The profundity of what takes place in a mirror is in perpetual danger of being lost through familiarity," notes science writer Adrian Desmond. *Mirror Mirror* has been an attempt to untarnish our mirrors, allowing us to look into them with fresh wonder and to help us understand their extraordinary place in human history.

Sometimes, however, I admit that I have thought we might be better off without mirrors, especially when I read that 850,000 Americans a year pay for botox injections to smooth their facial wrinkles with a paralytic poison, or when I consider other such attempts to manipulate image and deny mortality. But without mirrors, we would still be human. It is not the blank slate of the mirror that I deplore—it is what we sometimes reflect from it.

Let me leave you with the parable of anthropologist Edmund Carpenter's encounter with the Biami, an isolated New Guinea tribe. "It was important to us to film the reactions of people totally innocent of mirrors," Carpenter wrote in 1975. "Such people exist in New Guinea, though they number only a handful and are disappearing like the morning mist." A few Biami men had mirror shards, but they were too small to show a face and were used only as light-reflectors. They lived near swift rivers but no standing bodies of water in which they could see themselves.

"Certainly their initial reaction to large mirrors suggested this was a wholly new experience for them," Carpenter observed. "They were paralyzed: after their first startled response—covering their mouths and ducking their heads—they stood transfixed, staring at their images, only their stomach muscles betraying great tension." Carpenter interpreted their reaction as the "terror of self-awareness," and he portrayed their looking into mirrors as something like Adam and Eve eating the apple in the Garden of Eden, then suddenly becoming self-conscious and covering themselves.

"Western man," Carpenter asserted, "values, above all else, the isolated, delimited, aware self," whereas for traditional New Guinea tribes "there was no isolating individualism, no private consciousness." As appealing as this romantic assessment may be, I don't buy it. Anthropologist William Mitchell, who did extensive fieldwork with New Guinea tribes, says, "I never met any 'primitive' male or female who didn't know who he or she was, and acted upon it."

Though the Biami did not have mirrors, they had the human capacity to recognize themselves in them, as well as the human need to consider and manipulate their image. After all, the men already applied elaborate face paint to one another in preparation for war. "In a matter of days," Carpenter was forced to report, "they groomed themselves openly before mirrors." When they first beheld the miracle of their own reflection, the Biami may have felt genuine terror, as Carpenter surmised, but perhaps they also felt awe, wonder, and dawning comprehension.

In the developed world, we would do well—as we look into the myriad mirrors that surround us daily and as we use innovative scientific mirrors to look ever farther into the reaches of space and time, to send messages ever more quickly over beams of light, and to direct deadly laser weapons—to learn from the Biami. Mirrors *should* inspire terror, wonder, and comprehension.

NOTES ON SOURCES

In writing *Mirror Mirror* I consulted around 700 written sources, including books, articles, dissertations, video and audio programs, and websites, as well as conducting some 250 interviews. In the interests of conserving space, I have written the bibliographic essay below. A work is usually cited only once, even though it may be a source for other sections as well. For the full bibliography and citations, email me at markp@nasw.org.

— MARK PENDERGRAST

GENERAL BOOKS ON MIRRORS

Benjamin Goldberg's *The Mirror and Man* (1985), provides a good overview but few characters. Richard Gregory's *Mirrors in Mind* (1997) is a smorgasbord of information, mostly from experimental psychology. Sabine Melchior-Bonnet's *The Mirror: A History* (1994, 2001), originally in French and now translated into English, provides a good history of Murano and French mirrors before veering into Jacques Lacan psychobabble. Bruno Schweig's *Mirrors: A Guide to the Manufacture of Mirrors and Reflecting Surfaces* (1973) offers some history and detailed (though dated) technical information. Serge Roche's *Mirrors* (1956) offers introductory text and many pictures, as does G. F. Hartlaub in *Zauber des Spiegels* (1951), though it is available only in German. Pamela Heyne's *Mirror by Design: Using Reflection to Transform Space* (1996) is more specialized. Johann Beckmann's *A History of Inventions, Discoveries, and Origins* (1846) includes a good section on mirrors.

CHAPTER 1

On Egypt: Christine Lilyquist's *Ancient Egyptian Mirrors* (1979) is invaluable, though there are many books on ancient Egyptian culture and articles on Egyptian mirrors. **On the Sumerians:** Leo A. Oppenheim's *Ancient Mesopotamia* (1964); Seton Lloyd's *The Archaeology of Mesopotamia* (1978); Samuel Noah Kramer's *The Sumerians* (1963); Harriet Crawford's *Sumer and the Sumerians* (1991); *Myths from*

Mesopotamia, translated by Stephanie Dalley (1989). **On the Jews**, aside from the Bible itself, see Thomas Cahill's *The Gifts of the Jews* (1998); Harry M. Orlinsky's *Ancient Israel* (1960); Joshua Trachtenberg's *Jewish Magic and Superstition* (1939). **For the Phoenicians**, the website *Phoenician Enterprising*, http://phoenicia.org/trade.html, was useful, as well as *A Soaring Spirit: TimeFrame 600–400 BC* (1987), which also covers the Etruscans. Nancy Thomson de Grummond edited *A Guide to Etruscan Mirrors* (1982), and the series covering Etruscan mirrors, *Corpus Speculorum Etruscorum*, published by Cambridge University Press in the 1990s, is essential. See Richard Zacks's amusing *History Laid Bare* (1994) on sexual practices. **On Greek** myths, Edith Hamilton and Robert Graves are standard sources, as is Ovid. The works of Pausanias and Plato are widely available. See also *What Life Was Like at the Dawn of Democracy* (1997); Lenor Congdon's "Greek Mirrors," in *Notes in the History of Art* (1985). **On the Celts and Romans,** see *A Soaring Spirit: TimeFrame 600–400 BC* (1987); *What Life Was Like When Rome Ruled the World* (1997); F. R. Cowell's *Everyday Life in Ancient Rome* (1961); Jan Kock and Torben Sode's "Medieval Glass Mirrors," *Journal of Glass Studies* (2002); Ingeborg Krueger's "Glass-Mirrors of Medieval Times," *Annales du 12th Congrés de l'Association Internationale pour l'Histoire du Verre* (1993). Seneca's and Pliny's works are available in a number of translations. **On India, China, and Japan,** see Stuart Piggott's *Prehistoric India to 1000 B.C.* (1962); *Dwellings of Eternity,* edited by Alberto Siliotti (2000); *Barbarian Tides: Time-Frame 1500–600 BC.* (1987); Ju-hsi Chou's *Circles of Reflection* (2000); A. Bulling's *The Decoration of Mirrors of the Han Period* (1960); Friedrich Hirth's "Chinese Metallic Mirrors," in *Boas Anniversary Volume* (1906); B. Karlgren's "Early Chinese Mirror Inscriptions," in the *Bulletin of the Museum of Far Eastern Antiquities* (1934); *Chinese Bronze Mirrors*, by Milan Rupert and O. J. Todd (1935); Doris M. Roger's "The Divine Mirror of Japan," in *Asia* (1936); Timon Screech's *The Western Scientific Gaze and Popular Imagery in Later Edo Japan* (1996). **On North American Indians,** see George Emmons's *Slate Mirrors of the Tsimshian* (1921). **On ancient Mesoamerican and Peruvian mirrors,** see Victor W. von Hagen's *The Ancient Sun Kingdoms of the Americas* (1961); Muriel Porter Weaver's *The Aztecs, Maya, and Their Predecessors* (1981); Linda Schele and David Freidel's *A Forest of Kings: The Untold Story of the Ancient Maya* (1990); Miguel Covarrubias's *Indian Art of Mexico and Central America* (1966); Richard L. Burger's *The Prehistoric Occupation of Chavín de Huántar, Peru* (1984); Nigel Davies's *The Ancient Kingdoms of Peru* (1997); Garcilaso de la Vega's *Royal Commentaries of the Incas*, translated by Harold V. Livermore (1966); Cottie Burland and Werner Forman's *Feathered Serpent and Smoking Mirror* (1975); *Archaeology of Ancient Mexico and Central America: An Encyclopedia* (2001); and articles by Gordon F. Ekholm, J. J. Lunazzi, Karl A. Taube, and Justin Kerr.

CHAPTER 2

On magic and demons: E. A. Wallis Budge's *Amulets and Talismans* (1930, 1961); James George Frazer's *The Golden Bough*, abridged edition (1960); Rosemary Ellen

Guiley's *Encyclopedia of Witches and Witchcraft* (1989); Gary Jennings's *Black Magic, White Magic* (1964); Agnes Reppelier's *In the Dozy Hours* (1894). **On religious metaphors:** Mircea Eliade's *Patterns in Comparative Religion*, translated by Rosemary Sheed (1963); *Upanishads* in various editions; Alex Wayman's "The Mirror as a Pan-Buddhist Metaphor-Simile," *History of Religions* (1974); Judith A. Berling's "Taoism, or the Way," *Focus on Asian Studies* (fall 1982); Masumeh Price's "Norooz," http://www.cais-soas.co.uk/norooz.htm; *The Penguin Dictionary of Religions* (1995); Dennis Tedlock's *Breath on the Mirror* (1993); *Popol Vuh*, translated by Dennis Tedlock (1985). **On early scrying:** Theodore Besterman's *Crystal Gazing: A Study in the History, Distribution, Theory, and Practice of Scrying* (1924); Lynn Thorndike's *A History of Magic and Experimental Science*, 8 volumes (1923–1958). Thorndike's classic work is cited throughout Chapters 2, 3, and 4; Hippolytus, *The Refutation of All Heresies* (1868); Richard Kieckhefer's *Forbidden Rites: A Necromancer's Manual of the Fifteenth Century* (1997); Edward Peters's *The Magician, the Witch, and the Law* (1978); **On scrying in folk belief:** *The Complete Grimm's Fairy Tales*, translated by Margaret Hunt (1944); George Lyman Kittredge's *Witchcraft in Old and New England* (1926, 1956); W. F. Ryan's *The Bathhouse at Midnight* (1999). **On the Inquisition and fifteenth century:** Mary E. Gekler's *Gutenberg: The Master Printer* (1991); Brayton Harris's *Johann Gutenberg and the Invention of Printing* (1972); Heinrich Kramer and James Sprenger's *Malleus Maleficarum*, translated by Montague Summers (1486, 1971); Colin Wilson's *The Occult: A History* (1971); Anthony F. Aveni's *Behind the Crystal Ball* (2002); Grillo de Givry's *Witchcraft Magic and Alchemy*, translated by J. Courtenay Locke (1931, 1971); Theodore K. Rabb's *Renaissance Lives* (1993); Keith Thomas's *Religion and the Decline of Magic* (1971). **On John Dee and his times:** Benjamin Woolley's *The Queen's Conjurer: The Science and Magic of Dr. John Dee* (2001) is an excellent biography. See also Peter French's *John Dee: The World of an Elizabethan Magus* (1972); Deborah E. Harkness's *John Dee's Talking with Angels* (1999); William H. Sherman's *John Dee: The Politics of Reading and Writing in the English Renaissance* (1995); E.G.R. Taylor's *Tudor Geography* (1930); Reginald Scot's *The Discoverie of Witchcraft* (1584, 1964); Francis Russell Hart's *Admirals of the Caribbean* (1922); plus John Dee's own works: *Autobiographical Tracts* in *Remains Historical and Literary Connected with the Palatine Counties of Lancaster and Chester*, vol. 24 (1851); *The Diaries of John Dee*, edited by Edward Fenton (1998); *John Dee on Astronomy*, edited and translated with notes by Wayne Shumaker (1978); *The Mathematical Preface to the Elements of Geometrie of Euclid of Megara* (1570, 1975); *True and Faithful Relation of What Passed For Many Years Between Dr. John Dee and Some Spirits*, edited by Meric Casaubon (1659, reprint).

CHAPTER 3

On Greek science: Thomas Heath's two-volume *History of Greek Mathematics* (1921, 1981); Robert Temple's *The Crystal Sun* (2000), which argues that the ancients invented sophisticated optics; Edith Hamilton's *The Greek Way to Western*

Civilization (1942); works by Plato, Aristotle, Archimedes, Apollonius, Diocles, and Hero; David Park's *The Fire Within the Eye* (1997), an entertaining history of light; Carl B. Boyer's *The Rainbow from Myth to Mathematics* (1959), which covers other scientists as well; David C. Lindberg's *Beginnings of Western Science* (1992) and *Theories of Vision from Al-Kindi to Kepler* (1976), excellent resources for much material in Chapters 3 and 4; George Sarton's *Introduction to the History of Science* (1927); Jeanne Bendick's *Archimedes and the Door of Science* (1995); *Source Book in Greek Science* (1948); "Hypatia of Alexandria," http://www-gap.dcs.st-and.ac.uk/~history/ Mathematicians, one of many useful mathematical biographies on this University of St. Andrews website; Mark A. Smith's *Ptolemy and the Foundations of Ancient Mathematical Optics* (1999) and *Ptolemy's Theory of Visual Perception* (1996); G. L. Huxley's *Anthemius of Tralles* (1959). **On Chinese science:** Joseph Needham's four-volume *Science and Civilization in China* (1962). **On Arab science:** David Lindberg's books; *Ibn Al-Haitham: Proceedings*, edited by Hakim Mohammed Said (1969); Muhammad Saud's *The Scientific Method of Ibn al-Haytham* (1990); Alhazen and Witelo's *Opticae Thesaurus* (1972); K. Ajram's *The Miracle of Islamic Science* (1992). **On Grosseteste:** A. C. Crombie's *Robert Grosseteste and the Origins of Experimental Science* (1953, 1962); Robert Grosseteste's *On Light*, translated by Clare R. Riedl (2000); R. W. Southern's *Robert Grosseteste* (1986). **On Roger Bacon:** Roger Bacon's *Opus Majus*, with an introduction by John Henry Bridges (1964) and *[Perspectiva]: Roger Bacon and the Origins of Perspectiva in the Middle Ages*, translated and with an introduction by David C. Lindberg (1996); David C. Lindberg's *John Pecham and the Science of Optics* (1970). **On della Porta:** Giambattista della Porta's *Natural Magic*, a CD from Natural Magick Books [1589]; Michael John Gorman's "Science in Culture," *Nature* (2000). **On Digges:** Leonard Digges's *Prognotication Everlastinge Corrected and Augmented by Thomas Digges* (1576, 1975); F. R. Johnson's "The Influence of Thomas Digges," *Osiris* (1936); Albert van Helden's "Invention of the Telescope," *Transactions of the American Philosophical Society* (1977).

CHAPTER 4

On Thomas Harriot: John W. Shirley's *Thomas Harriot* (1983). **On Kepler:** *Dictionary of Scientific Biography*, ed. Charles Coulston Gillispie (1970–1990), which has articles on many other scientists in *Mirror Mirror* as well; Timothy Ferris's *Coming of Age in the Milky Way* (1988), an excellent overview of astronomy and cosmology; Johannes Kepler's *Optics: Paralipomena to Witelo*, translated by William H. Donahue (1604, 2000). **On Galileo:** Stillman Drake's *Galileo at Work* (1978) and *Galileo Studies* (1970); Galileo Galilei's *Discoveries and Opinions of Galileo*, translated by Stillman Drake (1957); Piero E. Ariotti's "Bonaventura Cavalieri, Marin Mersenne, and the Reflecting Telescope," *Isis* (1975); Henry C. King's *The History of the Telescope* (1955, 1979), a classic that covers many astronomers in *Mirror Mirror*. **On Descartes:** René Descartes' *Discourse on Method, Optics, Geometry, and Metrology*, translated by Paul J. Olscamp (2001); D. J. Lovell's *Optical Anecdotes* (1981). **On**

Kircher: Joscelyn Godwin's *Athanasius Kircher: A Renaissance Man and the Quest for Lost Knowledge* (1979); *The Great Art of Knowing: The Baroque Encyclopedia of Athanasius Kircher*, ed. Daniel Stolzenberg (2001); Ingrid D. Rowland's *The Ecstatic Journey: Athanasius Kircher in Baroque Rome* (2000). **On Huygens and Hevelius and Gregory**: *Dictionary of Scientific Biography*; A. E. Bell's *Christian Huygens and the Development of Science in the Seventeenth Century* (1947); Christiaan Huygens's *Treatise on Light* (1690, 1945). **On Hooke**: Ellen Tan Drake's *Restless Genius: Robert Hooke and His Earthly Thoughts* (1996); *Robert Hooke: New Studies*, edited by Michael Hunter and Simon Schaffer (1989); Robert Hooke's *Micrographia* (1665, 1961). **On Newton**: Gale E. Christianson's *In the Presence of the Creator: Isaac Newton and His Times* (1984); Dennis L. Sepper's *Newton's Optical Writings: A Guided Study* (1994); Michael White's *Isaac Newton: The Last Sorcerer* (1997); Isaac Newton's *Opticks* (1704, 1979).

CHAPTER 5

On glass mirrors: William S. Ellis's *Glass: From the First Mirror to Fiber Optics, the Story of the Substance That Changed the World* (1998); Alan MacFarlane and Gerry Martin's *Glass: A World History* (2002); Jaroslav R. Vávra's *5000 Years of Glass-Making: The History of Glass* (1954); Chloe Zerwick's *A Short History of Glass* (1980); Tom Grundy's *The Global Miracle of Float Glass* (1990), an insider's account of the invention of float glass; Heniz G. Pfaender's *Schott Guide to Glass* (1996); John E. Crowley's *The Invention of Comfort* (2001); Kock and Sode's "Medieval Glass Mirrors" (2002); Ingeborg Krueger's "Glass-Mirrors of Medieval Times" (1993); L. Y. Rahmani's "Mirror-Plaques from a Fifth-Century A.D. Tomb," *Israel Exploration Journal* (1964). **On Venice and Murano**: Jacques Barzun's *From Dawn to Decadence: 500 Years of Western Cultural Life* (2000); Dora Thornton's *The Scholar in His Study* (1997); Melchior-Bonnet's *The Mirror* (2001); Luigi Zecchin's *Vetro e Vetrai di Murano* (1990). **On literary specula**: Herbert Grabes's *The Mutable Glass: Mirror-Imagery in Titles and Texts of the Middle Ages and English Renaissance* (1973, 1982); RitaMary Bradley's "Backgrounds of the Title *Speculum* in Mediaeval Literature," *Speculum* (January 1954); James Williams's "Mirror and Speculum in Book Titles," *Law Magazine and Review* (1901); Guillaume de Lorris and Jean de Meun's *The Romance of the Rose*, translated by Charles Dahlberg (1995); Alan Gunn's *Mirror of Love* (1952); Dante Alighieri's *The Divine Comedy*, translated by Lawrence Grant White (1958); James L. Miller's "Three Mirrors of Dante's *Paradiso*," *University of Toronto Quarterly* (spring 1977). **On Elizabethan mirrors**: Anna Torti's *The Glass of Form: Mirroring Structures from Chaucer to Skelton* (1991); William Shakespeare's *The Riverside Shakespeare* (1974).

CHAPTER 6

On mirrors and lenses used in art: Norbert Schneider's *The Art of the Portrait* (1999); Heinrich Schwarz's "The Mirror in Art," *Art Quarterly* (1952); *Studies in the History* *of Art* (1959); H. W. Janson's *History of Art* (1962); Lawrence Weschler's "The Looking Glass," *New Yorker* (January 31, 2000); David Hockney's *Secret Knowledge* (2001); Jonathan Miller's *On Reflection* (1998); Martin Kemp's *The Science of Art* (1990); Giorgio Vasari's *Lives of the Artists: A Selection* (1965). **On Brunelleschi:** Ross King's *Brunelleschi's Dome* (2000); Antonio di Tuccio Manetti's *Life of Brunelleschi*, translated by Catherine Enggass (1970); Leon Battista Alberti's *On Painting*, translated by John R. Spencer (1956). **On van Eyck:** Elisabeth Dahnens's *Hubert and Jan Van Eyck* (1980); *The Complete Paintings of the Van Eycks* (1970); Craig Harbison's *Jan van Eyck* (1991); Edwin Hall's *Arnolfini Betrothal* (1994); Linda Seidel's *Jan van Eyck's Arnolfini Portrait* (1993); Jonathan Jones's "Arnolfini Portrait," *Guardian* (April 15, 2000); David Hockney's *Secret Knowledge* (2001). **On Leonardo:** Leonardo da Vinci's *The Notebooks of Leonardo da Vinci*, translated by Edward MacCurdy (1958); Anna Røsstad's *Leonardo da Vinci*, translated by Ann Zwick (1995); Robert Zwijnenberg's *The Writings and Drawings of Leonardo da Vinci*, translated by Caroline A. van Eck (1999). **On Dürer:** Mrs. Charles Heaton's *Life of Albrecht Dürer of Nürnberg* (1881); Erwin Panofsky's *Life and Work of Albrecht Dürer* (1955); Peter Streider's *Albrecht Dürer: Paintings, Prints, Drawings*, translated by Nancy M. Gordon and Walter L. Strauss (1989); Jan Bialostocki's *The Message of Images* (1988). **On anamorphic art:** Jurgis Baltrusaitis's *Anamorphic Art*, translated by W. J. Strachan (1977); Fred Leeman's *Hidden Images*, translated by Ellyn Childs Allison and Margaret L. Kaplan (1976); Martin Gardner's *Time Travel and Other Mathematical Bewilderments* (1988). **On Louis XIV mirrors:** Melchior-Bonnet's *Mirror*; Goldberg, *Mirror and Man*; Will and Ariel Durant's *Age of Louis XIV* (1963); Maurice Hamon's *From Sun to Earth, 1665–1999: A History of Saint-Gobain* (1999). **On mirrors in the North American fur trade:** Charles E. Hanson Jr.'s "Trade Mirrors," *Museum of the Fur Trade Quarterly* (winter 1986); **On James Graham:** Roy Porter's *Quacks* (2000); C.J.S. Thompson's *Mysteries of History* (1928).

CHAPTER 7

On William Herschel: Agnes M. Clerke's *The Herschels and Modern Astronomy* (1895); Caroline Herschel's *Memoir and Correspondence* (1879); William Herschel's *The Scientific Papers of Sir William Herschel* (1912); Michael Hoskin's *William Herschel and the Construction of the Heavens* (1964) and "William Herschel and the Making of Modern Astronomy," *Scientific American* (February 1986); Constance A. Lubbock's *The Herschel Chronicle* (1933); Patrick Moore's *William Herschel: Astronomer and Musician* (2000); Simon Schaffer's "Herschel in Bedlam," *British Journal for the History of Science* (November 1980); J. B. Sidgwick's *William Herschel* (1953); **On Hadley and Short:** John Hadley's "An Account of a Catadioptrick Telescope," *Philosophical Transactions* (1723); Rolf Willach's "James Short and the Development of the Reflecting Telescope," *Antique Telescope Society* (Winter 2001). **On King George:** John Brooke's *King George III* (1972). **On John Herschel:** Gunther Buttmann's *The Shadow of the Telescope: A Biography of John Herschel*, translated by E. J. Pagel (1965); Allan

Chapman's "An Occupation for an Independent Gentleman: Astronomy in the Life of John Herschel," *Vistas in Astronomy* (1993); John F. W. Herschel's *Poems and Pastimes* (1938); Patrick Moore's *Sir John Herschel: Explorer of the Southern Sky* (1992). **On Fraunhofer:** Wolfgang Jahn's *Historic Fraunhofer Glass-works: a Permanent Exhibition in Kloster Benediktbeuern* (1990). **On reflecting microscopes:** Thomas E. Jones's "History of the Light Microscope," http://www.utmem.edu/~thjones/hist/c2.htm; S. Bradbury's *Evolution of the Microscope* (1967). **On Birr's Leviathan:** Patrick Moore's *Astronomy of Birr Castle* (1991); Allan Chapman's *Victorian Amateur Astronomer* (1998); Robert Grant's *History of Physical Astronomy* (1825, 1966); J. P. Nichol's *Architecture of the Heavens* (1851); Agnes M. Clerke's *Popular History of Astronomy During the Nineteenth Century* (1902); *Science and Society in Ireland* (1999); Earl of Rosse's "Observations on the Nebulae," *Philosophical Transactions* (1850). **On Foucault, Grubb, Clark, Great Melbourne Telescope, Lick Observatory:** David Leverington's *History of Astronomy from 1890 to the Present* (1996); S.C.B. Gascoigne's "Great Melbourne Telescope," *Quarterly Journal of the Royal Astronomical Society* (1996); J. L. Perdrix's "Last Great Speculum," *Australian Journal of Astronomy* (April 1992); John Tallis's *Tallis's History and Description of the Crystal Palace* (1852); William Tobin's "Foucault's Invention of the Silvered-Glass Reflecting Telescope," *Vistas in Astronomy* (1987); J. A. Bennett's *Church, State and Astronomy in Ireland* (1990); Alvan Clark, "Great Telescopes of the Future," *Astronomy and Astro-Physics* (October 1893); Donald E. Osterbrock's *Eye on the Sky* (1988); John F. W. Herschel's *Telescope* (1861); Deborah Jean Warner's *Alvan Clark and Sons* (1995); I. S. Glass's *Victorian Telescope Makers* (1997); Howard Grubb's "On Great Telescopes of the Future," *Scientific Transactions of the Royal Dublin Society* (1877).

CHAPTER 8

On light, Young, Fresnel, Wollaston: David Park's *Fire Within the Eye* (1997); Vasco Ronchi's *Nature of Light* (1939, 1970); Jacob Abbott's *Light* (1871); Thomas D. Rossing's *Light Science* (1999); *Encyclopedia of Physics* (1990); *History of Astronomy: An Encyclopedia* (1997); *Dictionary of Scientific Biography*. **On photography invention:** Larry J. Schaaf's *Out of the Shadows* (1992); Beaumont Newhall's *History of Photography from 1839 to the Present Day* (1949). **On Brewster and Wheatstone:** *Brewster and Wheatstone on Vision*, edited by Nicholas J. Wade (1983); David Brewster's *The Kaleidoscope* (1858, 1987) and *Treatise on Optics* (1841); *'Martyr of Science': Sir David Brewster, 1781–1868*, edited by A. D. Morrison-Low (1984); Brian Bowers's *Sir Charles Wheatstone* (1975). **On Joseph Plateau:** Brian Coe's *History of Movie Photography* (1981). **On Bell's photophone:** Edwin S. Grosvenor's *Alexander Graham Bell* (1997). **On Feynman:** Richard P. Feynman's *QED: The Strange Theory of Light and Matter* (1985). **On Huggins:** William Huggins's "The New Astronomy," *Nineteenth Century* (June 1897) and *Scientific Papers* (1909); *Book of the Cosmos*, edited by Dennis Richard Danielson (2000). **On Rowland:** Henry Rowland's "Preliminary Notice," *Philosophical Magazine* (1882). **On nineteenth-century magic:** John Melville's *Crystal*

Gazing and Clairvoyance (1896, 1970); Eusebe Salverte's *Philosophy of Magic, Prodigies, and Apparent Miracles* (1847); James Steinmeyer's *Two Lectures on Theatrical Illusions* (2001); J. H. Pepper's *Cyclopaedic Science Simplified* (1869); Abbott's *Light*; Milbourne Christopher's *Illustrated History of Magic* (1973, 1996); Albert A. Hopkins's *Magic: Stage Illusions and Scientific Diversions* (1897, 1967); Edwin A. Dawes's *Stodare* (1998); J. C. Cannell's *Secrets of Houdini* (1931, 1971). **On Stratton:** George M. Stratton's "Some Preliminary Experiments on Vision," *Psychological Review* (1896), "The Spatial Harmony of Touch and Sight," *Mind* (1899); *Experimental Psychology and Its Bearing Upon Culture* (1908), *Man: Creator or Destroyer* (1952); Nicholas J. Wade's "An Upright Man," *Perception* (2000).

CHAPTER 9

On Hale and Ritchey: Donald E. Osterbrock's *Pauper and Prince: Ritchey, Hale, and Big American Telescopes* (1993); Helen Wright's *Explorer of the Universe: A Biography of George Ellery Hale* (1966); Sheehan and Osterbrock's "Hale's 'Little Elf,'" *Journal for the History of Astronomy* (2000); George Ellery Hale's *Study of Stellar Evolution* (1908); articles by Hale and Ritchey. **On Keeler:** Osterbrock's *Eye on the Sky* (1988); James E. Keeler's "The Crossley Reflector of the Lick Observatory," *Astrophysical Journal* (June 1900). **On World War I optics:** Glass's *Victorian Telescope Makers* (1997); Halvorson and Hussey's "Evolution of Light Projection," *Transactions of the Illuminating Engineering Society* (August 30, 1917); F. H. Kohloss's "The Development of Military Searchlights," *Military Engineer* (1930); W. F. Tompkins's "Co-operation of Aeroplane Searchlights," *Military Engineer* (1920). **On Edwin Hubble:** Gale E. Christianson's *Edwin Hubble* (1995). **On 200-inch telescope:** Ronald Florence's *The Perfect Machine* (1994); David O. Woodbury's *Glass Giant of Palomar* (1939); Frederick A. Collins's, *Greatest Eye in the World* (1942); G. Edward Pendray's *Men, Mirrors, and Stars* (1935). **On Don Hendrix:** Osterbrock's "Don Hendrix, Mount Wilson and Palomar Observatories," *Journal of the Antique Telescope Society* (Summer 2001). **On Russell Porter:** Berton C. Willard's *Russell W. Porter* (1976); Russell W. Porter's "The Telescope Makers of Springfield, Vermont," *Popular Astronomy* (March 1923) and *Giants of Palomar* (1983); Webb Waldron's "One Really Happy Man," *American Magazine* (November 1931); Albert G. Ingalls's *Amateur Telescope Making* (1935, 1972) and "The Heavens Declare the Glory of God," *Scientific American* (November 1925); **On Bernhard Schmidt:** A. A. Wachmann's "From the Life of Bernhard Schmidt," *Sky & Telescope* (November 1955); Erik Schmidt's *Optical Illusions: The Life Story of Bernhard Schmidt* (1995); Milton Silverman's "The Eye That Exposes Secrets," *Saturday Evening Post* (April 22, 1950).

CHAPTER 10

On early twentieth-century mirrors: *1897 Sears Roebuck Catalog* (1897, 1976); William Leach's *Land of Desire* (1993); David Nasaw's *Going Out* (1993); Frederick

Lewis Allen's *Only Yesterday* (1931); *Guiness Book of Car Facts and Feats* (1994). **On amusement parks:** John F. Kasson's *Amusing the Million* (1978); Robert W. Rydell's *All the World's a Fair* (1984); Edo McCullough's *World's Fair Midways* (1976); Judith A. Adams's *The American Amusement Park Industry: A History of Technology and Thrills* (1991); Pilat and Ranson's *Sodom by the Sea* (1941); Gary Kyriazi's *The Great American Amusement Parks* (1976). **On movie palaces:** Margolies and Gwathmey's *Ticket to Paradise* (1991); Ben M. Hall's *Best Remaining Seats* (1961); Roger Brett's *Temples of Illusion* (1976). **On cosmetics:** Kathy Peiss's *Hope in a Jar* (1998); Fenja Gunn's *Artificial Face* (1983); Frida Kerner Furman's *Facing the Mirror* (1997); Richard Corson's *Fashions in Makeup* (1972) and *Fashions in Hair* (1969); Anne Hard's "The Beauty Business," *American Magazine* (November 1909); Warshaw Collection, National Museum of American History. **On mirrors in architecture:** William Lawrence Bottomley's "Mirrors in Interior Architecture," *Architectural Forum* (1932); Robert W. Swedberg's *Furniture of the Depression Era* (1992); Jeffrey Hogrefe's "Ascent of Glass," *Smithsonian* (July 2001).

CHAPTER 11

On radio astronomy: *Early Years of Radio Astronomy*, edited by W. T. Sullivan III (1984); J. S. Hey's *Evolution of Radio Astronomy* (1973) and *Radio Universe* (1975); John D. Kraus's *Radio Astronomy* (1966) and *Big Ear Two* (1995); Pausey and Bracewell's *Radio Astronomy* (1955); *Serendipitous Discoveries in Radio Astronomy*, edited by K. Kellermann and B. Sheets (1983); George C. Southworth's "Early History of Radio Astronomy," *Scientific Monthly* (February 1956); Gerrit L. Verschuur's *Invisible Universe Revealed* (1987); Bernard Lovell's *Astronomer by Chance* (1990) and *Voice of the Universe* (1987); Henbest and Marten's *The New Astronomy* (1983); John Pfeiffer's *The Changing Universe* (1956); Marshall H. Cohen's "The Owens Valley Radio Observatory," *Engineering & Science* (spring 1994). **On Penzias and Wilson:** Steven Weinberg's *The First Three Minutes* (1977); Timothy Ferris's *The Red Limit* (1977); John Mather's *The Very First Light* (1996); R. W. Wilson's "The Cosmic Microwave Background Radiation," *Science* (August 31, 1979). **On pulsars:** S. Jocelyn Bell Burnell's "Little Green Men, White Dwarfs, or Pulsars?" *Annals of the New York Academy of Science* (December 1977); W. J. Cocke's "Discovery of Optical Signals from Pulsar NP 0532," *Nature* (February 8, 1969); Richard N. Manchester's *Pulsars* (1977); F. G. Smith's *Pulsars* (1977); Joan Warnow's "Moments of Discovery: Optical Pulsars" (no date). **On Frank Drake:** Frank Drake's *Is Anyone Out There?* (1992). **On millimeter waves:** Mark A. Gordon's "Cold Heart of the Cosmos," *Mercury* (January–February 1997) and "A New Surface for an Old Scope," *Sky & Telescope* (April 1984). **On x-ray astronomy:** Wallace Tucker and Riccardo Giacconi's *The X-Ray Universe* (1985); Giacconi's "Grazing-Incidence Telescopes for X-Ray Astronomy," *Space Science Reviews* (1969) and "X-Ray Astronomy," *Physica Scripta* (1996). **On gamma rays:** John V. Jelley and Trevor C. Weekes's "Ground-Based Gamma-Ray Astronomy," *Sky & Telescope* (September 1995).

CHAPTER 12

On Meinel: Aden and Marjorie Meinel's "Telescopes on the Horizon," unpublished manuscript; Ginger Oppenheimer's "Defining the Golden Age of Science," *OE Magazine* (June 2002). **On MMT:** A. B. Meinel's "A Large Multiple Mirror Telescope (MMT) Project," *Optical Engineering* (March/April 1972); Nathaniel P. Carleton's "The Multiple-Mirror Telescope," *Physics Today* (September 1978); Smithsonian Videohistory Program, "Multiple Mirror Telescope" (1989); *Telescopes for the 1980s* (1981); Richard Learner's *Astronomy Through the Telescope* (1981). **On Kitt Peak:** James E. Kloeppel's *Realm of the Long Eyes* (1983). **On lasers:** Jeff Hecht's *Laser* (1982, 1998); "LAGEOS 1, 2," http://msl.jpl.nasa.gov/QuickLooks/lageosQL.html. **On infrared:** David A. Allen's *Infrared* (1975); F. J. Low's "Ground-based Observations at 34 Microns," *Astrophysical Journal* (1973); Gerry Neugebauer's "Observations of Extremely Cool Stars," *Astrophysical Journal* (1965); Habing and Neugebauer's "The Infrared Sky," *Scientific American* (November 1984); "Kuiper Airborne Observatory Fact Sheet," http://spacelink.nasa.gov/. **On military mirrors:** J. Kelly Beatty's "Up in the Sky!" *Sky & Telescope* (February 1999); William W. Ward's "Thirty Years of Research," *Lincoln Laboratory Journal* (spring 1989); "Project West Ford," *Proceedings of the IEEE* (May 1964); William E. Burrows's *Deep Black* (1986); Jeffrey T. Richelson's *Wizards of Langley* (2001); Jams Bamford's *Body of Secrets* (2001). **On adaptive optics:** John W. Hardy's *Adaptive Optics for Astronomical Telescopes* (1998), "Adaptive Optics" interview by Frederick Su, *OE Reports* (December 1994); Fugate and Wild's "Untwinkling the Stars," *Sky & Telescope* (May/June 1994); Roger H. Ressmeyer's "Robert Q. Fugate," *Sky & Telescope* (May 1994). **On Hubble Space Telescope:** Robert W. Smith's *The Space Telescope* (1993); Eric J. Chaisson's *The Hubble Wars* (1994, 1998); Lyman Spitzer Jr.'s "Astronomical Advantages of an Extra-Terrestrial Observatory," *Astronomy Quarterly* (1946, 1990); Eric J. Lerner's "What Happened to Hubble?" *Aerospace America* (February 1991); *Hubble Space Telescope Optical Systems Failure Report* (1990); Robert E. Fischer's *Optical System Design* (2000); Richard Tresch Fienberg's "Hubble's Road to Recovery," *Sky & Telescope* (November 1993); John N. Bahcall's "The Space Telescope," *Scientific American* (July 1982) and "Lyman Spitzer, Jr.," *Physics Today* (October 1997). **On Jerry Nelson:** John R. Gustafson's "The Keck Observatory," *Mercury* (March 1988); Jerry Nelson's "The Keck Telescope," *American Scientist* (March–April 1989). **On 1980 Tucson conference:** *Optical and Infrared Telescopes for the 1990s* (1980). **On liquid mercury telescopes:** Brad Gibson's "Liquid Mirror Telescopes: History," *Journal of the Royal Astronomical Society of Canada* (1991); Ermanno F. Borra's "The Liquid-Mirror Telescope as a Viable Astronomical Tool," *Journal of the Royal Astronomical Society of Canada* (1982) and "Liquid Mirrors," *Scientific American* (February 1994); **On Roger Angel:** J. Madeline Nash's "Shoot for the Stars," *Time* (April 27, 1992); Mirror Lab, http://medusa.as.arizona.edu/mlab/mlab.html; J.R.P. Angel articles; *The NOAO 8-M Telescopes* (September 1989). **On future space telescopes:** Alan Dressler's *HST and Beyond* (1996); "Next Generation Space Telescope," http://www.stsci.edu/ngst/overview; *Ter-*

restrial Planet Finder website, http://planetquest.jpl.nasa.gov/TPF/tpf_sample.html; *Ultra Lightweight Space Optics Challenge Workshop* (March 24–25, 1999) http:// origins.jpl.nasa.gov/meetings/ulsoc/. **On space advertising and Znamya:** "Orbiting-Billboard Proposal," *Sky & Telescope* (November 1993); Kathy Prentice's "Rocketing Your Message," *Media Life* (July 24, 2000); http://www.space-frontier.org; "Znamya Falls to Earth," *Sky & Telescope* (February 5, 1999). **On solar sails:** Colin Robert McInnes's *Solar Sailing* (1999); Planetary Society website, http://planetary.org/html. **On future ground telescopes:** 2002 SPIE Conference in Hawaii, v. 4840 (2003); Michael Lemonick's "Beyond Hubble," *Time* (November 13, 2000); OWL website, http://www. eso.org/projects/owl/. **On accelerating universe:** Donald Goldsmith's *Runaway Universe* (2000); Robert P. Kirshner's *Extravagant Universe* (2002).

CHAPTER 13

Much of this chapter is based on personal experience and interviews. **On modern magic mirrors:** Donald Tyson's *How to Make and Use a Magic Mirror* (1995); Kirsten Lagatree's "Feng Shui," *Los Angeles Times* (October 31, 1999, April 30, 2000); Richard Webster's *101 Feng Shui Tips for the Home* (1999); Kristine Nyhout's "Feng Shui Helps Me," *Toronto Star* (December 8, 1998); R. Yong's "Companies Corporate Feng Shui," *Malaysia Business Times* (May 3, 2000). **On solar mirrors:** Goldberg, *Mirror and Man* (1985); Aden B. Meinel's *Applied Solar Energy* (1976). **On Shoemaker** and **Helin:** David Levy's *Shoemaker by Levy* (2000). **On kaleidoscopes:** Cozy Baker's *Kaleidoscopes: Wonders of Wonder* (1999) and *Through the Kaleidoscope . . . and Beyond* (1987); William Novak's "Surprise Party," *Washingtonian* (June 1998); *Brewster Society News Scope* issues, 1999–2002. **On mirror symmetry:** Chris McManus's *Right Hand, Left Hand* (2002); Martin Gardner's *New Ambidextrous Universe* (1990). **On mirror universes:** Robert Foot's *Shadowlands* (2002); Ron Cowen's "Through the Looking Glass," *Science News* (September 9, 2000); Frank Close's "Fearful Symmetry," *New Scientist* (April 8, 2000); Martin Rees's *Before the Beginning* (1997). **On LIGO:** Gary H. Sanders's "LIGO," *Sky & Telescope* (October 2000). **On high-tech mirrors:** David J. Bishop's "The Rise of Optical Switching," *Scientific American* (January 2001); Barnaby J. Feder's "Big Step Forward in Tiny Technology," *New York Times* (May 8, 2000); Charles Platt's "Bright Switch," *Wired* (September 2000). **On mirror reversal:** Gregory, *Mirrors in Mind*; Gardner, *New Ambidextrous Universe*; N. J. Block's "Why Do Mirrors Reverse Right/Left," *Journal of Philosophy* (May 16, 1974). **On visual rays:** Gerald A. Winer's "Fundamentally Misunderstanding Visual Perception," *American Psychologist* (June/July 2002). **On True Mirror:** William E. Benton's "Duality Mirror," Bookfinder.com (1930); McManus, *Left Hand*; Richard O'Mara's "Here's Looking at You," *Baltimore Sun* (November 23, 1998); Cullen Murphy's "The Mirror of Dorian Gray," *Atlantic Monthly* (June 1999); Barbara Smith's "Splitting Hairs," *Syracuse Herald American* (January 17, 1999); Alison Roberts's "You Might Flip," *Sacramento Bee* (November 18, 1998); R. Andrew Hicks's "Catadioptric Sensors," http://www.mcs.drexel.edu/~ahicks/mirrors.html. **On** psychology and mirrors: Mark Pendergrast's *Victims of Memory* (1996); Elizabeth Loftus's "The Most Dangerous Book," *Psychology Today* (November/December 2000); Marlene Steinberg's *Stranger in the Mirror* (2000); Katharine A. Phillips's *Broken Mirror* (1996); Simone Nilsson's "Heartbreak of Bigorexia," *Insight on the News* (April 24, 2000); Arthur C. Traub's "Psychophysical Studies of Body-Image," *Archives of General Psychiatry* (1964); Bill Choisser's *Face Blind!* http://www.choisser.com/faceblind!; Martin J. Tovée's "Neural Substrates of Face Processing Models," *Cognitive Neuropsychology* (1993); *Signal and Sense*, edited by Gerald M. Edelman (1990); Julian Paul Keenan's "Self-Awareness and the Right Prefrontal Cortex," *Trends in Cognitive Science* (2000); Robert Mayer's "Does a Mirror Deter Wandering?" *International Journal of Geriatric Pschiatry* (1991); Nora Breen's "Mirrored-Self Misidentification," *Neurocase* (2001); V. S. Ramachandran's "Synaesthesia in Phantom Limbs," *Proc. Royal Society of London* (1996); Luis Schwarz's "Illusions Induced by the Self-Reflected Image," *Journal of Nervous and Mental Disease* (April 1968); Suzanne Barry Osborn's "Little Things," *Chain Store Age* (June 1, 2000); Donna Greene's "Looking Glass for Telemarketers' Image," *New York Times* (December 5, 1999). On animal self-recognition: Gordon G. Gallup Jr.'s "Chimpanzees: Self-Recognition," *Science* (January 2, 1970) and many other Gallup articles; Adrian J. Desmond's *The Ape's Reflexion* (1979); Bernd Heinrich's *Mind of the Raven* (1999); Gretel H. Schueller's "Hey! Good Looking," *New Scientist* (June 17, 2000); *Self-Awareness in Animals and Humans*, edited by Sue Taylor Parker (1996); Marc Hauser's *Wild Minds* (2000); Daniel J. Povinelli's "Failure to Find Self-Recognition in Asian Elephants," *Journal of Comparative Psychology* (1989); Patricia Simonet's "Social and Cognitive Factors in Asian Elephant Mirror Behavior," presentation (August 2000); Steven M. Wise's *Drawing the Line* (2002); Diana Reiss and Lori Marino's "Mirror Self-Recognition in the Bottlenose Dolphin," *PNAS* (May 8, 2001). On babies and mirrors: Beulah Amsterdam's "Mirror Self-Image Reactions Before Age Two," *Developmental Psychobiology* (1972); Michael Lewis's *Social Cognition and the Acquisition of Self* (1979); Jerome Kagan's *Nature of the Child* (1984); Richard Langton Gregory's "Recovery from Early Blindness," *Experimental Psychology Monograph No. 2* (1963); Frans de Waal's *Good Natured* (1996) and *Bonobo* (1997); Maser and Gallup, "Theism as a By-Product," *Journal of Religion* (1990); Jacques Lacan's *Écrits: A Selection* (1977). On Biami: Edmund Carpenter's "Tribal Terror of Self-Awareness," in *Principles of Visual Anthropology* (1975).

LIST OF INTERVIEWS, CONDUCTED 1999–2003.

Alan Adler	Al Baez
Torben Andersen	Mark Bailey
Dave Anderson	Dan Bajuk
Roger Angel	Cozy Baker
Maryann Arrien	Guido Barbini
Eva Baboula	Chris Barry

Jacques Beckers

Charles Beichman

Pierre Bely

Tim Bentley

Jean Bernard

John Biretta

Pascal Bordé

Amanda Bosch

Edward Bowell

Jim Brase

Joachim Bretschneider

Dan Brocious

Lucio Bullo

Jim Burge

John Butler

Richard Capps

Webster Cash

Bob Cassanova

Veronique Cassel

Peter Chen

Harry Chocolate

Ju-hsi Chou

Mark Clark

Moe Cloutier

Mark Colavita

Doug Cooke

Chris Corbally

Dan Coulter

Bill Craig

Dave Crawford

Ray Creager

Dave Crowe

Larry D'Addario

Mario d'Alpaos

Frans de Waal

Joe Derek

George Djorgovski

Don Doak

John Dobson

Robert Dolland

Ewan Douglass

Jerry Doyle

Mark Dragovan

Alan Dressler

Georges Dreyer

Ted Dunham

Junie Esslinger

Emilio Falco

Timothy Ferris

Doug Finkbeiner

Adrian Fisher

Vincent Coudé du Foresto

Jason Fournier

Ellie Fowler

Gordon Gallup

Fritz Garrison

Elinor Gates

Daniel Gauthier

Riccardo Giacconi

Bruce Gillespie

Roberto Gilmozzi

Joseph Giral

Howard Glatter

Paul Glenn

Dan Goldin

Mark Gordon

Tom Gorka

Wes Grammer

David Graves

Richard Gregory

Tom Grundy

Craig Gullixson

Bob Gurwicz

Bob Hall

John Hamilton

Maurice Hamon

John Hardy

Tim Hawarden

Steve Hegwer

Eleanor "Glo" Helin

Jeff Hendrickson

Andy Hicks

Paul Hickson

John Hill

Michael Hill

David Hilyard

Allan Hobson
Xavier Hodges
David Hogg
Olga Horakova
Bob Horton
Rex Hunter
Rick Hunter
Ralph Jacobs
Phil Jewell
Anders Jorgensen
John Joyce
Charles Karadimos
Steve Keil
Mike Kendall
Dean Ketelsen
Jim Kilcrease
Robert Kirshner
Atsuko Nitta Kleinman
Ingeborg Krueger
Antoine Labeyrie
Marc Lacasse
French Leger
John Lennhoff
David Levy
Chuck Lillie
Frank Low
Real Manseau
Lori Marino
Bob Marriott
Buddy Martin
Marc Martinez
Terry Mast
John Mather
Gary Matthews
Bob May
Drew Mayberry
Don McCarthy
John McFarland
Ramsey Melugin
Marilyn Michalski
Jason Midiri
Robert Midiri
Richard Miles

Doug Miller
Jonathan Miller
Steve Miller
Bob Millis
Tony Misch
William E. Mitchell
Rabindra Mohapatra
Michael Molitz
Mel Montemerlo
Sherry Moser
Mark Mulrooney
Bill Napier
Jerry Nelson
Gerry Neugebauer
Bob Neville
Bruce Newman
Peter Newman
Don Nicholson
John Noble
Gary Nowak
Kay O'Connor
Gabrielle Olalde
Don Osterbrock
Brian Otekur
Frank Pakulski
C. J. Park
Dave Parker
Alecia Parsons
John Payne
John Pearce
Terry Pearce
Irene Pepperberg
Larry Phillips
Tim Pickering
Gary Poczulp
Gwen Pollock
Terry Pope
V. Radhakrishnan
Joanna Rankin
John Ratje
Dave Redding
Dan Reichart
Paul Reid

George Rieke
Thomas Rimmele
Bruno Rivoire
Jean-Francois Robert
Phil Rounseville
Vojtech Saman
Jeff Sauer
David Schlegel
Erik Schmidt
Bill Scott (Atmavidyamanda)
Bernard Seery
Roberto Sella
Tony Semeraro
Jean-Jacques Serrat
Robert Shannon
John Shelton
Carolyn Shoemaker
Julian Shull
Mike Shull
Bob Silversteen
Patricia Somonet
David Sinden
Patrick Slane
Al Slomba
Ray Smartt
Paul Smith
Steph Snedden
Gary Sommergren
John Spencer
Keith Spittler
Phil Stahl
Larry Stepp
Bill Stoeger
Remington Stone
Pierre Suzanne
Judith Swaddling
Don Sweeney
Lino Taglipietra

Don Taylor
Robert Temple
Charles Thayer
Bob Thicksten
Aaron Thomas
Norm Thomas
Harley Thronson
Ellis Tinios
Luigi Toso
John Trauger
Virginie Trillas
Wallace Tucker
Bob Tull
Paul Valleli
Erik Van Cort
Leon Van Speybroeck
John Vogt
Catherine Walter
John Walter
Philip Webb
Larry Webster
Trevor Weekes
Anke Weidenhaff
Alycia Weinberger
Martin Weisskopf
Michel Wenger
Mike Werner
Edward Whitehead
Henry Wildey
Bert Willard
J. T. Williams
Richard Williams
Robert Wilson
Laura Woodley
David Woody
Neville Woolf
Simon "Pete" Worden
Richard Wortley

ACKNOWLEDGMENTS

In some measure I can blame Lisa Bankoff, my ICM agent, for this three-year mirror project. We were celebrating the publication of *Uncommon Grounds*, my coffee history, over lunch in a fine New York restaurant. She asked what I would do next, and I said I wasn't sure, but I didn't want to write about anything as complex or time-consuming as coffee. So Lisa suggested writing on an everyday object. I looked around and saw a mirror on the wall. I've always been fascinated by mirrors and said, "Mirrors might be interesting." But like Alice, I stepped into more than I bargained for. It has been a many-mirrored journey. Thanks, Lisa, for sparking the idea and helping me to sell the concept.

Tim Bartlett, my then-editor at Basic Books, loved the proposal and championed it before leaving for Oxford University Press. Fortunately, Bill Frucht, who inherited the book at Basic, is an imaginative, encouraging editor with a broad scientific background. He has provided invaluable direction and suggestion. Regina Hersey is the freelance editorial wizard who helped me reduce my oversized manuscript to manageable proportions. Graphic artist Eric Fuenticilla produced a gorgeous, evocative cover.

In my travels, I usually stayed in inexpensive youth hostels or inns, but some kind people took me into their homes. In Paris, Nathalie McKinley let me use her apartment. In Portsmouth, England, Adrian and Marie Fisher invited me to stay in their home while I interviewed Adrian about mirror mazes, then Roger and Guri Scotford housed me in Bradford-on-Avon, near Bath. In Los Angeles, where the monks of the Vedanta monastery put me up, Bill Scott, also known as Atmavidyamanda, was particularly helpful. In Tucson, Arizona, grad student Doug Miller and his housemates shared their home, as did optician David Hilyard and his wife, Darrie, in Santa Cruz, California. In Fairfax, Virginia, Sue Taylor gave me a place to stay while working in Washington, D.C., as did Barbara Benjamin in White Plains for my research in New York City. Brent and Janie Cohen lodged me in their guest room in Oakland, California. I also discovered that isolated mountaintop observatories are hospitable to researchers. Thanks particularly to Bob Millis and his staff at Lowell Observatory; Bruce Gillespie at Apache Point Observatory; Don Nicholson for the tour of Mount Wilson; Bob Thicksten for taking me around Mount Palomar, Dan Brocious at Mount Hopkins, John Ratje at Mount Graham, Phil Jewell at Green Bank. Infrared astronomer Don McCarthy was a fantastic host and teacher during Astronomy Camp atop Mount Lemmon near Tucson. In Ireland, Mark Bailey and colleagues welcomed me at Armagh Observatory, and John Joyce and Alecia Parsons at Birr Castle. Closer to home, the members of my local Vermont Astronomical Society welcomed this neophyte observer.

I am indebted to the librarians and archivists at the British Library, British Museum, Library of Congress, Archives Center of the National Museum of American History, Hartman Center of Duke University, New York Public Library, and Mary Lea Shane Archives of the Lick Observatory. I couldn't have survived without interlibrary loan librarians Linda Willis-Pendo, Norma Lemieux, and Mara Siegel at the Midstate Regional Library in Berlin, Vermont, and Erika Trudeau at the Burnham Memorial Library in Colchester, Vermont. Peter Hingley of the Royal Astronomical Society Library in London helped in person and then sent requested material I couldn't find in the United States. Maurice Hamon, the Saint-Gobain historian, gave me his book and his time, and Clementine Albano, librarian at Stazione Sperimentale de Vetro in Murano, was very helpful. Christine Kleinegger of the New York State Museum in Albany sent me an invaluable script from the museum's mirror exhibit.

I am grateful to everyone I interviewed for sharing their time, knowledge, and expertise, but I must single out Harley Thronson, who spent many hours with me, patiently explaining NASA projects; Doris Tucker, Roger Angel's assistant, who helped so much, and Roger Angel himself for putting up with multiple questions and interviews; Larry Stepp of the NOAO for all his time; Bert Willard for articles and information; Cozy Baker, not only for much help but also for donating kaleidoscopes to my wife's hospice program. Thanks to Dave Barnes of Arbor Scientific for the Mirage and Virtual Reality Mirrors, and to John and Catherine Walter for a True Mirror.

I am fortunate to know unselfish, intellectually curious people who helped by sending mirror information periodically, including Chris Dodge, Diane Foulds, Peter Freyd, Chris Hadley, Connie Kite, Henry Lilienheim, Jack Malinowski, Loren Pankratz, and James Harvey Young.

Many people read parts of the manuscript and, though I am solely responsible for the content, I value comments from Irene Angelico, Jacques Beckers, Steve Carlson, Jim DeFilippi, Margaret Edwards, Diane Foulds, Riccardo Giacconi, Tim Hawarden, Andy Hicks, Linda Rice Lorenzetti, Chris Miller, Bill Mitchell, Frank Pakulski, Loren Pankratz, Nan and Britt Pendergrast, John Pendergrast, Larry Ribbecke, Robert Sharer, Joe Sherman, Mark Smith, Benjamin Woolley, and Steve Young. Various members of the online narrative nonfiction group, WriterL, were also good critics.

ILLUSTRATION PERMISSIONS
AND CREDITS

COVER PHOTO

"The Baleful Head," by Edward Burne-Jones, 1886–87, courtesy Staatsgalerie, Stuttgart, Germany.

AUTHOR PHOTO

Abbey Neidik photo.

TEXT ILLUSTRATIONS

page

59   Law of reflection drawing: from Thomas D. Rossing, Christopher J. Chiaverina, *Light Science,* Springer, 1999.

62   Conic sections: from Henry C. King, *History of the Telescope.* Dover, 1955, 1979.

63   Spherical aberration: from Richard Berry, *Build Your Own Telescope*, Scribner, 1985.

65   Boy in mirror: from Thomas D. Rossing, Christopher J. Chiaverina, *Light Science,* Springer, 1999.

75   Rainbow drawing: courtesy Adelaide Tyrol, Plainfield, VT, from Gale Lawrence, *Field Guide to the Familiar*, Prentice-Hall, 1984.

88   Mersenne designs: from Henry C. King, *History of the Telescope.* Dover, 1955, 1979.

99, 105, 108, 242   Gregorian, Newtonian, Cassegrain, & Schmidt telescope drawings: from *Eyes on the Universe: A History of the Telescope* by Isaac Asimov. Copyright © 1975 by Isaac Asimov. Reprinted by permission of Houghton Mifflin Company. All rights reserved.

134   Brunelleschi drawing: from Ross King, *Brunelleschi's Dome*. Pimlico, 2001, drawing by Sarah Challis.

188 double slit experiment: from Conrad G. Mueller, Mae Rudolph, *Light and Vision*, Time-Life, 1966, drawing by Otto van Eersel.

195 Stereoscope drawing: public domain.

198 Foucault speed of light: http://home.istar.ca/~jbogan/speedc.htm.

202 Spectrum of light: from Donald Goldsmith, *The Evolving Universe*, Benjamin-Cummings, 1981.

214, 215 Stratton drawings: from George Malcolm Stratton, *Experimental Psychology*, Macmillan, 1908.

285 Grazing incidence x-ray drawing: from Wallace Tucker and Riccardo Giacconi, *The X-Ray Universe*, Cambridge, MA: Harvard U. Pr., 1985.

INSERT PHOTOS

INSERT 1

Egyptian mirror: courtesy Pierre Devinoy photography collection, Paris, France, mirror in Louvre collection.

Etruscan mirror: courtesy Corpus Christi College & Fitzwilliam Museum, UK. Drawing by Garth Denning.

Greek vase: courtesy Bibliothèque Nationale de France, Paris.

Chinese magic mirror: from David Brewster, *A Treatise on Optics*. rev. ed. Philadelphia: Lea & Blanchard, 1841.

Tezcatlipoca drawing: courtesy The Board of Trustees of the National Museums and Galleries on Merseyside (Liverpool Museum), Liverpool, UK.

John Dee engraving: frontispiece of Meric Casaubon's *A True and Faithful Relation of What Passed for Many Yeers . . .* , London: D. Maxwell for T. Garthwait, 1659. By permission of the Syndics of Cambridge University Library.

John Dee's Aztec mirror: Mark Pendergrast photo.

Archimedes drawing: public domain.

Dietrich of Freiberg drawing, circa 1300: public domain.

Convex self-portrait: Mark Pendergrast photo.

Concave mirror handshake: courtesy Arbor Scientific.

Camera obscura: courtesy Gernsheim Collection, Harry Ransom Humanities Research Center, The University of Texas at Austin.

Descartes tennis player: René Descartes, *Vie & Oeuvres de Descartes*, par Charles Adam. Paris: Leopold Cerf, 1910.

Kircher engraving: Athanasius Kircher, *Ars Magna Lucis et Umbrae*, 2d ed., Amsterdam, 1671.

Hevelius telescope: courtesy Science Museum, London, UK.

Newton portrait: Godfrey Kneller 1689 portrait, by kind permission of the Trustees of the Portsmouth Estates, Basingstoke, Hampshire, UK.

Newton prism drawing: courtesy Bausch & Lomb.

Ivory mirror case: from Fenja Gunn, *The Artificial Face*, New York: Hippocrene, 1983.

Dante illustration: Dante Alighieri, *Purgatory and Paradise*, trans. by Henry Francis Cary, illustrated by Gustave Doré. New York: P. F. Collier, 1892.

Richard II: Photo of Richard Thomas by T. Charles Erickson, 1993 production, courtesy The Shakespeare Theatre, Washington, DC.

Marcia the nun: courtesy Bibliothèque Nationale de France, Paris.

Arnolfini Portrait: courtesy National Gallery, London, UK.

Dürer self-portrait, age 13: courtesy Albertina, Vienna, Austria.

Dürer self-portrait, age 21: courtesy Universitatsbibliothek, Graphische Sammlung der Universitat, Erlangen, Germany.

Baldung painting: courtesy Kunsthistorisches Museum, Vienna, Austria.

1493 engraving: from *Ritter von Turn*, Basel, 1493 book, from Wilhelm Worringer, *Die Altdeutsche Buchillustration*, Munich: Piper Verlag, 1919.

Bellini painting: courtesy Kunsthistorisches Museum, Vienna, Austria.

Parmigianino self-portrait: courtesy Kunsthistorisches Museum, Vienna, Austria.

Chinese anamorphic art: from Fred Leeman, *Hidden Images: Games of Perception, Anamorphic Art, Illusion*, trans. Ellyn Childs Allison and Margaret L. Kaplan. NY: Harry N. Abrams, 1976, p. 140, Collection Jean Vlug, Brussels.

Charles I anamorphosis: courtesy Nationalmuseum, Stockholm, Sweden.

French glass casting: from Denis Diderot and Jean le Rond d'Alembert. *Encyclopédie, ou, Dictionnaire Raisonné des Sciences, des Arts et des Métiers*. 33 vol. 3rd ed. Livourne, 1770–1775.

Hall of Mirrors at Versailles: courtesy Pierre Devinoy photography collection, Paris, France.

Rowlandson drawing: © Copyright The British Museum.

Male dandy caricature: courtesy New York Public Library, Prints Division.

William Herschel portrait: from Constance A. Lubbock, *The Herschel Chronicle*, Cambridge: Cambridge U. Pr., 1933, frontispiece, from a crayon copy of the oil painting by L. T. Abbott in the National Portrait Gallery.

Herschel's 40-foot telescope: engraving dedicated to King George III, from Constance A. Lubbock, *The Herschel Chronicle*, Cambridge: Cambridge U. Pr., 1933, p. 166.

John Herschel: photo by Julia Margaret Cameron, 1867, courtesy Gernsheim Collection, Harry Ransom Humanities Research Center, The University of Texas at Austin.

Leviathan of Parsonstown: from Robert S. Ball, *The Story of the Heavens*, London: Cassell and Co., 1900.

Concave lighthouse mirrors: from Jacob Abbott, *Light*. NY: Harper & Brothers, 1871.

Fresnel engraving: from John Tallis, *Tallis's History and Description of the Crystal Palace*, 3 vol. London: John Tallis & Co., 1852.

Fresnel drawing: from Jacob Abbott, *Light*. New York: Harper & Brothers, 1871.

Photophone engravings: courtesy Alexander Graham Bell collection, Library of Congress.

X-ray photo by Röntgen: courtesy Deutsches Museum, Munich, Germany.

Henry Rowland photo: The Ferdinand Hamburger Archives of The Johns Hopkins University.

John Melville book: from John Melville, *Crystal Gazing and Clairvoyance*. London: Nichols & Co, 1897.

Pepper's Ghost: from Jacob Abbott, *Light*. NY: Harper & Brothers, 1871.

Sphinx illusion: *Illustrated Times*, Oct. 18, 1865.

Mystic Maze: from Albert Allis Hopkins, *Magic: Stage Illusions and Scientific Diversions*. New York: Munn & Co, 1897.

Alice through mirror: John Tenniel illustration from Lewis Carroll, *Alice's Adventures in Wonderland*, and *Through the Looking-glass*. New York: Macmillan, 1923.

Dodgson photo: Morris L. Parrish Collection of Victorian Novelists, Princeton University Library, Image No. 897.

Stratton composite photo: courtesy Nicholas Wade, Dept. Psychology, University of Dundee, UK.

INSERT 2

George Ellery Hale photo: courtesy Palomar Observatories/Caltech.

George Ritchey photo: courtesy Astronomical Society of the Pacific.

Mobile searchlight: photo courtesy of The Society of American Military Engineers.

Bernhard Schmidt photo: from Erik Schmidt, *Optical Illusions: The Life Story of Bernhard Schmidt*, Estonian Academy Publishers, 1995.

Russell Porter embracing Springfield Mount telescope: courtesy Springfield Telescope Makers.

Turret telescope: Mark Pendergrast photo.

Stellafane convention: Mark Pendergrast photo.

Pyrex picture: courtesy Palomar Observatories/Caltech.

Marcus Brown et al.: courtesy Palomar Observatories/Caltech.

Final touches on 200-inch: courtesy Palomar Observatories/Caltech.

Shop windows: 1925 photo by Eugene Atget, Museum of Modern Art, NY, Abbott-Levy Collection, partial gift of Shirley C. Burden, Art Resource.

Coney Island photo: courtesy Museum of the City of New York.

Gillette ad: Warshaw Collection of Business Americana, Barbering, Archives Center, National Museum of American History, Smithsonian Institution.

Pesky boyfriend cartoon: "Make-Up," by W. E. Hill, copyright 1934 Chicago Tribune New York News Syndicate, from Rutgers University Library.

"Are You Afraid" ad: courtesy Hartman Center for Sales, Advertising and Marketing History, Duke University, Durham, NC.

1924 movie still: from 1924 *Toilet Requisites*, public domain.

Rosie the Riveter: from "Is Beauty Worth Half a Billion?" by Mary-Elizabeth Parker, Dec. 20, 1942, *New York Times Magazine*, p. 25.

Norman Rockwell's "Girl at the Mirror": courtesy Norman Rockwell Family Agency.

Gerald Brockhurst's "Adolescence": courtesy British Museum.

Karl Jansky: courtesy Bell Labs/Lucent Technologies.

Jodrell Bank radio dish: courtesy Jodrell Bank Observatory.

Arecibo dish: courtesy David Parker (1997), Science Photo Library, Arecibo Observatory.

Horn reflector photo and graphic: courtesy Bell Labs/Lucent Technologies.

Jocelyn Bell Burnell photo: courtesy Jocelyn Bell Burnell.

Very Large Array photo: Dave Finley NRAO/AUI/NSF.

Riccardo Giacconi photo: from Wallace Tucker and Riccardo Giacconi, *The X-Ray Universe*, Cambridge, MA: Harvard U. Pr., 1985.

Chandra mirrors: courtesy Harvard-Smithsonian Center for Astrophysics, Cambridge, MA.

MMT photo: courtesy MMT, photo by Howard Lester. The MMT is a joint facility of the Smithsonian Institution and the University of Arizona.

Laser diagram: from Jeff Hecht, and Dick Teresi, *Laser: Light of a Million Uses*. Mineola, NY: Dover, 1982, 1998.

Corner cube reflector: courtesy 3M Corporation.

Adaptive optics diagram: from John W. Hardy, *Adaptive Optics for Astronomical Telescopes*, New York: Oxford U. Pr., 1998.

Hubble mirror: courtesy Goodrich Corporation.

Fuzzy star image: courtesy Space Telescope Science Institute, Baltimore, MD.

Hubble cartoon: courtesy *Journal Gazette*, Fort Wayne, Indiana, cartoon by Dan Lyons.

Roger Angel photo and 8.4 meter mirror: courtesy Steward Observatory Mirror Laboratory, Tucson, AZ. Photos by Lori Stiles/John Florence.

Schott 8.2 meter mirror: courtesy Schott Glas, Mainz, Germany.

Three mirror methods: courtesy Andy Christie/Slim Films, New York, NY.

OWL telescope: courtesy Manchu/Ciel et Espace, Paris, France.

Two girls: Mark Pendergrast photo.

Musée Grevin: Mark Pendergrast photo.

INDEX

CPSIA information can be obtained at www.ICGtesting.com
Printed in the USA
LVOW05s1656100214

373092LV00004B/963/P